T0332923

Systems & Control: Foundations & Applications
Volume 3

Series Editor
Christopher I. Byrnes, Washington University

Harold J. Kushner

Weak Convergence Methods and Singularly Perturbed Stochastic Control and Filtering Problems

1990

Birkhäuser
Boston · Basel · Berlin

Harold J. Kushner
Division of Applied Mathematics
Brown University
Providence, RI 02912
U.S.A.

Library of Congress Cataloging-in-Publication Data
Kushner, Harold J. (Harold Joseph), 1933–
    Weak convergence methods and singularly perturbed stochastic
control and filtering problems / Harold J. Kushner.
        p.   cm. — (Systems & control ; v. 3)
    Includes bibliographical references.
    ISBN 0-8176-3437-1 (alk. paper)
        1. Control theory.   2. Stochastic processes.   3. Filters
(Mathematics)   4. Convergence.   I. Title.   II. Series.
    QA402.3.K83   1990
    629.8′312—dc20   90-31373

Printed on acid-free paper.

ISBN 0-8176-3437-1
ISBN 3-7643-3437-1

Camera-ready copy supplied by the author using LaTeX.
Printed and bound by Edwards Brothers, Inc., Ann Arbor, Michigan.
Printed in the U.S.A.

9 8 7 6 5 4 3 2 1

This book is dedicated to Linda for her strong support and wonderful inspiration.

Preface

The book deals with several closely related topics concerning approximations and perturbations of random processes and their applications to some important and fascinating classes of problems in the analysis and design of stochastic control systems and nonlinear filters. The basic mathematical methods which are used and developed are those of the theory of weak convergence. The techniques are quite powerful for getting weak convergence or functional limit theorems for broad classes of problems and many of the techniques are new. The original need for some of the techniques which are developed here arose in connection with our study of the particular applications in this book, and related problems of approximation in control theory, but it will be clear that they have numerous applications elsewhere in weak convergence and process approximation theory. The book is a continuation of the author's long term interest in problems of the approximation of stochastic processes and its applications to problems arising in control and communication theory and related areas. In fact, the techniques used here can be fruitfully applied to many other areas.

The basic random processes of interest can be described by solutions to either (multiple time scale) Itô differential equations driven by wide band or state dependent wide band noise or which are singularly perturbed. They might be controlled or not, and their state values might be fully observable or not (e.g., as in the nonlinear filtering problem). Such processes occur in many applications in engineering and science. The systems of interest are usually much too hard to analyze or use directly for purposes of obtaining qualitative properties, good controls or linear or nonlinear filters, etc. For the purpose of both analysis and design, it is desirable to find approximate models of the systems (whether controlled or not) which are simpler and which retain the essential properties of the physical systems, and which can be more conveniently used. Generally, such "approximate" methods are of the diffusion or jump diffusion type. Many of the methods which we develop to show that sequence of physical processes converges (weakly) to the desired approximation as some "approximation parameter" goes to, say, zero, are essentially extensions of those in [K20], but with the adaptations which are necessary to handle the more complex problems here. The methods are well adapted to the types of noise processes and approximation and limit problems which occur in control and communication theory, and at the moment seem to be the methods of choice for such cases. These methods can be used for a wide variety of problems as in [K20], and either over a finite or infinite time interval. The reference [K20] did not treat the control problem or the types of state dependent noise which occur in the singularly perturbed system, and some new types of averaging methods are introduced here for that purpose.

The particular applications with which this book is concerned are control and filtering problems for the so-called singularly perturbed system. The methods can be used for the study and approximation of other classes of systems or parametrized systems. In our case, the parameters are the singular perturbation parameter, and the "bandwidth" or time scale parameter of the driving noise, where appropriate. Such singularly perturbed systems are also known as two (or multiple) time scale systems. In a sense which will be made precise later in the book, they are essentially systems modelled by state dependent wide band noise. There is a vast literature on the deterministic problem [K4, K5], but work on the stochastic case has been much more limited (e.g., [B2, B4, B5]). There are stochastic analogues of most of the deterministic problems. Also, there are stochastic problems which have no deterministic counterpart; for example, the ergodic cost problems, the wide band noise problems, and the nonlinear filtering problems. The techniques are entirely different from those used in the deterministic case.

We can view the state variable of the physical singularly perturbed system as being divided into two sets, called the "slow" variable and the "fast" variable. Such two time scale systems occur frequently in applications. For example the motions of some states of an airplane are much faster than others, and certain state variables of a power system change much faster than others do. In many tracking systems, some of the components of the object to be tracked change slowly relative to others. Many concrete and interesting examples are in [K5]. Sometimes in the deterministic case, the "fast" dynamics are known as transients, in the sense that after a short time their values essentially stabilize about some steady state function of the slow variables. The stochastic analog is that the probability distribution of the fast variable essentially quickly stabilizes about some "steady state distribution," which is a function of the "slow" variable.

For most engineering systems that are too complicated to allow a simple analysis, it is a common practice to simplify the system by eliminating or averaging out some of the "transient" or "quickly stabilized" dynamical terms or state variables in some way. This is facilitated by exploiting the two time scale property of the system. One procedure for deterministic systems is roughly as follows. One first proves an appropriate type of stability of the "fast" components of the system. Then the two time scale property is used. Consider the dynamical system for the "fast" variable, and fix the value of the slow variable in that equation at some constant value. Then, it is often the case that the fast variable converges to some steady state value which depends on the value at which the other variable was fixed. In practice, one often approximates the total system by the reduced order system which is obtained by using the differential equation for the slow variable, but where the fast variable is fixed at its steady state value as a function of the slow variable. This procedure does indeed simplify the system, and is valid under quite broad conditions in the sense that the resulting system can be effectively used for the analysis of the original system, and the controls which are good for the approximating system are also good for the original system [K5]. Methods for the deterministic problem are relatively well developed and widely used, either explicitly or implicitly.

The stochastic problem is much more difficult and relatively little had been done. Many more types of problems arise in the stochastic case and powerful deep probabilistic methods need to be used. The basic physical problem is usually described by two sets of coupled differential equations, which are either driven by wide band noise or are of the diffusion or jump diffusion type. These systems might be of interest for purposes of control or we might simply want to analyze their properties. Also, if only noise corrupted data on the states is available, then we might want to construct a filter, linear or nonlinear. These systems are very difficult to analyze, and it makes sense to try to simplify the models as much as possible while retaining a structure which preserved the basic properties, and which can be used to get good controls and filters for the original physical system. The problems which are involved are those of the approximation of random processes, controlled or not, by simpler processes. One of the most powerful methods of treating such classes of problems is the theory of weak convergence.

Chapters 1 to 3 contain introductory material from the theory of weak convergence and stochastic differential equations. These chapters are not intended to be complete, but serve to collect most of the particular background results which will be needed in the sequel, for ease of reference in the form in which they will be needed. Chapter 1 contains many of the basic definitions and concepts of the theory of weak convergence. The basic path spaces are discussed together with the general questions of compactness and convergence, and useful criteria for tightness are given. For the reader with little prior acquaintance with the theory of weak convergence, the material might seem somewhat abstract. But the methods are powerful and are quite convenient to use in the applications. Chapter 2 contains a survey of the theory of stochastic differential equations, both of the diffusion and jump diffusion type. Several approaches to the construction of solutions are discussed and the solutions are related to the solution of the so-called martingale problem. Much of the book is concerned with sequences of random processes (say, parametrized by the singular perturbation parameter) and to get the desired system simplifications or approximations, we need to know whether either the original sequence or a suitable subsequence converges in an appropriate sense, and whether the limit process is a solution of a controlled stochastic differential equation. The relation between the solution of a stochastic differential equation and the martingale problem provides the basic tool that we will use to show that a limit of a sequence of physical processes is, indeed, a solution to a stochastic differential equation, since it is relatively easy to show that the types of limit processes which arise here actually solve the martingale problem. This method has been used for a long time. Controlled stochastic differential equations are introduced in Chapter 3. The methods of the book are usable for all of the cost functionals that are of use in control theory. For the sake of simplicity, we confine our attention to a few standard cases. The chapter introduces the important concept of "relaxed control," and shows how to prove the existence of an optimal stochastic control. The basic technique involves starting with a minimizing sequence of control functions, and shows that the sequence of processes which correspond to them actually converges to an optimal process. The idea is to show first that the weak limit solves a certain "controlled" martingale problem; hence, it is the solution to a controlled stochastic differential equation. We then show that the limit is indeed the optimally controlled process. The basic technique will be often used in the sequel for the more complicated singularly perturbed problems. It is a standard method for such approximation and limit problems and has applications much beyond those in this book.

The controlled singularly perturbed system is introduced in Chapter 4. The basic model which is used here is the two time scale diffusion or jump diffusion process. The fast system is studied first (in a stretched out time scale) under the assumption that the slow variable is treated as a parameter. Under the assumptions used, this so-called fixed-x system has a unique invariant measure. This invariant measure plays an important role in the simplification of the basic control problem. An averaged system and cost function is defined by integrating the fast variable out of both the dynamical equation for the slow variable and the cost function, by use of the cited invariant measure. This averaged system has the dimension of the "slow" variable, and doesn't involve the singular perturbation parameter. It is just what one would use in practice for the approximate model for the singularly perturbed problem. This use is justified. We show that optimal value functions for the original singularly perturbed problem converge to the optimal value function for the averaged problem. Also, good controls for the averaged problem are good controls for the original problem. Also, good controls for the averaged problem are good controls for the original problem. The methods used are basically relatively straightforward, considering the generality of the results. The average cost per unit time problem is also treated with the same type of results, except that the approximating averaged problem is now an ergodic cost problem. The methods which are used in the chapter appear to be quite efficient and robust.

In Chapter 5, we are concerned with a powerful new method for dealing with the pathwise average cost per unit time problem. The pathwise average cost per unit time arises in applications where the system is of interest over a long time interval, and only one realization might be taken, so that the use of the mathematical expectation in the cost function might not be justified. The functional occupation measure method concerns limits of normalized sample occupation measures of the paths of the system. The limits of these occupation measures are measure valued random variables whose sample values induce processes which are stationary controlled "averaged" processes. This result is then used to show that the best limit value for the pathwise average cost for the singularly perturbed problem (as the time interval and singular perturbation parameter go to their limits in any way at all) is the minimum value for the ergodic cost problem for the averaged system, and that this value can actually be attained arbitrarily closely for small values of the singular perturbation parameter. The basic ideas and results of the functional occupation measure method are given, since at this time there is no other source for them. The method which is used is purely probabilistic. Although it involves the abstract ideas concerning occupation measures over the Borel sets in the path space, it will be seen that the techniques are not hard to use on actual problems. There are many other applications of the functional occupation measure approach to functional limit theorems for pathwise averages for many types of sequences, and some of these are indicated in the chapter. Another indication of the potential of the functional occupation measure approach is shown in [K27] for a different type of limit problem. In that reference, the pathwise average limits of the average cost per unit time problem for a queueing system in heavy traffic is obtained, as the time interval and the heavy traffic parameter go to their limits in any way at all. It is shown that the limit values are (w.p.1) the ergodic cost for a certain stationary reflected diffusion.

The nonlinear filtering problem is dealt with in Chapter 6. Under appropriate conditions, it is shown that the output of the optimal filter for the singularly perturbed problem is close to that of the output of the filter for the averaged system, but where the actual physical observations are used as inputs (we call this latter filter the averaged filter). If our conditions are violated, then this might not be the case, as shown by a counterexample. But even if our conditions are violated, the averaged filter can still be shown to be nearly optimal with respect to a large and natural class of alternative filters. The averaged filters are usually of much lower dimension than the optimal filters for the original problem, and do not involve the high gains of that optimal filter. Thus, they are of considerable interest in practice. An example is given of the use of the averaged filter for purposes of control under partial information, and it is shown that it provides a nearly optimal control. The chapter also gives a useful computational approximation to the averaged filter, and proves its robustness, uniformly in both the singular perturbation and the approximation parameters. We also deal with the averaged filter on the infinite time interval.

So far, all the work in the book has concerned singularly perturbed diffusion or jump diffusion processes. Practical systems tend to be driven by wide bandwidth noise, and a substantial advantage of the weak convergence methods is that they can be used for this case, as opposed to, say, methods based on partial differential equations or classical semigroup techniques. Basic methods for proving weak convergence of wide band noise driven systems are given in Chapter 7. The methods are extensions of the techniques in the reference [K20]. In particular, we develop further the so-called "perturbed test function-direct averaging" method. It generally uses only the relatively simple first order perturbed test functions. Some of the ideas of this method were introduced in [K20] and used in many of the examples there. Here some of the ideas are simplified and carried further, and then applied to the singular perturbation problem. The methods uses a combination of the simple form of the perturbed test function method with ergodic type theorems to get a fairly powerful approach to obtaining the operator of the (controlled or uncontrolled) limit systems. The word "bandwidth" is used in a loose sense. The processes involved need not have a bandwidth in the usual sense. The word simply denotes a "time scale separation." In Chapter 8, we treat the singularly perturbed wide bandwidth noise driven system and obtain results analogous to those of Chapters 4 and 5. Note that there are several parameters here which must be dealt with in the weak convergence analysis: the bandwidth and the singular perturbation parameter; also, in the average cost per unit time problem, we need to consider the time as a parameter as well. The functional occupation measure method works well here also, and this problem well illustrates the power of that method. For the wide band noise driven case, the correct averaged system is not obvious, but it comes naturally out of the analysis. Here, the limit and approximation results provide even more of a simplification and averaging than for the case of Chapter 4, since we obtain the limit "approximating" averaged problem by averaging out both the driving noise (in the sense of getting the appropriate white noise driven system) and the fast variable.

Many of the problems of the book involve stability of either the fast or the slow system, particularly if the problem is of interest over an infinite time interval. Liapunov function methods which have been used to prove the stability of singularly perturbed deterministic systems are adapted to the stochastic case (for the diffusion or jump diffusion model) in Chapter 9. Stability methods for wide band noise driven systems were given in [K20]. These methods are combined with the above cited adaptations of the deterministic method to obtain a Liapunov function method which can be used to prove the stability properties which are needed elsewhere in the book for the wide band singularly perturbed system. Chapter 10 begins and quickly ends the study of parametric singularities, such as occur when a system contains a small or parasitic parameter, say, a small inductance or mass. This important topic involves too many details to be dealt with effectively here, but the chapter gives the flavor of some of the possible results.

The manuscript has been read and many suggestions for improvement given by my colleagues Paul Dupuis, Matthew James, Wolfgang Runggaldier and Jichuan Yang, and I gratefully acknowledge their considerable help. I would also like to thank Katherine MacDougall for her masterful typing work. For many years, the author's work in stochastic control has been supported by (not necessarily all at the same time) the Air Force Office of Scientific Research, the Army Research Office and the National Science Foundation.

Contents

1

Weak Convergence

0. Outline of the Chapter

The basic questions dealt with in this book concern approximations of relatively complex processes by simpler and more tractable processes. These processes might be controlled or not controlled. The approximation might be of interest for either the purposes of engineering design, or for other analytical or numerical work. The theory of weak convergence of measures seems to be the fundamental and most widely used and successful tool for such purposes. See, for example, the techniques, examples and references in the books [B6, E2, K20]. In this chapter, we will discuss the basic ideas in the subject which will be useful for our work in the sequel. The detailed examples will not be dealt with in this chapter, but are left to the rest of the book. The basic techniques discussed in this chapter will all be further developed and worked out in detail in connection with the applications dealt with in the succeeding chapters. The concepts and criteria have a rather abstract flavor as they are initially introduced, but when applied in the succeeding chapters, they take on concrete and readily usable forms.

Section 1 contains the basic definitions, and some of the fundamental equivalences. The theory of weak convergence is a powerful extension of the techniques and results associated with the classical central limit theorem. The examples in Section 2 further illustrate this point and provide an intuitive picture of some of the concepts introduced in Section 1 and raise basic questions concerning process approximation. In Section 3, we continue the discussion of Section 1, and give the very useful Skorohod representation. Since we are concerned with the approximation of random processes, we must be concerned with the particular space of paths for the processes and work with the spaces of paths which allow the most convenient analysis. The two most widely used spaces of paths are discussed in Sections 4 and 5; i.e., the space of continuous functions with the sup norm, and the space of functions which are right continuous and have left hand limits, with the so-called Skorohod topology (to be defined in Section 5).

The classical central limit theorem is concerned with the convergence in distribution of a sequence of vector-valued random processes. We take an analogous point of view to the problem of process approximations or the problem of convergence of a sequence of processes. We view a stochastic process as a random variable with values in some appropriate space of paths. Given a sequence of processes $X^n(\cdot)$, we study the "weak" convergence of the associated probability measures. From this, we can get the convergence in distribution or expectation of a very large class of functionals of $X^n(\cdot)$ to the same functional of the process $x(\cdot)$ induced by the limit measure. I.e., $X^n(\cdot)$ will converge in a generalized sense of distribution to $x(\cdot)$. This is a very powerful idea, and is the basis of process approximation theory.

Some of the most interesting applications in the book make use of measure-valued random variables and processes. The use of such random variables and processes provides a unifying approach to many important problems, and actually simplifies their development considerably. Some of the appropriate concepts are introduced in Section 6, and are further developed and filled out in the succeeding chapters. Section 6 might seem to be somewhat abstract to the reader whose main interest is in the applications, but it will be seen that the ideas can often be used with facility in concrete applications, and yield strong results. We cite, in particular, the results on approximation and convergence over an infinite time interval in Chapter 6.

1. Basic Properties and Definitions

Let S denote a metric space with metric d and $C(S)$ the space of real valued and continuous functions defined on S. The symbol $\mathcal{B}(S)$ is used to denote the collection of Borel subsets of S. Let $C_b(S)$ and $C_0(S)$ denote the subspaces of $C(S)$ which are bounded and which have compact support, resp. Now, let $S = R^r$, Euclidean r-space. Let $C^k(S)$, $C_b^k(S)$ and $C_0^k(S)$ denote the further subspaces (of $C(S)$, $C_b(S)$, $C_0(S)$, resp.) whose mixed partial derivatives up to order k are continuous, bounded and continuous, and bounded and continuous with compact support, resp. Unless otherwise specified, we use the norm $|f| = \sup_{x \in S} |f(x)|$ on $C(S)$ and its subspaces.

If P is a measure on $(S, \mathcal{B}(S))$ and $f(\cdot)$ is a real valued integrable function, then we sometimes write the integral as $(P, f) = \int f(x)P(dx)$. For $A \in S$, define the set $A^\epsilon = \{x \in S : d(x, x') < \epsilon$ some $x' \in A\}$. For P and P' being probability measures on $(S, \mathcal{B}(S))$, the *Prohorov metric* is [E2, p. 96] defined by

$$\pi(P, P') = \inf\{\epsilon > 0 : P(A^\epsilon) \leq P'(A) + \epsilon, \text{all closed sets } A \in \mathcal{B}(S)\}.$$

Let $\mathcal{P}(S)$ denote the space of probability measures on $(S, \mathcal{B}(S))$ with the Prohorov metric. If the space $\mathcal{P}(S)$ is obvious, then we might omit reference to it in the notation. One of the most useful properties of the Prohorov metric is given in the next theorem.

Theorem 1.1 [E2, p. 101]. *If S is complete and separable, then $\mathcal{P}(S)$ is complete and separable.*

We say that a sequence $\{P_n\}$ in $\mathcal{P}(S)$ converges to P in $\mathcal{P}(S)$ in the *weak topology* if and only if $(P_n, f) \to (P, f)$ for all $f(\cdot)$ in $C_b(S)$. If $\{P_n\}$

converges to P in this topology, then we say that it *converges weakly* and use the notation $P_n \Rightarrow P$. The relationship between convergence in the Prohorov metric and weak convergence is covered in Theorem 1.3 below.

Let P_n and P be in $\mathcal{P}(S)$ and let X_n and X denote S-valued random variables which induce the measures P_n and P on $(S, \mathcal{B}(S))$, resp. We often say that X_n and X are the random variables *associated with P_n and P*, resp. The weak convergence $P_n \Rightarrow P$ is equivalent to the convergence $X_n \to X$ in distribution; i.e., it is equivalent to $Ef(X_n) \to Ef(X)$ for each $f(\cdot) \in C_b(S)$. If P_n is in $\mathcal{P}(S)$ and X_n is the associated random variable, then for a set A in $\mathcal{B}(S)$ we use the notation $P_n(A)$ and $P(X_n \in A)$ interchangeably, depending on convenience. Let P_n and P be in $\mathcal{P}(S)$ with $P_n \Rightarrow P$. Let X_n and X be the associated random variables. Then we will frequently abuse terminology by saying that X_n *converges weakly* to X and writing $X_n \Rightarrow X$.

The reader should keep in mind that (in order to reduce the notational burden) we use the term *weak convergence* and the symbol $\Rightarrow$ in two different senses. If the X_n and X are random variables and we say that $X_n \Rightarrow X$ or that "X_n converges weakly to X," we mean that $Ef(X_n)$ converges to $Ef(X)$ for each $f(\cdot)$ in $C_b(S)$, or, equivalently, $f(X_n) \to f(X)$ in distribution for $f(\cdot) \in C(S)$. If P_n and P are in $\mathcal{P}(S)$, then $P_n \Rightarrow P$ or "P_n converges weakly to P" implies convergence in the Prohorov metric. By Theorem 1.3 below, convergence in the Prohorov metric is (usually) equivalent to convergence in the weak topology. The only places where care needs to be exercised is where the random variables of interest are measure-valued themselves.

A set $\{P_\alpha\} \subset \mathcal{P}(S)$ is said to be *tight* (or *tight in $\mathcal{P}(S)$*) if for each $\epsilon > 0$ there is a compact set $K_\epsilon \subset S$ such that

$$\sup_\alpha P_\alpha(x : x \notin K_\epsilon) \le \epsilon.$$

If $\{P_n\}$ is tight in $\mathcal{P}(S)$, and there is some P in $\mathcal{P}(S)$ such that $(P_n, f) \to (P, f)$ for each $f(\cdot) \in C_0(S)$, then $P_n \Rightarrow P$ in $\mathcal{P}(S)$. If X_n are random variables which induce the probability measures P_n, we use the term "tightness of $\{P_n\}$" interchangably with "tightness of $\{X_n\}$". The reason for the importance of tightness is given in the next theorem: It is essentially equivalent to the fact that each sequence has a subsequence which converges weakly. A set of probability measures is said to be relatively compact if its closure is compact and all elements of the closure have unit mass.

Theorem 1.2 (Prohorov's Theorem) [E2, Theorem 3.2.2]. *Let S be complete and separable. Then a set $M \in \mathcal{P}(S)$ is relatively compact if and only if M is tight.*

Sometimes we work with measures whose total mass is not unity. Let $\mathcal{M}(S)$ denote the space of measures on $(S, \mathcal{B}(S))$ each with finite total mass and with the Prohorov topology. Then Theorem 1.2 continues to hold with $\mathcal{M}(S)$ replacing $\mathcal{P}(S)$ provided that $\sup_\alpha P_\alpha(S) < \infty$.

Let $S = R^k$, Euclidean k-dimensional space. If $k = 1$, we will simply write R instead of R^1. For each n, let X_n be an R^k-valued random variable and P_n the measure induced by X_n. Then the classical Heine–Borel Theorem says that $\{P_n\}$ (a sequence in $\mathcal{P}(R^k)$) has a weakly convergent subsequence if and only if $\{P_n\}$ is tight. Of course, in this case tightness of $\{P_n\}$ is equivalent to

$$\lim_N \sup_n P_n(x : |x| \geq N) = 0.$$

For any real valued random variable X, define the distribution function $F(\cdot)$ by $F(x) = P(X < x)$. If P_n and P are in $\mathcal{P}(R)$, then the weak convergence $P_n \Rightarrow P$ is equivalent to $P(X_n < x) \to F(x)$ for each x which is a continuity point of $F(\cdot)$. This is a special case of Theorem 1.4. below.

Let $\overline{R}^k$ denote the one-point compactification of R^k. If $k = 1$, then we simply write $\overline{R}$. The compact sets of $\overline{R}^k$ contain those of R^k, but they also include the "infinite sets" of the form $\{x : |x| \geq N\}$ for each N. Thus any sequence of probability measures on $\overline{R}^k$ is tight.

A measurable set B is said to be a P-continuity set if $P(\partial B) = 0$, where $\partial B = \overline{B} - B^0 = $ closure of $B - $ interior of $B = $ boundary of B. We then have the important result of Theorem 1.3. This result will be illustrated in an example in the next section.

Theorem 1.3 [E2, Theorem 3.3.1]. *For any metric space S and a sequence of measures P_n and P in $\mathcal{M}(S)$ of uniformly bounded total mass, (i)–(iv) are equivalent. and are implied by (v). If S is separable, then (v) is equivalent to any of (i)–(iv).*

(i) $P_n \Rightarrow P$

(ii) $\limsup_n P_n(B) \leq P(B)$ *for closed sets* B

(iii) $\liminf_n P_n(D) \geq P(D)$ *for open sets* D

(iv) $P_n(F) \to P(F)$ *for all P-continuity sets* F

(v) $\pi(P_n, P) \to 0$.

Often, in a weak convergence study, we cannot restrict ourselves to working with functions in $C_b(S)$ only. For example, certain classes of functions which are not continuous at some points in S do appear very naturally as cost functions in control problems when the control is applied or which are of interest otherwise until some boundary or target set is reached. For such cases, we have the following useful corollary of Theorem 1.3.

Theorem 1.4 [B6, Theorem 5.1]. *For a metric space S, let P_n and P be in $\mathcal{P}(S)$ and let X_n and X, resp., be associated random variables. Let $P_n \Rightarrow P$. Let $f(\cdot)$ be a real-valued measurable function on S and let D_f denote the set of points in S at which $f(\cdot)$ is not continuous. Then D_f is measurable. If $P(D_f) = 0$, then $f(X_n) \Rightarrow f(X)$.*

2. Examples

Perhaps the best known and most classical use of weak convergence is the central limit theorem. Many "physical" phenomena are the result of the composition of numerous small effects, where the cumulative effect of the small effects is significant. Consider the following canonical example of such a situation. Let u_n denote a sequence of mutually independent and identically distributed real valued random variables with mean zero and variance σ^2 and with $E|u_n|^{2+\alpha} < \infty$ for some $\alpha > 0$. Define $X_n = \sum_{i=1}^{n} u_i/(\sigma\sqrt{n})$. Then the classical central limit theorem states that $X_n \Rightarrow X$ where X is normally distributed with mean zero and unit variance [B6, B9, E2].

We now give an illustration of Theorem 1.4. The sequence $\{X_n\}$ of real valued random variables defined above is tight. This can be seen from the fact that $EX_n^2 = 1$ and use of the Chebychev inequality

$$P(|X_n| \geq N) \leq EX_n^2/N^2.$$

Let $u_n = \pm 1$, each with probability $1/2$. Then $\sigma = 1$ and X_n takes values in the set

$$\left\{ \sum_1^n \alpha_i/\sqrt{n}, \alpha_i = \pm 1 \right\} \equiv V_n.$$

Define $V = \cup_n V_n$, a countable set. Let $F_1(\cdot)$ denote the distribution function of a normally distributed random variable with mean zero and unit variance, and let $P_1(\cdot)$ be the associated probability measure. Let $f(\cdot)$ denote the indicator function of the interval $(a, b] \equiv B$, where $a < b$. Then $\partial B = \{a\} \cup \{b\}$. The function $f(\cdot)$ is not continuous at the points a and b, but since $P_1(\partial B) = 0$, we have $f(X_n) \Rightarrow f(X)$. If $f(\cdot)$ were the indicator function for the set V, then $f(X_n) \not\Rightarrow f(X)$, since $f(X_n) = 1$, $f(x) = 0$ for $x \notin V$ and $P_1(V) = 0$. Here $\partial V = R$, so that V is not a P_1-continuity set.

The example which was described above can be extended to cover a class of problems that are very important in applications to stochastic dynamical systems. We will introduce some of the ideas and main points of the extensions by means of a relatively simple example. In fact, the weak convergence ideas in general are extensions of the concepts of convergence in distribution of a sequence of vector valued random variables. In this book, we are concerned with the convergence of expectations of functionals of a sequence of random processes which are defined on some finite or infinite time interval to the expectation of a related functional of a "limit" random process which is defined on the same time interval. Consider one particular example. For each $\epsilon > 0$, let $x^\epsilon(\cdot)$ denote a real-valued process with continuous paths and which is defined on the time interval $[0, \infty)$. We then would like to know if there is a process with continuous paths and defined on $[0, \infty)$ and which we can conveniently characterize, such that $Ef(x^\epsilon(\cdot)) \to Ef(x(\cdot))$ for a "large enough" class of functions $f(\cdot)$. In this case, the space S of values of the "random variables" $x^\epsilon(\cdot)$ is just the space of continuous functions on $[0, \infty)$ with an appropriate topology used.

We now give a concrete and standard example in order to illustrate the above remarks; namely, an approximation to the Wiener process. For each n, let $\{u_n^i, i < \infty\}$ be a sequence of mutually independent and identically distributed random variables, with mean zero and variance σ_n^2, where $\sigma_n^2 \to 0$ as $n \to \infty$. Suppose that for each n we have a sequence of random variables τ_i^n such that $\{\tau_i^n - \tau_{i-1}^n, u_i^n, i < \infty\}$ are mutually independent and

$$P\big(\text{some } \tau_j^n \in [t, t+\delta] \,|\, \tau_i^n, u_i^n, i : \tau_i^n < t\big) = \lambda_n \delta + o(\delta),$$

$$P\big(\text{more than one } \tau_j^n \in [t, t+\delta] \,|\, \tau_i^n, u_i^n, i : \tau_i^n < t\big) = o(\delta).$$

Let $\lambda_n \to \infty$, $\lambda_n \sigma_n^2 \to \sigma^2 > 0$, and $\lambda_n E|u_i^n|^3 \to 0$ as $n \to \infty$. Such processes are widely used to model the underlying "atomic level" noise in electrical circuits and for many other applications [D1, P2].

Define $W_n(t) = \sum_{\tau_i^n \leq t} u_i^n$. The $W_n(\cdot)$ do not have continuous paths; they are piecewise constant and right continuous. Let $D[0, T]$ denote the space of paths which are right continuous and have left hand limits. The topology which is used on this space will be discussed in Section 5 below. By a classical central limit theorem [B6, B9], for each t $W_n(t)$ converges weakly to a random variable $W(t)$ which is normally distributed with mean zero and variance $\sigma^2 t$. Also if $t_{i+1} > t_i$, for all i, then $\{W_n(t_{i+1}) - W_n(t_i), i = 1, \ldots, k\} \equiv \{Y_i^n, i \leq k\}$ are mutually independent random variables and the $\{Y_i^n, i \leq k\}$ converge weakly to a set of mutually independent random variables $\{Y_i, i \leq k\}$, where Y_i is normally distributed with mean zero and variance $\sigma^2(t_{i+1} - t_i)$. Thus the classical central limit theorem yields the convergence of the *multivariate distributions* of $W_n(t)$, $t < \infty$, to those of a Wiener process $W(\cdot)$ with variance parameter σ^2.

Let us now ask a more interesting question. Do we have

$$\int_0^T f(W_n(s))ds \Rightarrow \int_0^T f(W(s))ds \tag{2.1}$$

for all $f(\cdot) \in C_b(R)$ and $T < \infty$? To get (2.1), we need more than convergence of the multivariate distributions. The following discussion of this question is for motivational purposes only and is somewhat loose. Roughly speaking, we can see that the convergence of the multivariate distributions would be enough if the paths of the $W_n(\cdot)$ were "smooth", uniformly in n, since then we could approximate (uniformly in n) the left side of (2.1) by a function which depends continuously on the samples $\{W_n(i\delta), i\delta \leq T\}$, for small enough δ. By "smooth, uniformly in n," we mean that in some appropriate statistical sense, the $W_n(\cdot)$ do not "vary much" in the intervals $\{[i\delta, i\delta + \delta), i \leq T/\delta\}$ for small δ, uniformly in n. In order to get such "smoothness" results, we need to study the structure of the entire set of *processes* $\{W_n(\cdot), n < \infty\}$ to see how the "smoothness" of the paths depends on n (in an appropriate statistical sense to be discussed below). As will be seen, the path space $D[0, T]$ with the topology used in Section 5 will be very convenient for our purposes, and is the most widely used path space in applications of the weak convergence theory. In this particular example, we could also use the path space $C[0, T]$ of continuous real valued functions on $[0, T]$ with the sup norm topology. Then we would use the continuous piecewise linear approximation $\overline{W}_n(\cdot)$ of $W_n(\cdot)$, where we define $\overline{W}_n(\tau_i^n) = W_n(\tau_i^n)$ and for $t \in (\tau_i^n, \tau_{i+1}^n)$, $\overline{W}_n(\cdot)$ is defined to be the linear interpolation of $W_n(\tau_i^n)$ and $W_n(\tau_{i+1}^n)$. If there were a compact set S_1 in $C[0, T]$ such that the paths of each of the processes $\overline{W}_n(\cdot)$ were in that set for all n, then the processes would be smooth enough to get the desired approximation result (2.1) simply from the convergence of the finite dimensional distributions, by using an easy interpolation. A similar result holds when working with the space $D[0, T]$. Unfortunately, in the most interesting cases and including the example above, there is no such set S_1. But this is precisely the point where the value of the concept of tightness appears.

Tightness of $\{W_n(\cdot)\}$ (or of $\{\overline{W}_n(\cdot)$, resp.) implies that for each $\delta > 0$ there is a compact set K_δ in the path space of the processes $W_n(\cdot)$ such that the paths of the $W_n(\cdot)$ (or of $\overline{W}_n(\cdot)$, resp.) are in K_δ with a probability of at least $1 - \delta$ for each n. The compactness of a set of paths implies enough smoothness on the paths (or smoothness between the finite set of jumps), so that we can get limit results such as (2.1) for a class of bounded and continuous functions, just from the convergence of the finite dimensional distributions and a suitable interpolation argument. Indeed, it follows from Theorems 5.1 and 5.2 below that $\{W_n(\cdot)\}$ is tight in $D[0, T]$ and that (2.1) does hold. Similarly, it follows from Theorems 4.1 and 4.2 below that $\{\overline{W}_n(\cdot)\}$ is tight in $C[0, T]$ and (2.1) holds if $\overline{W}_n(\cdot)$ is used there. The convergence holds for any continuous functional replacing the integral. These results are a special case of what is known as Donsker's Theorem [B6].

Next, consider the somewhat more complicated problem, where for each n the "impulses" $\{u_i^n, i < \infty\}$ are the driving forces for a dynamical system. Let us represent the system in the form

$$X_n(t) = X(0) + \int_0^t b(X_n(s))ds + W_n(t), \qquad (2.2)$$

where $b(\cdot)$ is a bounded and smooth function. The process defined by (2.2) is Markov, but it can be a nearly impossible object to study directly. Actually the situation is much worse if the $\{u_i^n\}$ are correlated or if the $\{\tau_{i+1}^n - \tau_i^n\}$ were not exponentially distributed. It would be very convenient for both numerical as well as analytical purposes, if it could be shown that the solution to (2.2) could be approximated by the solution to the simpler process defined by the Itô equation (2.3):

$$x(t) = x(0) + \int_0^t b(x(s))ds + W(t). \tag{2.3}$$

where $W(\cdot)$ is a Wiener process with variance parameter σ^2. This seems to be an appropriate approximation, since $W_n(\cdot)$ converges weakly to $W(\cdot)$ (in $D[0,T)$, in the topology defined in Section 5). If there were a continuous function $F_0(\cdot)$ on $D[0,T]$ such that $X_n(\cdot) = F_0(W_n(\cdot))$ for all n, then the weak convergence $X_n(\cdot) \Rightarrow x(\cdot)$ would follow from the weak convergence $W_n(\cdot) \Rightarrow W(\cdot)$. Unfortunately, it is not usually possible to show that the solutions to equations such as (2.2) are continuous functions of the driving processes (or even almost everywhere continuous with respect to Wiener measure so that we could use Theorem 1.4).

By the use of the word "approximation" above, we mean that the distribution of path functionals $F(X_n(\cdot), W_n(\cdot))$ could be approximated by the distributions of the functionals $F((x(\cdot), W(\cdot))$ for large n and a large enough class of functionals $F(\cdot, \cdot)$. But this is just a question of weak convergence of $(X_n(\cdot), W_n(\cdot))$ to $(x(\cdot), W(\cdot))$. Again, in order to get the desired approximation or convergence results we must look at $(X_n(\cdot), W_n(\cdot))$ as being random variables with values in a space of paths, and study the associated measures on the path spaces. The theory of weak convergence is motivated by such approximation problems. We will describe the usual path spaces below, and we will discuss basic techniques which have been found to be very useful for getting the desired weak convergence. In applications, the abstract flavor of this discussion rapidly reduces to concrete and readily usable techniques. In fact, the tightness proofs and the characterization of the limit processes is all accomplished via "local" and checkable properties of the $X_n(\cdot)$ and $W_n(\cdot)$ (or of their analogs in other problems) which are obtainable from the defining equations.

Our motivation arises in problems associated with singular perturbations problems and with other modelling and approximation problems in stochastic control and systems theory. The techniques which we develop are particularly useful for such problems, and for many other classes of problems as well. The methods are, indeed, arguably the most efficient currently available for problems arising in control and communication theory. See also [E20] for lots of applications to these subjects. The field of weak convergence covers many other problem areas as well, and there is a wealth of results available [B6, E2, K20].

3. The Skorohod Representation

Suppose that $\{X_n, n < \infty\}$ is a sequence of S-valued random variables and $\{P_n,\ n < \infty\}$ the associated measures. Generally, the first step in the application of the ideas of weak convergence theory involves proving tightness of $\{P_n,\ n < \infty\}$. Once this tightness (i.e., relative compactness) is proved, one extracts a weakly convergent subsequence $\{P_{n_i}, i < \infty\}$ and tries to characterize the S-valued random variable X associated with the limit P. With our usual abuse of notation, we also say that we first prove tightness of $\{X_n\}$, then extract a weakly convergent subsequence $\{X_{n_i}, i < \infty\}$ and try to characterize or "identify" the limit X. If S is a "path space", so that X is actually a random process, then we wish to characterize the law of evolution of that process. Of course, strictly speaking X is not a limit of $\{X_{n_i}, i < \infty\}$ but is the random variable associated with the "weak limit" P. In the cases which are most important to us in this book, the X_n and X take their values in a path space; e.g., they might be the processes which are the solutions to differential equations which are "driven" by some sequence of stochastic processes indexed by n. The entire procedure can often be reduced to a relatively routine verification of conditions on the distributions of the noise driving the system and on the structure of the system. A great deal of information on this procedure, as well as a useful introduction to the ideas of weak convergence theory as it applies to the models and problems which are common in control and communications theory can be found in [K20].

We are concerned only with weak convergence and not with convergence with probability one (w.p.1). Because of this, the actual probability space on which the X_n and X are defined is not important and we can choose it in any convenient way, provided only that the distributions of the random variables of interest are preserved. In fact, the X_n might originally be defined on different probability spaces. We can always define them on a common probability space such that the probability law of each X_n is preserved. But there are many choices for such a space, since we can "correlate" the $\{X_n\}$ in many ways. Let $X_n \Rightarrow X$. For purposes of *characterizing* the limit X, it often simplifies the calculations if the X_n and X are defined on a common probability space with the additional property that the weakly convergent sequence $\{X_n, n < \infty\}$ actually converged to its limit X w.p.1 in the topology of the space S. We will see later in the book just how useful this can be.

It turns out that, in all applications of interest in this book, we can choose the common probability space so that w.p.1 convergence on that chosen space will be implied by the weak convergence. Such a choice of probability space is referred to as *Skorohod Representation*. The formal statement is given by the next theorem.

Theorem 3.1 [E2, Theorem 3.1.8 and Theorem 1.3 above]. *Let S be separable and let $P_n \Rightarrow P$ with X_n and X, resp., being the associated S-valued random variables. Then there is a probability space $(\tilde{\Omega}, \tilde{B}, \tilde{P})$ with S-valued random variables $\tilde{X}_n$, $\tilde{X}$ defined on it and such that (3.1) holds for each Borel set B:*

$$\tilde{P}(\tilde{X}_n \in B) = P(X_n \in B)$$

$$\tilde{P}(\tilde{X} \in B) = P(X \in B) \tag{3.1}$$

$$d(\tilde{X}_n, \tilde{X}) \to 0 \quad w.p.1.$$

Remark. Suppose that $X_n \Rightarrow X$, that S is separable and that the Skorohod representation is used for the common probability space. For notational simplicity, and without loss of generality, we will usually *not* use the tilde $\sim$ notation, and simply assume that the probability space was chosen such that $X_n \to X$ w.p.1 in the topology of S.

4. The Function Space $C^k[0,T]$

Let $C^k[0,T]$ denote the space of continuous R^k-valued functions on the interval $[0,T]$, with the sup norm used. Let $C^k[0,\infty)$ denote the space of continuous functions on the infinite interval $[0,\infty)$ with the metric $\rho(\cdot)$ defined by

$$\rho(x(\cdot), y(\cdot)) = \int_0^\infty (\exp -t) \min[1, \sup_{s \le t} |x(s) - y(s)|] dt.$$

If $k = 1$, then we write simply $C[0,T]$ or $C[0,\infty)$.

The next theorem gives a widely used criterion for tightness. In that theorem, $S = C^k[0,T]$ or $C^k[0,\infty)$, and the random variables X_n are actually random processes with paths in S. The proof of the theorem and other properties of the spaces $C^k[0,T]$ are to be found in [B6, Chapter 2].

Theorem 4.1. *For each integer n, let $x_n(\cdot)$ be a process with paths in $C^k[0,T]$. Then $\{x_n(\cdot), n < \infty\}$ (or, equivalently in our notation, the set of measures on $\mathcal{B}(C^k[0,T])$ induced by the $X_n(\cdot)$) is tight in $C^k[0,T]$ if and only if (i) and (ii) hold.*

(i) $\sup_n P(|x_n(0)| \ge N) \to 0$ *as* $N \to \infty$.

(ii) *For each $\epsilon > 0$ and $\eta > 0$ there are $\delta > 0$ and $N_0 < \infty$ such that*

$$P\left(\sup_{|t-s| \le \delta} |x_n(t) - x_n(s)| \ge \epsilon \right) \le \eta$$

for all $n \ge N_0$. A sufficient condition for (ii) *is that:*

(iii) *There are $a > 0$, $b > 0$, and $K < \infty$ such that*

$$E|x_n(t) - x_n(s)|^a \leq K|t-s|^{1+b}, \quad all\ n.$$

A set of processes $\{x_n(\cdot), n < \infty\}$ with paths in $C^k[0,\infty)$ is tight if and only if its restriction to the interval $[0,T]$ is tight for each $T < \infty$.

Theorem 4.2. *Let the sequence of processes $\{x_n(\cdot), n < \infty\}$ be tight in $C^k[0,\infty)$. Suppose that there is a continuous process $x(\cdot)$ such that the finite dimensional distributions of $\{x_n(\cdot), n = 1,2,\ldots\}$ converge to those of $x(\cdot)$. Then $x_n(\cdot) \Rightarrow x(\cdot)$.*

Remarks. Let us note the following. In the space $C^k[0,T]$ any bounded and equicontinuous set of functions is precompact (i.e., the closure of any such set is compact). A set B of equicontinuous functions is bounded if and only if the set $\{\phi(0) : \phi(\cdot) \in B\}$ is bounded. A set B in $C^k[0,T]$ is equicontinuous if and only if for each $\epsilon > 0$ there is a $\delta > 0$ such that for all $\phi(\cdot) \in B$, and $t, s \leq T$,

$$\sup_{|t-s|\leq\delta} |\phi(t) - \phi(s)| \leq \epsilon.$$

These remarks essentially imply the necessity and sufficiency of (i) and (ii) of the theorem. The inequality (iii) is called the Kolmogorov criterion. Suppose that it is satisfied for a fixed separable process $x(\cdot)$. Then the proof in [N1, Chapter III.5] shows that there is a continuous version of the process. In fact, that proof shows that for each $T < \infty$ there is a $K_T(\omega) < \infty$ w.p.1 such that for $t + \delta \leq T$ and $\beta < b/a$ we have

$$|x(t + \delta) - x(t)| \leq K_T(\omega)\delta^\beta.$$

When that same proof is applied to the sequence $\{x_n(\cdot), n < \infty\}$, it implies that there are $K_{T,n}(\omega)$ such that as $N \to \infty$ we have

$$\sup_n P(K_{T,n} \geq N) \to 0 \quad as \quad N \to \infty,$$

and

$$|x_n(t) - x_n(s)| \leq K_{T,n}(\omega)|t - s|^\beta$$

5. The Function Space $D^k[0,T]$

Let $D^k[0,T]$ (resp., $D^k[0,\infty)$) denote the space of R^k-valued functions on the interval $[0,T]$ (resp., on the interval $[0,\infty)$) which are right continuous and have left-hand limits.

In many applications, the processes of interest have their paths in $D^k[0,T]$ even if their "weak limits" have continuous paths. The processes

$W_n(\cdot)$ defined in Section 2 provide one standard example. In proving the weak convergence of the $\{W_n(\cdot),\ n < \infty\}$, it is possible to work with the piecewise linear interpolations, which are continuous processes. The weak limits are the same in both cases. Introducing the piecewise linear interpolation can be notationally awkward. But much more importantly, it is usually much easier to prove tightness in the space $D^k[0,T]$ than in the space $C^k[0,T]$. In fact (as will be commented on further below) if a sequence of processes $\{x_n(\cdot),\ n < \infty\}$ converges weakly to a process $x(\cdot)$ in $D^k[0,T]$, and if the discontinuities of the $x_n(\cdot)$ go to zero in some uniform sense as $n \to 0$, then the paths of $x(\cdot)$ must be continuous w.p.1. The usual metric which is put on $D^k[0,T]$ and the one which we will always use is known as the *Skorohod metric*.

The Skorohod metric [E2, Chapter 3.5], [B6, Chapter 3]. Let Λ denote the space of continuous and strictly increasing functions from the interval $[0,T]$ onto the interval $[0,T]$. The functions in this set will be "allowable time scale distortions". Actually, there are two equivalent (in the sense of inducing the same topology on $D^k[0,\infty)$) metrics which are known as the Skorohod metric. The first Skorohod metric $d_T(\cdot)$ is defined by

$$d_T(f(\cdot), g(\cdot)) = \inf\{\epsilon : \sup_{0 \leq s \leq T} |s - \lambda(s)| \leq \epsilon \text{ and }$$

$$\sup_{0 \leq s \leq T} |f(s) - g(\lambda(s))| \leq \epsilon \text{ for some } \lambda(\cdot) \in \Lambda\}.$$

This Skorohod metric $d_T(\cdot)$ has the following properties. If $f_n(\cdot) \to f(\cdot)$ in $d_T(\cdot)$ where $f(\cdot)$ is continuous, then the convergence must be uniform on $[0,T]$. If there are $\eta_n \to 0$ such that the discontinuities of $f_n(\cdot)$ are less than η_n in magnitude and if $f_n(\cdot) \to f(\cdot)$ in $d_T(\cdot)$, then the convergence is uniform on $[0,T]$ and $f(\cdot)$ must be continuous. Because of the "time scale distortion" which is involved in the definition of the metric $d_T(\cdot)$, we can have (loosely speaking) convergence of a sequence of discontinuous functions where there are only a finite number of discontinuities, where both the locations and the values of the discontinuities converge, and a type of "equicontinuity" condition holds between the discontinuities. For example, define $f(\cdot)$ by: $f(t) = 1$ for $t < 1$ and $f(t) = 0$ for $t \geq 1$. Define the function $f_n(\cdot)$ by $f_n(t) = f(t + 1/n)$. Then $f_n(\cdot)$ converges to $f(\cdot)$ in the Skorohod topology, but not in the sup norm. These and many other properties of the Skorohod metric can be found in [B6, E2].

Under the metric $d_T(\cdot)$, the space $D^k[0,T]$ is separable but not complete [B6, Chapter 3]. Owing to Theorem 1.2, it is important to have the completeness property. There is an equivalent metric $d'_T(\cdot)$ under which the space is both complete and separable. We will refer to both $d'_T(\cdot)$ and $d'(\cdot)$ as the Skorohod metric. The $d'_T(\cdot)$ weights the "derivative" of the time scale changes $\lambda(t)$ as well as its deviation from t. For $\lambda(\cdot) \in \Lambda$ define

$$|\lambda| = \sup_{s<t<T} \left| \log\left\{ \frac{\lambda(t) - \lambda(s)}{t - s} \right\} \right|.$$

The metric $d'_T(\cdot)$ is defined by

$$d'_T(f(\cdot), g(\cdot)) = \inf\{\epsilon : |\lambda| \le \epsilon \text{ and } \sup_{0 \le s \le T} |f(s) - g(\lambda(s))| \le \epsilon,$$

$$\text{for some } \lambda(\cdot) \in \Lambda\}$$

On the space $D^k[0,\infty)$, the Skorohod metric is defined by

$$d'(f(\cdot), g(\cdot)) = \int_0^\infty (\exp -t) \min(1, d'_t(f(\cdot), g(\cdot))) dt.$$

We will also have occasion to use the fact that if $x(\cdot)$ is a process with paths in $D^k[0,T]$, then there are at most a countable number of points $t \in (0,\infty)$ at which there is a positive probability of a discontinuity. I.e., where $P(x(t) \ne x(t^-)) > 0$. Let T_P denote the complement of this countable set. Fix $t > 0$ and let $F(\cdot)$ be the real valued function on $D[0,T]$ which takes the values $F(f) = f(t)$. Then $F(\cdot)$ is not continuous at the point $f(\cdot)$ in the Skorohod topology unless $f(\cdot)$ is continuous at t. This remark together with Theorem 1.4 implies the following fact. Let $x_n(\cdot) \Rightarrow x(\cdot)$ in $D^k[0,T]$, and let $x(\cdot)$ induce the measure P on $D^k[0,T]$. Let $\{t_i, i \le q\} \in T_P$. Then we have the convergence of the finite dimensional distributions: $\{x_n(t_i), i \le q\} \Rightarrow \{x(t_i), i \le q\}$. A partial converse to this statement is contained in Theorem 5.2 below. In applications, T_P is not usually known until $x(\cdot)$ is characterized. But this is usually unimportant in the analysis, since the $x(\cdot)$ with paths in $D^k[0,\infty)$ are characterized by the distribution of $\{x(t), t \in$ any countable dense set$\}$. See Chapter 3.4 for a typical application. The set T_P is only of technical use; for use in proofs. Sometimes, we need to exclude points not in T_P from an argument. The set T_P exists whether known or not. In the applications in this book $T_P = (0,\infty)$, although that will not be known until the limit processes are actually characterized.

The spaces $D[S_0; 0, T]$ and $D[S_0; 0, \infty)$. Let S_0 denote a metric space with metric $\rho(\cdot)$. We will occasionally need to work with processes $x(\cdot)$ with values in some spaces S_0 which are not Euclidean. Let $D[S_0; 0, T]$ denote the space of S_0-valued functions on the interval $[0, T]$ which are right continuous, have left-hand limits and with the Skorohod metric defined by the $d'_T(\cdot)$ above, but with $\rho(f(s), g(\lambda(s)))$ used in place of $|f(s) - g(\lambda(s))|$, where both $f(\cdot)$ and $g(\cdot)$ are now points in $D[S_0; 0, T]$. Define the space $D[S_0; 0, \infty)$ analogously. If $S_0 = R^k$, then we return to the original notation and write $D^k[0, T]$ or $D^k[0, \infty)$. If S_0 is complete and separable then so are $D[S_0; 0, T]$ and $D[S_0; 0, \infty)$.

The following criterion for tightness, due to Aldous and Kurtz [K9, Theorem 2.7b], seems to be the most convenient one for our purposes.

Theorem 5.1. *Let $x_n(\cdot)$ be processes with paths in $D[S_0; 0, \infty)$, where S_0 is separable and complete. For each $\delta > 0$ and rational $t < \infty$, let there be a compact set $S_{\delta,t} \subset S_0$ such that $\sup_n P(x_n(t) \notin S_{\delta,t}) \le \delta$. Let $\mathcal{F}_t^n$ be the σ-algebra determined by $\{x_n(s), s \le t\}$ and $T_n(T)$ the set of $\mathcal{F}_t^n$-stopping times which are no bigger than T. Suppose that*

$$\lim_{\delta \to 0} \limsup_n \sup_{\tau \in T_n(T)} E \min\{1, \rho(x_n(\tau + \delta), x(\tau))\} = 0$$

for each $T < \infty$. Then $\{x_n(\cdot), n < \infty\}$ is tight in $D[S_0; 0, \infty)$.

The following theorem is analogous to Theorem 4.2.

Theorem 5.2 [B6, Theorem 15.1]. *Let $\{x_n(\cdot), n < \infty\}$ be a tight sequence of processes with paths in $D^k[0, \infty)$. Let $x(\cdot)$ be a process with paths in $D^k[0, \infty)$ and which induces the measure P. Suppose that for arbitrary q and $t_1, \ldots, t_q \in T_P$,*

$$\{x_n(t_i), i \le q\} \Rightarrow \{x(t_i), i \le q\}.$$

Then $x_n(\cdot) \Rightarrow x(\cdot)$.

6. Measure Valued Random Variables and Processes

Measure valued random variables. We will have a number of occasions to work with random variables whose sample values are measures, particularly in Chapters 5 and 6. Let S_0 denote a complete and separable metric space. If S_0 is compact, then $\mathcal{P}(S_0)$ is also compact in the Prohorov topology (or, equivalently, in the weak topology) and any sequence in $\mathcal{P}(S_0)$ has a weakly convergent subsequence. Let $\{v_\alpha\}$ be a set measures on $(\mathcal{B}(S_0), S_0)$ with uniformly bounded total mass. Suppose that for each $\delta > 0$ there is a compact set $S_\delta \subset S_0$ such that $\sup_\alpha v_\alpha(S_0 - S_\delta) < \delta$. Then $\{v_\alpha\}$ is tight. Hence, by Prohorov's Theorem it is precompact in the Prohorov topology, and has a subsequence which converges in the Prohorov topology (therefore, also in the weak topology by Theorem 1.3).

Let $\{Q_n\}$ denote a sequence of $\mathcal{P}(S_0)$-valued *random variables*. By definition, $Q_n \Rightarrow Q$ (here, we speak of weak convergence of the sequence of $\mathcal{P}(S_0)$-valued random variables) if and only if for each $F(\cdot) \in C_b(\mathcal{P}(S_0))$, $EF(Q_n) \to EF(Q)$. The following assertion is a consequence of Prohorov's Theorem (Theorem 1.2). Note that EQ_n is a measure on $(\mathcal{B}(S_0), S_0)$.

Theorem 6.1. *Suppose that there are nondecreasing (as $m \to \infty$) compact sets $S_m \subset S_0$ such that*

$$\limsup_m \sup_n P(Q_n(S_0 - S_m) > 1/m) = 0, \tag{6.1}$$

or equivalently that there are compact S_m such that

$$\limsup_{m} \, _n EQ_n(S_0 - S_m) = 0. \qquad (6.2)$$

Then $\{Q_n\}$ is tight (in the sense of a sequence of random variables) and has a weakly convergent subsequence.

Remark. According to the theorem, tightness of the sequence of measure valued random variables $\{Q_n\}$ is equivalent to tightness of the sequence of measures $\{EQ_n\}$.

Proof. Note that (6.1) and (6.2) are equivalent statements. Given any $\delta > 0$, we need to find a compact set $K_\delta \in \mathcal{P}(S_0)$ such that $\sup_n P(Q_n \notin K_\delta) \leq \delta$. By (6.1), for each $\epsilon > 0$ there is a compact S_ϵ such that $P(Q_n(S_0 - S_\epsilon) \geq \epsilon) \leq \epsilon$, for all n. Choose $\epsilon_i \to 0$ such that $\Sigma_i \epsilon_i < \infty$. Given $\delta > 0$, choose $m_\delta < \infty$ such that $\sum_{m_\delta}^{\infty} \epsilon_i \leq \delta$. Define the set $K_\delta = \{\hat{P} \in \mathcal{P}(S_0): \hat{P}(S_0 - S_{\epsilon_i}) \leq \epsilon_i, \; i \geq m_\delta\}$. By construction, K_δ is precompact, since it is tight. In particular, for each $\rho > 0$ there is a compact set S_ρ such that $\hat{P}(S_0 - S_\rho) < \rho$ all $\hat{P} \in K_\rho$. Since $P(Q_n \notin K_\delta) \leq \delta$ for all n, the proof is complete.   Q.E.D.

The following examples illustrate some of the particular applications which will be of interest later in the book.

Example 1. Let $\mathcal{U}$ be a compact set in some Euclidean space and define $S_0 = \mathcal{U} \times [0, T]$, $T < \infty$, with the Euclidean metric used on S_0. Then S_0 is compact and any set $\{Q_\alpha\}$ of $\mathcal{M}(S_0)$-valued random variables has a weakly convergent subsequence if

$$\limsup_{k} \, _\alpha P(Q_\alpha(S_0) \geq k) = 0.$$

Example 2. Consider $S_0 = \mathcal{U} \times [0, \infty)$, where $\mathcal{U}$ is as in Example 1 but where we use the "compactified" metric on $\mathcal{M}(S_0)$ defined by $v_n \to v$ if and only if $(v_n, f) \to (v, f)$ for all $f(\cdot) \in C_0(S_0)$. Then $\mathcal{M}(S_0)$ is a complete and separable metric space. Any set $\{Q_\alpha\}$ of $\mathcal{M}(S_0)$-valued random variables has a weakly convergent subsequence if for each $T < \infty$

$$\limsup_{k} \, _\alpha P(Q_\alpha(\mathcal{U} \times [0, T]) \geq k) = 0.$$

Example 3. Let $S_0 = D^k[0, \infty)$. Let $\{Q_\alpha\}$ be a set of random variables with values in $\mathcal{P}(S_0)$. By Theorem 6.1, if for each $\delta > 0$ there is a compact set $S_\delta \in S_0$ with

$$\limsup_{\delta} \, _\alpha EQ_\alpha(S_0 - S_\delta) = 0,$$

then Q_α will have a weakly convergent subsequence in $\mathcal{P}(S_0)$.

Let us examine this example in more detail to see how tightness might be
proved. Each $\overline{Q}_\alpha \equiv EQ_\alpha$ is a probability measure on S_0. Let $\overline{X}_\alpha(\cdot)$ denote
the process induced on S_0 by $\overline{Q}_\alpha$. Then tightness of $\{Q_\alpha\}$ is equivalent to
tightness of $\{EQ_\alpha\}$ which, in turn, is equivalent to tightness of $\{\overline{X}_\alpha(\cdot)\}$.

Measure valued random processes. In Chapter 6 on nonlinear filtering theory, we will have occasion to deal with measure-valued random processes, where the measures of the entire space are not necessarily normalized
to be unity. The measures of concern will be related to the unnormalized
conditional distributions which occur in nonlinear filtering theory. We will
now state some of the results which will be needed there.

Let S be a compact metric space and let $\mathcal{M}(S)$ denote the set of measures
on $(S, \mathcal{B}(S))$ with finite total mass and with the Prohorov topology used.
In the book, we will use only the compact set $S = \overline{R}^k$. The space $\overline{R}^k$
will be used in lieu of R^k in Chapter 6 because it allows some notational
simplification in dealing with a particular convergence problem there. It
is possible that a measure $P_0 \in \mathcal{M}(\overline{R}^k)$ have "mass at infinity," since the
"point of infinity" is in $\overline{R}^k$, i.e., it might be that $\lim_N P_0(x: |x| \geq N) > 0$.
But such an occurence will be ruled out in the applications.

The space $\mathcal{M}(S)$ is also complete and separable. All the results of Section
1 continue to hold. In particular, the sets in $\mathcal{M}(S)$ which have uniformly
bounded total mass are tight and relatively compact, and weak convergence and convergence in the Prohorov topology are equivalent. Thus the
topology of weak convergence and the Prohorov topology are equivalent,
and the compact sets are the same in both topologies. The topology of
weak convergence on $\mathcal{M}(S)$ can be metrized as follows. Let $\{\phi_i(\cdot)\}$ be a
countable dense set in $C(S)$. For v and v' in $\mathcal{M}(S)$, use the metric

$$d(v, v') = \sum_i 2^{-i}|(v - v', \phi_i)|/(1 + |(v - v', \phi_i)|).$$

This equivalence between the topologies implies the following result.

Theorem 6.2. *Let $\{Z_\alpha\}$ be a set of random processes with values in*
$D[\mathcal{M}(\overline{R}^k); 0, \infty)$. *Suppose that $\{(Z_\alpha(\cdot), \phi)\}$ is tight in $D[0, \infty)$ for each*
$\phi(\cdot) \in C(\overline{R}^k)$. *Then $\{Z_\alpha(\cdot)\}$ has a weakly convergent subsequence.*

Remark. By Theorem 5.1, the set $\{Z_\alpha(\cdot)\}$ is tight if for each $T < \infty$,
$\sup_{\alpha, t \leq T} EZ_\alpha(t, \overline{R}^k) < \infty$ holds, and for each $T < \infty$ and $\phi(\cdot) \in C(\overline{R}^k)$ we
have

$$\lim_{\delta \to 0} \sup_\alpha \sup_{\tau \leq T} E|(Z_\alpha(\tau + \delta) - Z_\alpha(\tau), \phi)| = 0. \qquad (6.3)$$

Here, τ denotes a stopping time.

2

Stochastic Processes: Background

0. Outline of the Chapter

In this chapter, we survey some of the concepts in the theory of stochastic processes which will be used in the sequel. The material is intended to be only a discussion of some of the main ideas which will be needed and to serve as a convenient reference. It is not intended to be complete in any sense. Additional details, motivation and background can be found in the references. The reader familiar with the basic facts concerning stochastic differential equations can skip the chapter and refer to the results as needed. Section 1 gives some definitions and facts concerning martingales. In Section 2, we define stochastic integrals with respect to a Wiener process and state Itô's Formula, the fundamental tool in the stochastic calculus. Section 3 uses the stochastic integrals to define stochastic differential equations (SDE) of the diffusion type, and obtains bounds on their solutions which will be used later in the study of properties of the solution processes, as well as in the proofs of tightness of sequences of solutions to such equations. In Section 4, we discuss three standard methods for obtaining existence and uniqueness of solutions to SDE's. These sections give some flavor of a few of the basic ideas. But they just touch the surface of the subject. Sections 5 and 6 concern an alternative and very useful method for verifying whether a process satisfies an SDE of the diffusion type. The so-called "martingale problem" method which is discussed in these sections is perhaps the most useful approach to showing that the limit of a weakly convergent sequence of processes is, indeed, a solution to a SDE. This "martingale problem" characterization of the solutions will be used frequently. In Sections 7 and 8, we redo the concepts of Sections 5 and 6 for jump-diffusion processes. The results of Sections 5, 6 and 8 will be used mainly to prove that certain processes which appear in the analysis are, indeed, solutions to SDE's.

1. Martingales

Let $(\Omega, \mathcal{F}, P)$ denote a probability space and $\{\mathcal{F}_t, t < \infty\}$ a family of σ-algebras satisfying $\mathcal{F}_t \subset \mathcal{F}_{t+s} \subset \mathcal{F}$ for all $t, s > 0$. The family $\{\mathcal{F}_t, t < \infty\}$ is called a *filtration*. All the random processes considered in this book will have values in some metric space S_0 and have versions which are right continuous and have left hand limits (w.p.1). I.e., their paths will always be in $D[S_0; 0, \infty)$. (See Chapter 1.5.) We always suppose that we are working with such a version. Unless otherwise specified, we suppose that $S_0 = R^k$, Euclidean k-space for some $k \geq 1$. Let $E_{\mathcal{F}_t}$ and $P_{\mathcal{F}_t}$ denote expectation and probability, conditioned on $\mathcal{F}_t$. Let $x(\cdot)$ be a vector-valued (i.e., R^k-valued) process such that $x(t)$ is $\mathcal{F}_t$-measurable, $E|x(t)| < \infty$, and

$$E_{\mathcal{F}_t} x(t + s) = x(s) \quad \text{w.p.1, all } t, s > 0. \tag{1.1}$$

Then $x(\cdot)$ is said to be an $\mathcal{F}_t$-*martingale*, or simply a *martingale* if the $\mathcal{F}_t$ is either obvious or unimportant. If the equality in (1.1) is replaced by the inequality $\leq$, then $x(\cdot)$ is said to be an $\mathcal{F}_t$-*supermartingale*. A random variable τ with values in $[0, \infty]$ is said to be an $\mathcal{F}_t$-*stopping time* if $\{\tau \leq t\} \in \mathcal{F}_t$ for each $t < \infty$. If τ is undefined at some ω, we always set it equal to infinity there.

Let $x(\cdot)$ be a vector-valued random process, such that $x(t)$ is $\mathcal{F}_t$-measurable for each t. Suppose that there is a sequence of $\mathcal{F}_t$-stopping times $\{\tau_n, n < \infty\}$ such that $\tau_n \to \infty$ w.p.1 as $n \to \infty$ and such that for each n the process $x(\cdot \wedge \tau_n)$ is an $\mathcal{F}_t$-martingale. Then $x(\cdot)$ is said to be a *local $\mathcal{F}_t$-martingale*, or simply a *local martingale*, if the $\mathcal{F}_t$ is obvious or unimportant.

The following inequalities will be used often in the book. Let $x(\cdot)$ be an $\mathcal{F}_t$-martingale. Then [D5] for any $\lambda > 0$ and $t \leq T$,

$$P_{\mathcal{F}_t} \left(\sup_{t \leq s \leq T} |(x(s)| \geq \lambda \right) \leq E_{\mathcal{F}_t} x^2(T)/\lambda^2 \quad \text{w.p.1,} \tag{1.2}$$

$$E_{\mathcal{F}_t} \sup_{t \leq s \leq T} |x(s)|^2 \leq 4 E_{\mathcal{F}_t} |x(T)|^2 \quad \text{w.p.1.} \tag{1.3}$$

If $x(\cdot)$ is a non-negative $\mathcal{F}_t$-supermartingale, then

$$P_{\mathcal{F}_t} \left(\sup_{t \leq s \leq T} x(s) \geq \lambda \right) \leq x(t)/\lambda, \quad \text{w.p.1.} \tag{1.4}$$

The Wiener process. Let $w(\cdot)$ be a real valued process such that $w(0) = 0$ and for each n and set of real numbers $0 \leq t_0 < t_1 < \ldots < t_n$, the random variables $\{w(t_{i+1}) - w(t_i), i \leq n\}$ are mutually independent. If $w(t_{i+1}) - w(t_i)$ is also normally distributed with mean zero and variance $\sigma^2(t_{i+1} - t_i)$, where $\sigma^2 > 0$, then there is a continuous version of the process $w(\cdot)$ and it is known as the *Wiener process* [B9]. If $\sigma = 1$, then $w(\cdot)$ is said to be a *standard Wiener process*. A finite set of mutually independent (standard) Wiener processes is called a *(standard) vector-valued Wiener process*.

For any random process $x(\cdot)$, let $\mathcal{B}(x(s), s \leq t)$ denote the minimal σ-algebra which measures $\{x(s), s \leq t\}$. If $w(\cdot)$ is a real-valued Wiener process, then it is easy to see that $w(\cdot)$ is an $\tilde{\mathcal{F}}_t$-martingale, where $\tilde{\mathcal{F}}_t = B(w(s),\ s \leq t)$. Also, the process defined by $M(t) = w^2(t) - t$ is an $\tilde{\mathcal{F}}_t$-martingale, since for all $t, s > 0$,

$$E_{\tilde{\mathcal{F}}_t}[w^2(t+s) - w^2(t)] = E_{\tilde{\mathcal{F}}_t}[2w(t)(w(t+s) - w(t)) + (w(t+s) - w(t))^2]$$
$$= s$$

by the independence of $w(t+s) - w(t)$ and $\{w(u),\ u \leq t\}$. Similarly, if $w(t)$ is a vector-valued Wiener process, then the process defined by $M(t) = w(t)w'(t) - tI$ is a matrix-valued $\tilde{\mathcal{F}}_t$-martingale.

There is a very important converse to these properties of the Wiener process. Let $\mathcal{F}_t$ denote a filtration and let both $w(\cdot)$ and $w^2(t) - t$ be continuous real-valued $\mathcal{F}_t$-local martingales. Then [I1, K8] $w(\cdot)$ is a standard Wiener process. Analogously, if $w(\cdot)$ is a continuous vector valued $\mathcal{F}_t$-local martingale and the process defined by $w(t)w'(t) - tI$ is a matrix valued $\mathcal{F}_t$-local martingale, then $w(\cdot)$ is a vector-valued standard Wiener process [I1, K8]. This is a widely used and very important characterization of a Wiener process. If the filtration $\mathcal{F}_t$ is important, we then refer to $w(\cdot)$ as a *standard* $\mathcal{F}_t$-Wiener process or as a standard *vector-valued* $\mathcal{F}_t$-Wiener process, as the case may be. Unless the adjective "vector" is used, it is understood that the Wiener process $w(\cdot)$ is real valued.

2. Stochastic Integrals and Itô's Lemma

Stochastic integrals with respect to a Wiener process

Definitions. Let the probability space be $(\Omega, \mathcal{F}, P)$. Let $B(T)$ denote the Borel σ-algebra over the interval $[0, T]$, and let $\mathcal{F} \times B(T)$ denote the minimal σ-algebra over the product sets in $\mathcal{F}$ and $B(T)$. Let $\mathcal{F}_t$ be a given filtration and $w(\cdot)$ a standard $\mathcal{F}_t$-Wiener process. A random process $x(\cdot)$ which is defined on the time interval $[0, T]$, $T \leq \infty$, is said to be *measurable* if, when considered as a function of (ω, t), it is $\mathcal{F} \times B(T)$-measurable. If $x(\cdot)$ is measurable and $X(t)$ is $\mathcal{F}_t$-measurable for each $t \geq 0$, then $x(\cdot)$ is said to be $\mathcal{F}_t$-*adapted*. If $w(\cdot)$ is an $\mathcal{F}_t$-Wiener process and $x(\cdot)$ is $\mathcal{F}_t$-adapted, then we say that $x(\cdot)$ is *non-anticipative with respect to* $w(\cdot)$, since $\{x(u),\ u \leq t\}$ and $\{w(s) - w(t),\ s \geq t\}$ are independent for all t. An $\mathcal{F}_t$-adapted random process on the interval $[0, T]$, $T \leq \infty$, is said to be *progressively measurable* if, when restricted to any interval $[0, T_1]$, $T_1 \leq T$, it is $\mathcal{F}_{T_1} \times B(T_1)$-measurable. Any $\mathcal{F}_t$-adapted continuous process is progressively measurable. Indeed, any $\mathcal{F}_t$-adapted process which is left continuous or is the limit of a sequence of left continuous processes is progressively measurable [I1, Chapter 5]. For any $\mathcal{F}_t$-adapted process $x(\cdot)$, there is a progressively measurable process $x'(\cdot)$ such that $x(t) = x'(t)$ for almost all (ω, t) [M1]. In any case, in the applications in this book, the functions that will be required to be progessively measurable will actually be continuous.

Let $\Phi_2(T)$ denote the set of $\mathcal{F}_t$-adapted progressively measurable real valued processes $\phi(\cdot)$ such that $E \int_0^T \phi^2(s)ds < \infty$. Let Φ_2 denote the set of processes which are in $\Phi_2(T)$ for each $T < \infty$, and Φ the set of $\mathcal{F}_t$-adapted progressively measurable processes for which $\int_0^T \phi^2(s)ds < \infty$ w.p.1 for each T. We say that a process $\phi(\cdot)$ in Φ_2 is a *step function* if there are $\mathcal{F}_t$-stopping times $\{\tau_n, n < \infty\}$ such that $\tau_0 = 0$, $\lim_n \tau_n = \infty$ w.p.1 and $\phi(t) = \phi(\tau_n)$ for $t \in [\tau_n, \tau_{n+1})$. Let Φ_{2s} denote the set of step functions in Φ_2.

Stochastic integrals. For $\phi(\cdot) \in \Phi_{2s}$, we *define* the *stochastic integral* of $\phi(\cdot)$ with respect to $w(\cdot)$ (to be simply referred to as the stochastic integral of $\phi(\cdot)$) to be the process $\psi(\cdot)$ with values

$$\psi(t) = \sum_{i=0}^{n-1} \phi(\tau_i)[w(\tau_{i+1}) - w(\tau_i)] + \phi(\tau_n)[w(t) - w(\tau_n)], \ t \in [\tau_n, \tau_{n+1}).$$

The stochastic integral of $\phi(\cdot)$ is often written in the 'integral' notation $\psi(t) = \int_0^t \phi(s)dw(s)$. The 'differential' notation of $d\psi = \phi(t)dw + f(t)dt$ is also used to indicate that $\psi(t)$ is the sum of the stochastic integral of $\phi(\cdot)$ and the Lebesgue integral of $f(\cdot)$.

Let $\phi(\cdot)$ and $\phi_i(\cdot)$, $i = 1, 2$, be in Φ_{2s} and let τ be a bounded $\mathcal{F}_t$-stopping time. By the definition of $\psi(\cdot)$, it is easily verified that it is a continuous $\mathcal{F}_t$-martingale and has the following properties: Let $t > s > \tau$. Then

$$E_{\mathcal{F}_s} \int_0^t \phi(u)dw(u) = \int_0^s \phi(u)dw(u) \ \text{w.p.1,} \tag{2.1}$$

$$E_{\mathcal{F}_\tau} \int_0^t \phi_1(u)dw(u) \int_0^s \phi_2(u)dw(u) = \int_0^s E_{\mathcal{F}_\tau}\phi_1(u)\phi_2(u)du$$

$$= \int_0^\tau \phi_1(u)\phi_2(u)du + \int_\tau^s E_{\mathcal{F}_\tau}\phi_1(u)\phi_2(u)du \ \text{w.p.1,} \tag{2.2}$$

$$E \left[\int_0^t \phi_1(u)dw(u) - \int_0^t \phi_2(u)dw(u) \right]^2 = \int_0^t E[\phi_1(u) - \phi_2(u)]^2 du, \tag{2.3}$$

$$\int_0^t \phi_1(u)dw(u) + \int_0^t \phi_2(u)dw(u) = \int_0^t [\phi_1(u) + \phi_2(u)]dw(u) \ \text{w.p.1,} \tag{2.4}$$

$$\int_0^{t\wedge\tau} \phi(u)dw(u) = \int_0^t I_{\{u\leq\tau\}}\phi(u)dw(u). \tag{2.5}$$

By (1.2), (1.3) and (2.3), we have

$$P\left(\sup_{t\leq T} |\psi(t)| \geq \lambda\right) \leq E|\psi(T)|^2/\lambda^2 = \int_0^T E\phi^2(u)du/\lambda^2, \tag{2.6a}$$

$$E \sup_{t \leq T} \psi^2(t) \leq 4 \int_0^T E\phi^2(u)du. \tag{2.6b}$$

With the use of the above properties we are now in a position to define the stochastic integral with respect to any $\phi(\cdot) \in \Phi_2$. Full details appear in [I1, Chapter 2], [G2, Chapter 8]. For each $\phi(\cdot) \in \Phi_2$, there is a sequence $\{\phi_n(\cdot), n < \infty\} \subset \Phi_2$, such that for each $T < \infty$

$$\int_0^T E|\phi_n(s) - \phi(s)|^2 ds \to 0 \quad \text{as} \quad n \to \infty.$$

Define $\psi_n(t) = \int_0^t \phi_n(s)dw(s)$. By (2.3), as n and m go to infinity,

$$E|\psi_n(t) - \psi_m(t)|^2 = \int_0^t E|\phi_n(s) - \phi_m(s)|^2 ds \to 0. \tag{2.7}$$

By (1.2) and (2.3)

$$P\left(\sup_{t \leq T} |\psi_n(t) - \psi_m(t)| \geq \lambda\right) \leq \int_0^T E|\phi_n(s) - \phi_m(s)|^2 ds/\lambda^2. \tag{2.8}$$

By (2.7), for each t, $\{\psi_n(t), n < \infty\}$ is a Cauchy sequence. Thus for each t there is a random variable $\psi(t)$ which is $\mathcal{F}_t$-measurable and is such that $E|\psi_n(t) - \psi(t)|^2 \to 0$. In fact, by (2.8) and the continuity of each $\psi_n(\cdot)$, we can choose $\psi(t)$ for each t such that $\psi(\cdot)$ is a continuous process w.p.1. The continuous limit $\psi(\cdot)$ does not depend on the approximating sequence $\{\phi_n(\cdot), n < \infty\}$ in the sense that the continuous limit processes associated with two different approximating sequences must be equal w.p.1. Using (1.3), it can also be shown that

$$E \sup_{t \leq T} |\psi_n(t) - \psi(t)|^2 \xrightarrow[n]{} 0 \quad \text{for each} \quad T < \infty.$$

We define the stochastic integral of $\phi(\cdot)$ to be the continuous limit process $\psi(\cdot)$, and write it either as $\psi(t) = \int_0^t \phi(u)dw(u)$ or as $d\psi = \phi\,dw$. The properties (2.1)–(2.6) continue to hold and $\psi(\cdot)$ is an $\mathcal{F}_t$-martingale.

By a "localization" method, the stochastic integral can be defined for all functions in Φ. This is done as follows. Let $\phi(\cdot) \in \Phi$. Then, by the definition of Φ, there is a sequence of $\mathcal{F}_t$-stopping times $\{\tau_n, n < \infty\}$ such that $\tau_n \to \infty$ w.p.1 and the processes with values $\phi(t)I_{\{t \leq \tau_n\}}$ are in Φ_2. For example, use

$$\tau_n = \min\left\{t: \int_0^t \phi^2(u)du = n\right\}.$$

Let $\psi_n(\cdot)$ denote the stochastic integral of $\phi_n(\cdot)$. Define the stochastic integral of $\phi(\cdot)$ by $\psi(\cdot) = \lim_n \psi_n(\cdot)$. The limit does not depend (w.p.1) on the particular $\{\tau_n, n < \infty\}$ which are used. The process $\phi(\cdot)$ is a local $\mathcal{F}_t$-martingale and Equations (2.4) and (2.5) continue to hold.

The vector case. Let $w(\cdot)$ be a standard R^k-valued $\mathcal{F}_t$-Wiener process, and let $\phi(\cdot) = \{\phi_{ij}(\cdot), i \le r, j \le k\}$ be a $(r \times k)$ matrix-valued process with components in Φ. We define the stochastic integral $\psi(\cdot)$ of $\phi(\cdot)$ to be the vector $\psi(t) = (\psi_i(t), i \le r)$, where

$$\psi_i(t) = \sum_{j=1}^{k} \int_0^t \phi_{ij}(u)dw_j(u) = \int_0^t \sum_{j=1}^{k} \phi_{ij}(u)dw_j(u).$$

Again, the notation $d\psi = \phi dw$ will also be used to denote $\psi(\cdot)$.

Itô's formula. Let $g(\cdot)$ be an integrable real-valued function on $[0, T]$ and define the integral $G(t) = \int_0^t g(s)ds$. For $f(\cdot) \in C^1(R)$, the fundamental theorem of the calculus states

$$F(t) \equiv f(G(t)) - f(G(0)) = \int_0^t f_x(G(s))g(s)ds \equiv \int_0^t dF(s). \qquad (2.9)$$

The integral-differential relationship (2.9) is often written in the "symbolic" differential form: $dF(t) = f_x(G(t))dG(t)$. The relationship (2.9) is one of the cornerstones of real analysis. Itô's Formula plays the same fundamental role in stochastic analysis. Let $f(\cdot) \in C^2(R)$. Let $x(0)$ be $\mathcal{F}_0$-measurable, let $\phi(\cdot) \in \hat{\Phi}$, and let $g(\cdot)$ be real valued and $\mathcal{F}_t$-adapted and satisfy $\int_0^T |g(s)|ds < \infty$ w.p.1 for each $T < \infty$. Define the process $x(\cdot)$ by

$$x(t) = x(0) + \int_0^t \phi(s)dw(s) + \int_0^t g(s)ds. \qquad (2.10)$$

Itô's Formula [I1, Chapter 2.5], [G2, p. 387] states that

$$f(x(t)) - f(x(0)) = \int_0^t f_x(x(s))dx(s) + \frac{1}{2}\int_0^t f_{xx}(x(s))\phi^2(s)ds, \qquad (2.11)$$

where we use the notation

$$\int_0^t f_x(x(s))dx(s) = \int_0^t f_x(x(s))\phi(s)dw(s) + \int_0^t f_x(x(s))g(s)ds. \qquad (2.12)$$

Since the function $f_x(x(\cdot))\phi(\cdot)$ is in Φ, the stochastic integral in (2.12) is well defined. (2.11) is an equality in that any continuous version of the two sides must be equal w.p.1 for all $t < \infty$. The formula (2.11) is one of the most powerful tools in stochastic analysis.

We next state the *vector version* of Itô's Formula (2.11). Let $w(\cdot)$ be a standard R^k-valued $\mathcal{F}_t$-Wiener process, let $\phi(\cdot) = \{\phi_{ij}(\cdot), i \le r, j \le k\}$

where $\phi_{ij}(\cdot) \in \Phi$. Let $g_i(\cdot)$, $i \leq r$, be real-valued and $\mathcal{F}_t$-adapted processes satisfying $\int_0^t |g_i(s)| ds < \infty$ w.p.1 for each $t < \infty$, and define $g(\cdot) = (g_i(\cdot), i \leq r)$. Define $x(\cdot)$ by (2.10). For $f(\cdot) \in C^2(R^r)$, define $F(t) = f(x(t))$. Then *Itô's Formula* states:

$$F(t) - F(0) = f(x(t)) - f(x(0)) = \int_0^t f'_x(x(s)) dx(s)$$

$$+ \frac{1}{2} \int_0^t \sum_{i,j} f_{x_i x_j}(x(s)) \left[\sum_\alpha \phi_{i\alpha}(s) \phi_{j\alpha}(s) \right] ds. \qquad (2.13)$$

The first integral on the right of (2.13) is defined to be

$$\int_0^t f'_x(x(s)) \phi(s) dw(s) + \int_0^t f'_x(x(s)) g(s) ds, \qquad (2.14)$$

(i.e., $dx \equiv \phi dw + g dt$). The last term on the right of (2.13) can also be written as

$$\frac{1}{2} \int_0^t \text{trace}[f_{xx}(x(s)) \cdot \phi(s)\phi'(s)] ds,$$

a representation which will be used often. The "differential form" notation

$$dF(t) = f'_x(x(s)) dx(s) + \frac{1}{2} \text{trace}[f_{xx}(x(s)) \cdot \phi(t)\phi'(t)] dt \qquad (2.15)$$

will also be used to denote that (2.13) holds.

3. Stochastic Differential Equations: Bounds

Introduction. Let $w(\cdot)$ be a vector valued Wiener process with respect to some filtration $\mathcal{F}_t$. From the point of view of applications to systems theory as well as to many problems in physics, biology and elsewhere, one of the most important classes of random processes is defined by processes $x(\cdot)$ which solve stochastic differential equations of the form:

$$x(t) = x + \int_0^t b(x(s)) ds + \int_0^t \sigma(x(s)) dw(s), \quad x \in R^r,$$

for appropriate $b(\cdot)$ and $\sigma(\cdot)$. When a control is added, the form

$$x(t) = x + \int_0^t b(x(s), u(s)) ds + \int_0^t \sigma(x(s)) dw(s)$$

is often used, where the control process $u(\cdot)$ is $\mathcal{F}_t$-adapted. (We will actually use a slightly more general form, called "relaxed controls" which will be introduced in the next chapter.) In this section, we are concerned with bounds which hold for arbitrary $u(\cdot)$. Because of this, it is convenient to work first with the solutions of the more general equations

$$x(t) = x + \int_0^t \tilde{b}(x(s), s)ds + \int_0^t \sigma(x(s))dw(s), \quad x \in R^r, \tag{3.1}$$

for functions $\tilde{b}(\cdot)$ and $\sigma(\cdot)$ satisfying (A3.1) below. For given $x(s) = x$ and s, $\tilde{b}(x, s)$ might be random. By a solution $x(\cdot)$ to (3.1), we mean a continuous process $x(\cdot)$ which is $\mathcal{F}_t$-adapted and such that (3.1) holds. Under (A3.1) below, the stochastic integral of $\sigma(x(\cdot))$ is well defined. We will also use the differential form:

$$dx = \tilde{b}(x, t)dt + \sigma(x)dw$$

to denote (3.1). There is an enormous literature concerning the mathematics and the applications of (3.1) and its solution processes [A1, I1, F4, G1, G2, F3, K13, S4].

Itô's Formula revisited. The differential operator. Suppose that $x(\cdot)$ is given and $\tilde{b}(x(\cdot), \cdot)$ and $\sigma(x(\cdot))$ are $\mathcal{F}_t$-adapted processes (vector and matrix-valued, resp.). Let the components of these processes satisfy the conditions imposed on the $g(\cdot)$ and $\phi(\cdot)$ which are used in (2.10). Define the operator A acting on functions $f(\cdot)$ in $C^2(R^r)$ by

$$(Af)(x, t) = f'_x(x)\tilde{b}(x, t) + \frac{1}{2}\text{trace } f_{xx}(x) \cdot \sigma(x)\sigma'(x). \tag{3.2}$$

A is called the differential operator of the process $x(\cdot)$ satisfying (3.1). Its value might depend on time. For notational convenience, we write its value when acting on $f(x)$ at time t by $(Af)(x, t)$. By Itô's Formula (2.13):

$$f(x(t)) = f(x) + \int_0^t (Af)(x(s), s)ds + \int_0^t f'_x(x(s))\sigma(x(s))dw(s). \tag{3.3}$$

If the function $\tilde{b}(\cdot)$ does not depend on either t or ω, then we write $Af(x)$ in lieu of $(Af)(x, t)$ for notational simplicity.

Bounds on the solutions. In this subsection, we obtain some a priori bounds on the solutions of (3.1) which will be useful later for proving the tightness of various sets of processes which are defined by solutions of (3.1), as well as for proving existence and uniqueness for the solutions of (3.1). Let E_x denote the expectation of functions of $x(\cdot)$, under the condition $x(0) = x$.

We will use the following assumption

A3.1. $\sigma(\cdot)$ *is continuous. If $x(\cdot)$ is $\mathcal{F}_t$-adapted, then so is $\tilde{b}(x(\cdot), \cdot)$. For some $K < \infty$, $|\tilde{b}(x, t)| \leq K(1 + |x|)$, $|\sigma(x)|^2 \leq K^2(1 + |x|^2)$.*

Theorem 3.1. *Assume* (A3.1). *For each* $T < \infty$, *there is a* $K_0(x,T)$ *which is bounded on each bounded x-set such that any solution to* (3.1) *satisfies*

$$E_x \sup_{t \leq T} |x(t)|^2 \leq K_0(x,T). \tag{3.4}$$

Proof. For notational simplicity, we do the scalar case only. By (2.6) and (A3.1),

$$E_x \sup_{s \leq T} \left| \int_0^s \sigma(x(u))dw(u) \right|^2 \leq 4E_x \int_0^T \sigma^2(x(u))du$$

$$\leq 4K^2 \int_0^T (1 + E_x |x(u)|^2)du. \tag{3.5}$$

For $t \leq T$, Schwarz's inequality and (A3.1) yield

$$E_x \left(\int_0^t |\tilde{b}(x(s),s)|ds \right)^2 \leq K^2 t \int_0^t (1 + E_x|x(s)|^2)ds. \tag{3.6}$$

The inequalities (3.5) and (3.6) hold whether the right sides are finite or infinite. Using (3.1), (3.5) and (3.6) yields for $t \leq T$

$$E_x \sup_{s \leq t} |x(s)|^2 \leq 3|x|^2 + 3E_x \sup_{s \leq t} \left| \int_0^t \tilde{b}(x(s),s)ds \right|^2$$

$$+ 3E_x \sup_{s \leq t} \left| \int_0^t \sigma(x(s))dw(s) \right|^2$$

$$\leq 3|x|^2 + 3K^2(4+T) \left(t + \int_0^t E_x|x(s)|^2 ds \right). \tag{3.7}$$

The theorem follows from the inequality (3.7) and Theorem 3.2 below. Q.E.D.

Theorem 3.2. *Assume* (A3.1) *and let* $x = x(0)$ *in* (3.1). *There is a* $K_p(x,T)$ *which is bounded on each compact x-set for each* $T < \infty$ *and such that*

$$\sup_{t \leq T} E_x|x(t)|^{2p} \leq K_p(x,T). \tag{3.8}$$

Proof. For notational simplicity, we do the scalar case and $p = 2$ only. Let n be an integer and let $f(\cdot)$ be in $C_0^2(R)$ with $f(y) = |y|^4$ for $|y| \leq n$. Define $\tau_n = \min\{t : |x(t)| \geq n\}$. Let $\tau_n = \infty$ if it is not otherwise defined. Let n be large enough so that $|x| \leq n$. By Itô's Formula (2.11)

$$df(x(t)) = f_x(x(t))[\tilde{b}(x(t),t)dt + \sigma(x(t))dw(t)] + \frac{f_{xx}(x(t))}{2}\sigma^2(x(t)).$$

Using the fact that $E_x \int_0^{t \wedge \tau_n} f'_x(x(s))\sigma(x(s))dw(s) = 0$, we have

$$E_x f(x(t \wedge \tau_n)) = f(x) + E_x \int_0^{t \wedge \tau_n} f_x(x(s))\tilde{b}(x(s), s)ds$$

$$+ \frac{1}{2} E_x \int_0^{t \wedge \tau_n} f_{xx}(x(s))\sigma^2(x(s))ds. \tag{3.9}$$

Thus, by (A3.1), (3.9) and the form of $f(y)$ for $|y| \leq n$, there is a $K' < \infty$ which depends only on K and is such that

$$E_x |x(t \wedge \tau_n)|^4 \leq |x|^4 + 4K E_x \int_0^{t \wedge \tau_n} |x(s)|^3(1 + |x(s)|)ds$$

$$+ 6K^2 E_x \int_0^{t \wedge \tau_n} |x(s)|^2(1 + |x(s)|^2)ds$$

$$\leq |x|^4 + K' E_x \int_0^{t \wedge \tau_n} (1 + |x(s)|^4)ds.$$

Thus, there is a $K'' < \infty$ such that for all n and $t \leq T$,

$$E|x(t \wedge \tau_n)|^4 \leq |x|^4 + K''T + K'' \int_0^t E_x |x(s \wedge \tau_n)|^4 ds. \tag{3.10}$$

Since $|x(s \wedge \tau_n)| \leq n$, (3.10) and the Bellman–Gronwall inequality imply that

$$\sup_{t \leq T} E_x |x(t \wedge \tau_n)|^4 \leq (|x|^4 + K''T) \exp K''T.$$

Since the right side doesn't depend on n, the theorem is proved.   Q.E.D.

Theorem 3.3. *Assume* (A3.1). *Then, for each $T < \infty$,*

$$\sup_{\tau \leq T} E_x |x(\tau + \delta) - x(\tau)|^2 \xrightarrow{\delta} 0,$$

uniformly in $x(0) = x$ in any compact set and where the sup is over all $\mathcal{F}_t$-stopping times $\tau \leq T$. The rate of convergence to zero depends only on the K of (A3.1) *and not on $\tilde{b}(\cdot)$ or $\sigma(\cdot)$ otherwise.*

Proof. We again do only the scalar case. Let τ be an $\mathcal{F}_t$-stopping time which satisfies $\tau \leq T$. Then by Itô's Formula (2.11),

$$[x(\tau + \delta) - x(\tau)]^2 = \int_\tau^{\tau + \delta} 2[x(s) - x(\tau)]dx(s) + \int_\tau^{\tau + \delta} \sigma^2(x(s))ds.$$

For some constants K_i,

$$
\begin{aligned}
E_x |x(\tau + \delta) - x(\tau)|^2 &= E_x \int_\tau^{\tau + \delta} [2(x(s) - x(\tau))\tilde{b}(x(s), s) + \sigma^2(x(s))]ds \\
&\leq K_1 E_x \int_\tau^{\tau + \delta} \{2|x(s) - x(\tau)| \cdot (|x(s)| + 1) \\
&\quad + 1 + |x(s)|^2\} \\
&\leq K_2 \delta E_x \max_{s \leq T} |x(s)|^2 + K_2 \delta.
\end{aligned}
$$

The last inequality and Theorem 3.1 yield the result.   Q.E.D.

Remark. Suppose that there is a sequence $\{x_n(\cdot), n < \infty\}$ of solutions to (3.1), associated with functions $\tilde{b}_n(\cdot)$ and $\sigma_n(\cdot)$, resp., all satisfying (A3.1) with the same constant K. Then by Theorem 3.3 and Theorem 1.5.1, $\{x_n(\cdot), n < \infty\}$ is tight. Also any limit process is continuous w.p.1 since each of the $x_n(\cdot)$ is continuous w.p.1.

4. Controlled Stochastic Differential Equations: Existence of Solutions

Controlled processes. Let $w(\cdot)$ be a standard vector-valued Wiener process on a probability space $(\Omega, \mathcal{F}, P)$ and with respect to the filtration $\mathcal{F}_t$. Let $\mathcal{U}$ be a compact set in some Euclidean space and $u(\cdot)$ a $\mathcal{U}$-valued process which is $\mathcal{F}_t$-adapted. Such processes $u(\cdot)$ will be called *admissible ordinary controls*. We say, equivalently, that $u(\cdot)$ is *admissible with respect to* $w(\cdot)$ or that $(u(\cdot), w(\cdot))$ is an *admissible pair*. In this section, we will discuss the existence and uniqueness of solutions of the SDE, where $u(\cdot)$ is admissible and $b(\cdot)$ and $\sigma(\cdot)$ satisfy A4.1 below:

$$
x(t) = x + \int_0^t b(x(s), u(s))ds + \int_0^t \sigma(x(s))dw(s). \tag{4.1}
$$

The differential operator. Relating the case (4.1) to (3.1), we see that $\tilde{b}(x, t) = b(x, u(t))$. For this case, it is useful to use a slightly different notation for the differential operator. For $\alpha \in \mathcal{U}$, and $f(x) \in C^2(R^r)$, define

$$
A^\alpha f(x) = f'_x(x)b(x, \alpha) + \frac{1}{2}\text{trace } f_{xx}(x) \cdot \sigma(x)\sigma'(x).
$$

When referring to the operator for the system (4.1) we use A^u, and write

$$
(A^u f)(x, t) = A^{u(t)}f(x).
$$

Three different methods will be discussed: a classical direct construction, an analytical method, and a "measure transformation" method. Each has its own advantages.

Definition. Let $(u(\cdot), w(\cdot))$ be an admissible pair. The solution to (4.1) is said to be *unique in the weak sense* if the probability law of $(u(\cdot), w(\cdot))$ determines that of $(x(\cdot), u(\cdot), w(\cdot))$. We say that the solution $x(\cdot)$ to (4.1) is *unique in the strong sense* if any two solutions to (4.1) defined on the same probability space are equal w.p.1. See [I1, Chapter 4.3] for a more complete discussion of uniqueness.

Existence of solutions under a Lipschitz condition; a direct construction. In this subsection, we will use the following condition.

A4.1. $b(\cdot)$ *and* $\sigma(\cdot)$ *are continuous functions on* $R^r \times U$ *and* R^r, *resp.* $b(\cdot, u)$ *and* $\sigma(\cdot)$ *are Lipschitz continuous (constant K) uniformly in* $u \in U$, *and* $b(0, u)$ *is uniformly bounded on* U.

The following theorem is the classical existence result, only slightly altered to incorporate the controls.

Theorem 4.1. *Assume (A4.1) and let* $(u(\cdot), w(\cdot))$ *be an admissible pair. Then there exists a solution to* (4.1) *for each initial condition* $x = x(0)$. *This solution is unique in both the strong and weak sense. The same result holds for* (3.1) *when* $\tilde{b}(\cdot)$ *and* $\sigma(\cdot)$ *are (uniformly in* ω, t *for* $\tilde{b}(\cdot)$) *Lipschitz continuous in* x, *and* $\tilde{b}(x, \cdot)$ *is uniformly bounded when* $x = 0$.

The proof is given in many places, but we provide some of the details, owing to the presence of the control (although the details are essentially the same whether there is a control or not). For more details on the proof see [I1, Chapter 4.2], [D5], [G2, Chapter 8.3]. We work with (4.1) only.

Proof. *Uniqueness.* Note that (A4.1) implies (A3.1). Let $\tilde{x}(\cdot)$ and $x(\cdot)$ be solutions to (4.1), and set $\delta x(\cdot) = x(\cdot) - \tilde{x}(\cdot)$. Then by (A4.1), (1.3) and Schwarz's inequality,

$$E_x \max_{s \leq t} |\delta x(s)|^2 \leq 2E_x \max_{s \leq t} \left| \int_0^s [b(x(s), u(s)) - b(\tilde{x}(s), u(s))]ds \right|^2$$

$$+ 2E_x \max_{s \leq t} \left| \int_0^t [\sigma(x(s)) - \sigma(\tilde{x}(s))]dw(s) \right|^2$$

$$\leq 2K^2 t \int_0^t E_x |\delta x(s)|^2 ds + 8K^2 \int_0^t E_x |\delta x(s)|^2 ds. \qquad (4.2)$$

By Theorem 3.1, $E_x \max_{s \leq t} |\delta x(s)|^2$ is bounded for each t. Hence (4.2) and the Bellman–Gronwall Lemma imply that $E_x \max_{s \leq t} |\delta x(s)|^2 = 0$ for all t. Hence, strong sense uniqueness.

Existence. The solution $x(\cdot)$ will be constructed recursively. Define the process $x_0(\cdot)$ by $x_0(t) = x$ and define $x_n(\cdot)$ recursively by

$$x_{n+1}(t) = x + \int_0^t b(x_n(s), u(s))ds + \int_0^t \sigma(x_n(s))dw(s).$$

Owing to the recursive definition and $\mathcal{F}_t$-adaptiveness of $u(\cdot)$, for each n the process $x_n(\cdot)$ is continuous w.p.1 and $\mathcal{F}_t$-adapted. Define $\delta x_n(\cdot) = x_{n+1}(\cdot) - x_n(\cdot)$, and let $T > 0$. Then for $t \leq T$,

$$E_x \max_{s \leq t} |\delta x_n(s)|^2 \leq 2E_x \max_{s \leq t} \left| \int_0^t [b(x_n(s), u(s)) - b(x_{n-1}(s), u(s))]ds \right|^2$$

$$+ 2E_x \max_{s \leq t} \left| \int_0^t [\sigma(x_n(s)) - \sigma(x_{n-1}(s))]dw(s) \right|^2$$

$$\leq 2K^2 T \int_0^t E_x |\delta x_{n-1}(s)|^2 ds + 8K^2 \int_0^t E_x |\delta x_{n-1}(s)|^2 ds$$

$$\leq K_1 \int_0^t E_x \max_{s \leq t} |\delta x_{n-1}(s)|^2 ds,$$

where $K_1 = 2K^2(T + 4)$. In the above, we used the Lipschitz condition, Schwarz's inequality and (2.6).

Iterating yields

$$E_x \max_{s \leq T} |\delta x_n(s)|^2 \leq \frac{K_1^n T^n}{n!} E_x \max_{s \leq T} |\delta x_0(s)|^2, \tag{4.3}$$

where the right hand expectation is finite. Inequality (4.3) implies that for some constant K_2,

$$P_x \left\{ \max_{s \leq T} |x_{n+1}(s) - x_n(s)| \geq 2^{-n} \right\} \leq K_2 \frac{K_1^n T^n 4^n}{n!},$$

an element of a summable sequence. It also implies that for $m > n$ the right side of

$$E_x |x_m(s) - x_n(s)|^2 \leq 2E_x |x_{n+1}(s) - x_n(s)|^2$$

$$+ \cdots + 2^{(m-n)} E_x |x_m(s) - x_{m-1}(s)|^2$$

goes to zero as m and n go to infinity. It can now be shown that the Cauchy sequence

$$x_n(\cdot) = x_0(\cdot) + (x_1(\cdot) - x_0(\cdot)) + \cdots + (x_n(\cdot) - x_{n-1}(\cdot))$$

converges uniformly w.p.1 on each $[0, T]$ and in mean square for each t to a continuous $\mathcal{F}_t$-adapted process $x(\cdot)$ which satisfies (4.1).

The constructive nature of the proof implies that the probability law of the constructed $x(\cdot)$ is uniquely determined by the probability law of $(w(\cdot),\ u(\cdot))$. This and the strong sense uniqueness implies the weak sense uniqueness.   Q.E.D.

Existence and uniqueness. The non-degenerate case. An analytic method. If the matrix $\sigma(\cdot)$ does not depend on u but is Hölder continuous and $\sigma(x,t)\sigma'(x,t)$ is 'uniformly' positive definite, then uniqueness and existence of the solutions to (4.1) can be shown by the analytic methods of [S9, Chapter 3]. The following result is typical, and will use the condition:

A4.2. $b(\cdot)$ *and* $\sigma(\cdot)$ *are bounded, Hölder continuous functions of* (x,t), *and there is a* $\delta > 0$ *such that* $\sigma(x,t)\sigma'(x,t) \geq \delta I$.

Theorem 4.2. *Assume* (A4.2). *Then the equation*

$$dx = b(x,t)dt + \sigma(x,t)dw \tag{4.4}$$

has a solution for each initial condition $x = x(0)$, *in the sense that there is a pair* $(x(\cdot),\ w(\cdot))$ *which solves* (4.4), *and* $x(\cdot)$ *is non-anticipative with respect to* $w(\cdot)$. *The solution is unique in the weak sense.*

Existence and uniqueness by a measure transformation method. We now sketch the outline of a very useful and popular method for the construction of solutions to (4.1) under minimal regularity conditions on the drift function $b(\cdot)$ and on the control process. In order to prove Theorem 4.1, we use a direct construction of the solution starting with given $(u(\cdot),\ w(\cdot))$. For Theorem 4.2, both partial differential equations and probabilistic techniques need to be used. For the method of this subsection, one starts with a solution $x(\cdot)$ to an SDE with drift term $b(\cdot)$ on some probability space $(\Omega, \mathcal{F}, P)$. Then a new measure P' on $(\Omega, \mathcal{F})$ is constructed, such that under P', the same $x(\cdot)$ satisfies an SDE with a different chosen drift term. Typically, $x(\cdot)$ under the measure P is an uncontrolled system, and the drift changes yield the control terms. This is further developed in Chapter 3.3. Full details are in [I1, Chapter 4.4].

To start the development, partition the system (4.1) into two parts by writing the state vector as $x = (x_1, x_2)$, where $x_1 \in R^{r_1}$, $x_2 \in R^{r_2}$ and $r = r_1 + r_2$. Attention will be restricted to processes on the bounded interval $[0, T]$. Let $\mathcal{F}_t$ be a filtration on the probability space. Let $\hat{w}_1(\cdot)$ and $w_2(\cdot)$ be mutually independent standard vector-valued $\mathcal{F}_t$-Wiener processes, and let $(x_1(\cdot), x_2(\cdot)) = x(\cdot)$ satisfy (4.5), where the $b_i(\cdot)$ and $\sigma_i(\cdot)$ are non-random and are such that (4.5) has a unique weak sense solution $x(\cdot)$ for each initial condition. Suppose that $\sigma_1^{-1}(x,t)$ exists and is bounded uniformly in (x,t), and let P be the measure on the probability space on which $x(\cdot)$ is defined.

$$dx_1 = b_1(x,t)dt + \sigma_1(x,t)d\hat{w}_1,$$
$$dx_2 = b_2(x,t)dt + \sigma_2(x,t)dw_2, \quad t \leq T. \tag{4.5}$$

Let $\hat{b}_1(\cdot)$ be a bounded random process which is $\mathcal{F}_t$-adapted and has values in R^{r_1}. We now define the measure transformation. Define the random variable

$$R(T) = \exp\left[\int_0^T (\sigma_1^{-1}(x(s), s)\hat{b}_1(s))'d\hat{w}_1(s) - \frac{1}{2}\int_0^T |\sigma_1^{-1}(x(s), s)\hat{b}_1(s)|^2 ds\right].$$

Then $ER(T) = 1$, and $R(T)$ is the Radon–Nikodyn derivative with respect to P of the measure $\hat{P}$ which is defined by

$$\hat{P}(A) = \int_A R(T)P(d\omega).$$

Define the "shifted" process $w_1(\cdot)$ by

$$w_1(t) = \hat{w}_1(t) - \int_0^t \sigma_1^{-1}(x(s), s)\hat{b}_1(s)ds.$$

Under $\hat{P}$, the processes $(w_1(\cdot), w_2(\cdot))$ are mutually independent standard vector-valued $\mathcal{F}_t$-Wiener processes ([I1, pp. 176–183]. Using the definition of $w_1(\cdot)$, we can rewrite (4.5) as

$$\begin{aligned} dx_1 &= [b_1(x,t) + \hat{b}_1(t)]dt + \sigma_1(x,t)dw_1 \\ dx_2 &= b_2(x,t)dt + \sigma_2(x,t)dw_2 \end{aligned} \tag{4.6}$$

Thus, starting with (4.5) under measure P, we constructed a solution to (4.6) under $\hat{P}$. The solution to (4.6) has the additional drift term $\hat{b}_1(\cdot)$. This method of constructing a solution with a general drift term $\hat{b}_1(\cdot)$ is called the *Girsanov transformation* method. The $\hat{b}_1(\cdot)$ is often considered to represent the effects of the control. See Chapter 3.3.

The fact that (4.5) has a unique weak sense solution implies that (4.6) does [I1, pp. 176–183].

Suppose that $\sigma(\cdot)$, $b(\cdot)$ satisfy (A4.1). Then, by Theorem 4.1, (4.5) does have the desired unique weak sense solution. Thus, if $\sigma_1^{-1}(\cdot)$ is bounded, as needed above, we have a unique weak sense solution to (4.6) for arbitrary bounded $\mathcal{F}_t$-adapted $\hat{b}_1(\cdot)$.

5. Representing a Martingale as a Stochastic Integral

One of the main concerns in the book is the characterization of the limit of a weakly convergent sequence of processes $\{x_n(\cdot), n < \infty\}$. We are usually interested in knowing that the limit process satisfies some uncontrolled or controlled SDE. The so-called *martingale method*, to be outlined in the next section, provides one of the most useful tools for that purpose. In applications, it is often possible to prove that certain functionals of the limit $x(\cdot)$ of a weakly convergent sequence are martingales, and to show that this implies that there is a Wiener process such that (4.1) holds. This approach will be used in Chapters 3, 4 and elsewhere. In this section, we give some background for the subsequent development. The results of this section are used only to motivate the method of the next section.

The quadratic variation process. Let $\mathcal{F}_t$ be a given filtration and let $M(\cdot)$ be a continuous and real valued $\mathcal{F}_t$-martingale. Then, there is a unique continuous non-decreasing $\mathcal{F}_t$-adapted process which we call $\langle M \rangle(\cdot)$ such that the process defined by $M_1(\cdot) = M^2(\cdot) - \langle M \rangle(\cdot)$ is a local martingale [I1, Chapter 2.2], [K8]. Even if $M(\cdot)$ is a martingale, it is not necessarily true that $M_1(\cdot)$ is a martingale, since $EM_1(t) < \infty$ might not hold. But $M_1(\cdot)$ will always be a local martingale. We always assume, w.l.o.g., that $\langle M \rangle(0) = 0$. The process $\langle M \rangle(\cdot)$ is called the *quadratic variation* of $M(\cdot)$.

It can be seen by a direct calculation that if $M(\cdot)$ is a real-valued standard Wiener process, then $\langle M \rangle(t) \equiv t$, and if $M(t) = \int_0^t \phi(s)dw(s)$ for $\phi(\cdot) \in \Phi_2$, then $\langle M \rangle(t) = \int_0^t \phi^2(s)ds$. It is an important fact that if $M(\cdot)$ is a continuous martingale with quadratic variation equal to t, then it is a real-valued standard Wiener process [I1, K8]. This property was mentioned at the end of Section 1. Furthermore, we have the "converse result" that any continuous martingale whose quadratic variation process is absolutely continuous w.p.1 can be represented as a stochastic integral with respect to some Wiener process. In particular, suppose that $\langle M \rangle(\cdot)$ can be written in the form $\langle M \rangle(t) = \int_0^t a(s)ds$ for some non-negative $\mathcal{F}_t$-adapted and locally integrable process $a(\cdot)$. Let $\phi(\cdot)$ be any $\mathcal{F}_t$-adapted process such that $\phi^2(\cdot) = a(\cdot)$ for almost all ω, t. We can always find a $\phi(\cdot)$ that is progressively measurable. If $a(t) > 0$ for almost all (ω, t), then there is a standard $\mathcal{F}_t$-Wiener process $w(\cdot)$ such that

$$M(t) = \int_0^t \phi(s)dw(s). \qquad (5.1)$$

Since the proof involves the concepts of stochastic integration with respect to certain martingales (and not only with respect to the Wiener process), we omit further details. For further discussion of the idea, see [K8, S9], or the formal development in [K20].

If $a(\cdot)$ is not positive for almost all (ω, t), then (5.1) continues to hold with a minor modification. To motivate the modification, consider the totally degenerate example where $a(t) \equiv 0$. Then we must have $M(t) \equiv 0$. Then (5.1) does hold with $\phi(\cdot) = 0$ and any $w(\cdot)$. But the probability space might not have any Wiener process definable on it. The minor modification which we need is just the augmentation of the original probability space by the addition of a Wiener process that is independent of all the other random variables defined on the original probability space. The representation (5.1) holds in general, even if $a(\omega, t) = 0$ on a non-null set, provided that we add such an "independent" Wiener process to the probability space. Loosely speaking, we note that the "increments dw" can be constructed from $a(\cdot)$ and $M(\cdot)$ on the set where $a(\cdot) > 0$. The added independent Wiener process simply provides "dw" on the set where $a(\cdot) = 0$.

The vector case. Now let $M(\cdot)$ be a continuous R^k-valued $\mathcal{F}_t$-martingale. There is a unique continuous non-decreasing (in the sense of non-negative definite matrices) $\mathcal{F}_t$-adapted matrix valued process which we also denote by $\langle M \rangle(\cdot)$, such that the process defined by $M(\cdot)M'(\cdot) - \langle M \rangle(\cdot)$ is a $k \times k$ matrix-valued $\mathcal{F}_t$-local martingale. The process $\langle M \rangle(\cdot)$ will also be called the *quadratic covariation* of $M(\cdot)$. We always assume, w.l.o.g., that $\langle M \rangle(0) = 0$. Suppose that $\langle M \rangle(t) = \int_0^t a(s)ds$ for some $\mathcal{F}_t$-adapted non-negative definite matrix-valued process $a(\cdot)$ which is locally integrable. Let $\phi(\cdot)$ be any matrix-valued $\mathcal{F}_t$-adapted process such that $a(\cdot) = \phi(\cdot)\phi'(\cdot)$ for almost all ω, t. Again, we can always find a version of $\phi(\cdot)$ that is progressively measurable. If $a(t)$ is positive definite for almost all (ω, t), then there is a standard R^k-valued $\mathcal{F}_t$-Wiener process $w(\cdot)$ such that (5.1) holds. If $a(t)$ is not positive definite for almost all (ω, t), then (5.1) still holds if we augment the probability space, as was done in the scalar case, by adding to the original probability space a R^k-valued standard Wiener process which is independent of $M(\cdot)$.

6. The Martingale Problem

Let $(x(\cdot), u(\cdot), w(\cdot))$ satisfy (4.1), where $(x(\cdot), u(\cdot))$ is non-anticipative with respect to $w(\cdot)$. In this section, we suppose:

A6.1. $u(t) \in \mathcal{U}$, *a compact set.* $b(\cdot)$ *and* $\sigma(\cdot)$ *are continuous and satisfy the linear growth condition*

$$|b(x, u)| + |\sigma(x)| \leq K(1 + |x|).$$

Let $f(\cdot) \in C_0^2(R^r)$ and let $\mathcal{F}_t$ denote the σ-algebra $\mathcal{B}(x(s), u(s), s \leq t)$. Recall the definition of the differential operator A^u given below (4.1). By Itô's Formula (3.3), the process defined by

$$M_f(t) = f(x(t)) - f(x) - \int_0^t (A^u f)(x(s), s)ds \qquad (6.1)$$

equals $\int_0^t f_x'(x(s))\sigma(x(s))dw(s)$. Since $f_x'(x(\cdot))\sigma(x(\cdot)) \in \Phi_2$, the stochastic integral is well defined and $M_f(\cdot)$ is an $\mathcal{F}_t$-martingale.

The martingale problem: definition. Let $x(\cdot)$, $u(\cdot)$ be given processes, assume (A6.1) and let $\mathcal{F}_t$ be a filtration satisfying $\mathcal{F}_t \supset \mathcal{B}(x(s), u(s), s \leq t)$. Let $x(\cdot)$ be continuous and suppose that (6.1) is a martingale for each $f(\cdot) \in C_0^2(R^r)$. Then $(x(\cdot), u(\cdot))$ is said to solve the *martingale problem for operator* A^α.

Now, consider the following "inverse" problem. Assume (A6.1). Let $(x(\cdot), u(\cdot))$ solve the martingale problem for operator A^α. Then, it turns out that there is a Wiener process $w(\cdot)$ such that $(x(s), u(s), s \leq t)$ is independent of $(w(u) - w(t), u \geq t)$ for all t and (4.1) holds. (See [S9], or see [K20] for a formal proof.) Thus, the set of solutions to the martingale problem associated with the operator A^α and the set of solutions to (4.1) are equivalent. This equivalence will be heavily exploited in the rest of the book. We will usually show that a "limit process" solves (4.1) by showing that it solves the martingale problem for appropriate A^α.

The description of the solution to (4.1) as a solution to a martingale problem is particularly convenient when using weak convergence methods, since it provides a relatively simple way of characterizing the process which is the limit of a "weakly convergent subsequence." Historically, the martingale problem characterization has been used heavily in this way, starting with the seminal work of Stroock and Varadhan, which was eventually summarized in [S9]. It will be much used in this book. We now formally illustrate how the "driving" Wiener process is obtained. First, we show how to obtain the quadratic variation of the martingale $M_f(\cdot)$.

The quadratic variation process $\langle M_f \rangle(\cdot)$. Suppose that $M_f(\cdot)$ is a martingale. For notational convenience, we will work with the scalar case and where $\sigma^{-1}(x)$ is bounded for all x. Define the filtration $\mathcal{F}_t = \mathcal{B}(x(s), u(s), s \leq t)$. By definition, the quadratic variation process $\langle M_f \rangle(\cdot)$ is the unique non-decreasing continuous $\mathcal{F}_t$-adapted process such that $M_f^2(\cdot) - \langle M_f \rangle(\cdot)$ is a martingale. Note that this process is a martingale here, not only a local martingale, since the $f(\cdot)$ in (6.1) has compact support. By the martingale property of $M_f(\cdot)$, we have

$$E_{\mathcal{F}_t}[M_f(t + s) - M_f(t)]^2 = E_{\mathcal{F}_t}\langle M_f \rangle(t + s) - \langle M_f \rangle(t), \qquad (6.2)$$

for $s \geq 0$. The uniqueness of the quadratic variation process $\langle M_f \rangle(\cdot)$ implies that any non-decreasing $\mathcal{F}_t$-adapted continuous function satisfying (6.2) is the quadratic variation. We now use this characterization as the unique solution to (6.2) in order to help us construct $\langle M_f \rangle(\cdot)$. The argument is formal and is intended for illustrative purposes only.

Divide the interval $[t, t + s]$ into n subintervals each of length Δ. By a direct calculation,

$$E_{\mathcal{F}_{t+i\Delta}}[M_f(t + i\Delta + \Delta) - M_f(t + i\Delta)]^2$$

$$= E_{\mathcal{F}_{t+i\Delta}}[f(x(t + i\Delta + \Delta)) - f(x(t + i\Delta))]^2 + o(\Delta)$$

$$= E_{\mathcal{F}_{t+i\Delta}}[f^2(x(t + i\Delta + \Delta)) - f^2(x(t + i\Delta))$$

$$- 2f(x(t + i\Delta))\{f(x(t + i\Delta + \Delta)) - f(x(t + i\Delta))\}] + o(\Delta) \qquad (6.3)$$

By the fact that (6.1) is a martingale for each $f(\cdot) \in C_0^2(R^r)$, for $j = 1$ or 2 we have

$$E_{\mathcal{F}_{t+i\Delta}} \int_{t+i\Delta}^{t+i\Delta+\Delta} (A^u f^j)(x(u), u)du$$

$$= E_{\mathcal{F}_{t+i\Delta}} f^j(x(t+i\Delta+\Delta)) - f^j(x(t+i\Delta)). \tag{6.4}$$

Now, summing (6.3) and using (6.4) yields

$$E_{\mathcal{F}_t}[M_f(t+s) - M_f(t)]^2 = E_{\mathcal{F}_t} \left(\sum_{i=0}^{n-1}[M_f(t+i\Delta+\Delta) - M_f(t+i\Delta)] \right)^2$$

$$= E_{\mathcal{F}_t} \sum_{i=0}^{n-1} E_{\mathcal{F}_{t+i\Delta}}[M_f(t+i\Delta+\Delta) - M_f(t+i\Delta)]^2$$

$$= E_{\mathcal{F}_t} \int_t^{t+s} [(A^u f^2)(x(\tau), \tau) - 2f(x(\tau))(A^u f)(x(\tau), \tau)]d\tau + \delta_\Delta$$

$$= E_{\mathcal{F}_t} \int_t^{t+s} |f_x(x(\tau))\sigma(x(\tau))|^2 d\tau + \delta_\Delta, \tag{6.5}$$

where $\delta_\Delta \to 0$ as $\Delta \to 0$. Thus, (6.2) and (6.5) imply that the quadratic variation process of $M_f(\cdot)$ is given by

$$\langle M_f \rangle(t) = \int_0^t |f_x(x(\tau))\sigma(x(\tau))|^2 d\tau. \tag{6.6}$$

The construction of the "driving" Wiener process $w(\cdot)$. We let $(x(\cdot), u(\cdot))$ solve the martingale problem for the operator A^α and show that there is a suitable $w(\cdot)$ such that (4.1) holds. We continue to do the scalar case only and suppose that $\sigma^{-1}(x)$ is uniformly bounded. The result is still true if $\sigma^{-1}(x)$ is not bounded, but then we might have to augment the probability space by the addition of an "independent" Wiener process, as discussed in Section 5. The vector case is treated similarly.

Fix n, and choose $f(\cdot) \in C_0^2(R)$ but with $f(x) = x$ for $|x| \leq n$. Define $\tau_n = \min\{t : |x(t)| \geq n\}$. By the absolute continuity of the quadratic variation process implied by (6.6) and the results in Section 5, there is a standard $\mathcal{F}_t$-Wiener process $w_n(\cdot)$ such that the martingale $M_f(\cdot)$ can be represented in the form

$$M_f(t) = \int_0^t \left[\frac{d}{dt} \langle M_f \rangle(\tau) \right]^{1/2} dw_n(\tau).$$

Hence by (6.6), for $t \leq \tau_n$,

$$M_f(t) = \int_0^t \sigma(x(\tau)) dw_n(\tau), \quad t \leq \tau_n. \tag{6.7}$$

Now, using the definition of $M_f(\cdot)$ in (6.1), for $t \leq \tau_n$ we have

$$x(t) = x + \int_0^t b(x(\tau), u(\tau))d\tau + \int_0^t \sigma(x(\tau))dw_n(\tau). \qquad (6.8)$$

Applying the estimate in Theorem 3.1 to (6.8) yields

$$E \max_{t \leq T} |x(t \wedge \tau_n)|^2 \leq K_1(x, T).$$

Since the right side does not depend on n, we have $E \max_{t \leq T} |x(t)|^2 < \infty$. Thus, $\lim_n \tau_n = \infty$ w.p.1. For $m > n$, we can suppose that $w_m(\cdot) = w_n(\cdot)$ on $[0, \tau_n]$. It can now be seen that the process defined by $w(\cdot) = \lim_n w_n(\cdot)$ satisfies our needs; i.e. that (6.8) holds for all $t \geq 0$ with $w(\cdot)$ replacing $w_n(\cdot)$.

7. Jump-Diffusion Processes

In this section, we consider jump-diffusions, an extension of the diffusion process (4.1) to a class of processes subject to "impulsive" driving terms as well as a diffusion term, hence discontinuous in t. Such models are used frequently when the system of interest is subject to sudden changes. They are diffusions of the type (4.1) between the jumps in the state values, and the jumps are of the "Markovian type." First, we give a simple example and then a general model of the impulsive forces which "drive" the SDE.

Example. Let $\{\tau_n\}$ be an increasing sequence of stopping times. Suppose that a sattelite is subject to micrometeorite impacts of "impulsive magnitude" g_n at τ_n. Due to the short duration of each impact, we suppose that the effect on the sattelite is simply an instantaneous change of g_n in the state. The process $g(\cdot)$ defined by $g(t) = 0$ unless $t = \tau_n$ for some n and $g(\tau_n) = g_n$ is called a *point process*, if $\{g_n, \tau_n\}$ is a random sequence and the $\{\tau_n\}$ have no finite accumulation point. For purposes of analysis it is useful to define certain measures concerning the number of impulses on each time interval which lie in given sets. We refer to the g_n as *impulses*.

Definition. Let Γ, a compact set (which does not include the origin) in some Euclidean space, be the range space of the "impulsive" forces g_n of a point process $g(\cdot)$. Suppose that the impulsive times are $\{\tau_n\}$, where $\tau_n \to \infty$ w.p.1. For each $H \in \mathcal{B}(\Gamma)$, define the process

$$N(t, H) = \text{number of impulses of } g(\cdot) \text{ on } [0, t] \text{ with values in } H.$$

Thus $N(\cdot, H)$ is right continuous. The measure valued process $N(\cdot)$ with value $N(t, \cdot)$ at time t is called a *counting measure*.

We will suppose that $EN(t, \Gamma) < \infty$ for each t. Our definition is not the most general, but it is adequate for our purposes. If $\mathcal{F}_t$ is a filtration such that $N(\cdot, H)$ is $\mathcal{F}_t$-adapted for each $H \in \mathcal{B}(\Gamma)$, then $g(\cdot)$ (resp., $N(\cdot)$) is said to be an $\mathcal{F}_t$-point process (resp., an $\mathcal{F}_t$-counting measure).

If the set $\{N(t + \cdot, H) - N(t, H), H \in \mathcal{B}(\Gamma)\}$ is independent of $\mathcal{F}_t$ for all t, then $g(\cdot)$ is said to be an $\mathcal{F}_t$-*Poisson point process*, and $N(\cdot)$ is called an $\mathcal{F}_t$-*Poisson measure*.

If $g(\cdot)$ is a Poisson point process and the distribution of the increments $\{N(t + s, H) - N(t, H), H \in \mathcal{B}(\Gamma)\}$ does not depend on t, then $g(\cdot)$ is said to be a *stationary Poisson point process*. The point processes of interest in this book will all be stationary *Poisson point processes*. In this case, there is a constant $\lambda > 0$ and a probability measure $\Pi(\cdot)$ on $\mathcal{B}(\Gamma)$ such that $E_{\mathcal{F}_t} N(t + s, H) - N(t, H) = s\Pi(H)\lambda$ [G2, Chapter 6]. The constant λ is interpreted as the "impulsive rate" of $g(\cdot)$ or the jump rate of $N(\cdot, \Gamma)$, since it follows from the properties of the Poisson point process that

$$P(g(s) \neq 0 \text{ on } [t, t + \Delta) \mid g(u), u < t) = \lambda\Delta + o(\Delta),$$

$$P(g(s) \neq 0 \text{ at more than one instant in } [t, t + \Delta) \mid g(u), u < t) = o(\Delta).$$

$\Pi(\cdot)$ is the jump distribution in the sense that

$$P(g(t) \in H \mid g(t) \neq 0, g(u), u < t) = \Pi(H).$$

The values and times of the impulses can be recovered from the integral (a pure jump process)

$$Q(t) \equiv \int_0^t \int_\Gamma \gamma N(ds d\gamma) = \sum_{s \leq t} g(s),$$

but the Poisson measure concept provides a convenient representation of the "driving forces" for use in constructing the solution of the stochastic differential equation.

Stochastic differential equations with jumps. Henceforth $g(\cdot)$ will be a *stationary Poisson point process*. $N(\cdot)$ will be the associated Poisson measure, with respect to some filtration $\mathcal{F}_t$ and with *compact* Γ, jump rate $\lambda < \infty$ and jump distribution $\Pi(\cdot)$. Also $w(\cdot)$ will be a standard vector-valued $\mathcal{F}_t$-Wiener process. Since the vector-valued process $(w(\cdot), N(\cdot))$ has independent increments, it follows that $w(\cdot)$ and $N(\cdot)$ are mutually independent [G2, Chapter 6].

The SDE. We will next be concerned with the existence, uniqueness and characterization of solutions to the "jump-diffusion" extension of the SDE (3.1), namely

$$x(t) = x + \int_0^t \tilde{b}(x(s), s)ds + \int_0^t \sigma(x(s))dw(s) + J(t) \qquad (7.1)$$

where $J(\cdot)$ has the form

$$J(t) = \int_0^t \int_\Gamma q(x(s^-), \gamma) N(ds\,d\gamma).$$

We will use the conditions (A7.1) and (A7.2).

A7.1. $q(\cdot)$ *is a bounded and continuous, R^r-valued function on $R^r \times R$.* $q(x,0) = 0$. *For each x, the value of γ can be uniquely determined from the value of $q(x,\gamma)$.*

A7.2. *For $\mathcal{F}_t$-adapted $x(\cdot)$ with paths in $D^r[0,\infty)$, $\tilde{b}(x(\cdot), \cdot)$ is $\mathcal{F}_t$-adapted. $\sigma(\cdot)$ is continuous. Also, for some $K < \infty$,*

$$|\tilde{b}(x,t)| \leq K(1 + |x|), \quad |\sigma(x)|^2 \leq K^2(1 + |x|^2).$$

The last part of (A7.1) is innocuous in general, since we can always enlarge the state space to include a component $dx_0 = \int_\Gamma \gamma N(dt\,d\gamma)$. The last integral on the right side of (7.1) is interpreted as a Stieltjes integral and equals

$$\sum_{s \leq t : g(s) \neq 0} q(x(s^-), g(s)).$$

By a solution to (7.1), we mean a process $x(\cdot)$ which is $\mathcal{F}_t$-adapted. The form (7.1) implies that $x(\cdot)$ will be right continuous and have left hand limits. The $x(s^-)$ appears in $J(\cdot)$ since, in order to have a Markov property, the magnitude of the jump can only depend on the value of the state *just before* the jump occurs.

Define the *centered Poisson measure* by $\tilde{N}(s, H) = N(s, H) - \lambda s \Pi(H)$. It is sometimes useful to write the last integral in (7.1) in the centered form

$$\int_0^t \int_\Gamma q(x(s^-), \gamma) \tilde{N}(ds\,d\gamma) + \lambda \int_0^t \int_\Gamma q(x(s), \gamma) \Pi(d\gamma) ds,$$

which is the sum of a martingale and an absolutely continuous process. Analogously to the differential form representation of (3.1), we will also write (7.1) symbolically in the form

$$dx = \tilde{b}(x,t)dt + \sigma(x)dw + \int_\Gamma q(x,\gamma) N(dt\,d\gamma). \tag{7.2}$$

Bounds on $x(\cdot)$.

Theorem 7.1. *Assume (A7.1) and (A7.2). Then Theorems 3.1 to 3.3 hold for (7.1).*

Remark. The proofs are essentially the same as those of Theorems 3.1 to 3.3, and are omitted. They are essentially the same since for each $p > 0$

$$E_x \sup_{s \le t} \left| \int_0^s \int_\Gamma q(x(s^-), \gamma) N(dsd\gamma) \right|^p$$

is bounded uniformly in $x(0) = x$, and for each $\mathcal{F}_t$-stopping time τ which is finite w.p.1,

$$E_x \left| \int_\tau^{\tau+\delta} \int_\Gamma q(x(s^-), \gamma) N(dsd\gamma) \right| = O(\delta),$$

where the $O(\delta)$ does not depend on either τ or x. Because of these bounds, the presence of the jump term in (7.1) does not alter the bounds or proofs in Theorems 3.1 to 3.3.

Itô's Formula. For the process $x(\cdot)$ satisfying (A7.1), we define the differential operator A by its value at t, acting on $f(\cdot) \in C^2(R^r)$ by

$$(Af)(x,t) = f'_x(x)\tilde{b}(x,t) + \frac{1}{2}\text{trace } f_{xx}(x)\sigma(x)\sigma'(x)$$

$$+ \lambda \int_\Gamma [f(x + q(x,\gamma)) - f(x)]\Pi(d\gamma). \tag{7.3}$$

If $\tilde{b}(x,t)$ does not depend explicitly on t or ω, then for notational convenience we write $Af(x)$ in lieu of $(Af)(x,t)$. The first two terms of (7.3) are just the $(Af)(x,t)$ of (3.2). The last term on the right is the 'mean instantaneous' rate of change of $f(x)$ due to the jump term.

For the process (7.1), Itô's Formula [11, Chapter 2.5] takes the form

$$f(x(t)) - f(x) = \int_0^t (Af)(x(s), s)ds + \int_0^t f'_x(x(s))\sigma(x(s))dw(s)$$

$$+ \int_0^t \int_\Gamma [f(x(s^-) + q(x(s^-), \gamma)) - f(x(s^-))]\tilde{N}(dsd\gamma). \tag{7.4}$$

The last two terms in (7.4) are $\mathcal{F}_t$-local martingales. Eqn. (7.4) can be rewritten as

$$f(x(t)) - f(x) = \int_0^t f'_x(x(s))\tilde{b}(x(s), s)ds + \int_0^t f'_x(x(s))\sigma(x(s))dw(s)$$

$$+ \frac{1}{2}\int_0^t \text{trace } f_{xx}(x(s)) \cdot \sigma(x(s))\sigma'(x(s))ds + J_f(t), \tag{7.5}$$

where

$$J_f(t) = \sum_{s \leq t}[f(x(s)) - f(x(s^-))]$$

$$= \int_0^t \int_\Gamma [f(x(s^-) + q(x(s^-),\gamma)) - f(x(s^-))]N(ds d\gamma). \quad (7.6)$$

Existence and uniqueness of solutions. We next turn our attention to the jump-diffusion extension of (4.1), and consider equations of the type

$$x(t) = x + \int_0^t b(x(s),u(s))ds + \int_0^t \sigma(x(s))dw(s) + J(t), \quad (7.7)$$

where

$$J(t) = \sum_{s \leq t}[x(s) - x(s^-)] = \int_0^t \int_\Gamma q(x(s^-),\gamma)N(ds d\gamma).$$

Analogously to what was done below (4.1), the differential operator for the $x(\cdot)$ solving (7.7) is defined as follows: Define A^α by

$$A^\alpha f(x) = f_x'(x)b(x,\alpha) + \frac{1}{2}\text{trace } f_{xx}(x)\sigma(x)\sigma'(x)$$

$$+ \lambda \int_\Gamma [f(x + q(x,\gamma)) - f(x)]\Pi(d\gamma). \quad (7.8)$$

Define the differential operator A^u of (7.7) by its values $(A^u f)(x,t) = A^{u(t)}f(x)$, as was done below (4.1). We say that $u(\cdot)$ is admissible with respect to $(w(\cdot), N(\cdot))$ or that $(u(\cdot), w(\cdot), N(\cdot))$ is admissible if $u(\cdot)$ is non-anticipative with respect to $(w(\cdot), N(\cdot))$. Theorem 4.1 can readily be carried over to the present case (7.7).

Definition. We say that the solution to (7.7) is *unique in the strong sense* if any two solutions on the same probability space are equal w.p.1. The solution is *unique in the weak sense* if the distribution of $(w(\cdot), u(\cdot), N(\cdot))$ determines that of $x(\cdot)$. If the solution of (4.1) is unique in the weak sense for all initial conditions and $\mathcal{F}_t$-adapted controls, then so is the solution of (7.7).

Theorem 7.2. *Assume (A4.1), (A7.1) and let $u(\cdot)$ be $\mathcal{U}$-valued and $\mathcal{F}_t$-adapted. Then there exists a solution to (7.7) for each initial condition $x(0) = x$: The solution is unique in both the strong and the weak sense. The same result holds for (7.1) if $\tilde{b}(\cdot)$ is Lipschitz continuous in x, uniformly in (ω, t) and is uniformly bounded at $x = 0$.*

Remark on the Proof. Since $N(t + \cdot) - N(t)$ is independent of $\{N(s), w(s), u(s), s \leq t\}$ and since there are only a countable number of jumps and they are isolated (w.p.1), the iterative method of Theorem 4.1 can be used to construct the solution as follows. Let $\{\tau_n, \, n < \infty\}$ denote the times at which $g(\cdot)$ is non-zero. Suppose that we have solved (7.7) until τ_n^-. At τ_n, we use $x(\tau_n) - x(\tau_n^-) = q(x(\tau_n^-), g(\tau_n))$. Then with the new initial condition $x(\tau_n^-) + q(x(\tau_n^-), g(\tau_n))$ at τ_n, solve (7.7) up to τ_{n+1} by the method of Theorem 4.1, add the next jump at τ_{n+1} and continue. Similarly for (7.1).

The solution to (7.1) or (7.7) has the properties

$$P_{\mathcal{F}_t}(\text{one jump occurs on } (t, t + \Delta])$$

$$= \Delta \lambda \Pi \{\gamma : q(x(t), \gamma) \neq 0\} + o(\Delta), \tag{7.9}$$

$$P_{\mathcal{F}_t}(\text{more than one jump occurs on } (t, t + \Delta]) = o(\Delta).$$

$$P_{\mathcal{F}_{t-}}(\text{jump of magnitude in } H \text{ occurs at } t \mid \text{jump at } t)$$

$$= \Pi\{\gamma : q(x(t^-), \gamma) \in H\}. \tag{7.10}$$

The measure transformation method. The "Girsanov measure transformation" method of Section 4 can be used in the present case and in exactly the same way as used in Section 4.

8. Jump-Diffusion Processes: The Martingale Problem Formulation

In Section 6, we found that there was an equivalence between the solutions of (4.1) and those of a certain "martingale problem." There is an analogous result for the jump-diffusion process. Just as for the diffusion case, this equivalence provides a very convenient way of showing that a particular process (which, e.g., might be obtained as a limit of a weakly convergent sequence) is actually a jump-diffusion process.

Assume (A6.1), (A7.1) and let $x(\cdot)$ solve (7.7), with $(u(\cdot), w(\cdot), N(\cdot))$ admissible. Recall the definitions of A^u given below (7.8). Define $\mathcal{F}_t = B(x(s), u(s), s \leq t)$. For $f(\cdot) \in C_0^2(R^r)$, Itô's Formula (7.4) implies that the process defined by

$$M_f(t) = f(x(t)) - f(x) - \int_0^t (Af)(x(s), s)ds \tag{8.1}$$

is an $\mathcal{F}_t$-martingale. As for the case of Section 6, there is a converse to this result.

Definition. The martingale problem. Let $u(\cdot)$ be a $\mathcal{U}$-valued measurable process, and $x(\cdot)$ a process with paths in $D^r[0, \infty)$. Let $\mathcal{F}_t$ be a filtration such that

$$\mathcal{F}_t \supset B(x(s), u(s), s \leq t) \tag{8.2}$$

and let $M_f(\cdot)$ be an $\mathcal{F}_t$-martingale for each $f(\cdot) \in C_0^2(R^r)$. Then we say that $(x(\cdot), u(\cdot))$ solves the martingale problem for the operator A^α.

Theorem 8.1. *Let* (A6.1) *and* (A7.1) *hold. For a filtration $\mathcal{F}_t$ satisfying* (8.2), *let $M_f(\cdot)$ be an $\mathcal{F}_t$-martingale for each $f(\cdot) \in C_0^2(R^r)$, where $u(\cdot)$ is a $\mathcal{U}$-valued measurable process and $x(\cdot)$ takes values in $D^r[0, \infty)$. Then there is a standard vector-valued $\mathcal{F}_t$-Wiener process such that* (7.7) *holds, where the jump term $J(\cdot)$ satisfies* (7.9), (7.10). *There is a Poisson measure $N(\cdot)$ such that $J(t) = \int_0^t \int_\Gamma q(x(s^-), \gamma) N(ds d\gamma)$.*

Remark. As in the pure diffusion case, it might be necessary to augment the probability space by adding an "independent" Wiener process.

Outline of Proof. The proof is very close to that for the case in Section 6. We first make a few remarks concerning the properties of the mean "jump term" in $M_f(\cdot)$, then we show how to get the $w(\cdot)$.

By a judicious selection of $f(\cdot)$ and use of the martingale property of $M_f(\cdot)$, the following properties can be shown (all w.p.1):

$$P_{\mathcal{F}_t}(x(\cdot) \text{ jumps on } (t, t + \Delta]) = \lambda \Delta \Pi \{\gamma : q(x(t), \gamma) \neq 0\} + o(\Delta) \quad (8.3)$$

$$P_{\mathcal{F}_t}(x(\cdot) \text{ has more than one jump on } (t, t + \Delta]) = o(\Delta) \quad (8.4)$$

$$P_{\mathcal{F}_{t-}}(x(t) - x(t^-) \in H \,|\, \text{jump at } t) = \Pi \{\gamma : q(x(t^-), \gamma) \in H\}, \quad (8.5)$$

where the $o(\Delta)$ terms are uniform in ω for $x(t)$ in any compact set.

We now illustrate some of the details of the proof of (8.3)–(8.5). Fix $\epsilon > 0$ and $z \in R^r$. For large $n > 0$, let $f_n(\cdot) \in C_0^2(R^r)$ satisfy $f_n(x) = 0$ for $|x - z| < 3\epsilon$, $0 \leq f_n(x) \leq 1$, $f_n(x) = 1$ for $3\epsilon + \frac{1}{n} < |x - z| < $ large number. Fix t and define the stopping time $\tau = \inf\{s \geq t : |x(s) - z| \geq 2\epsilon\}$. Let $|x(t) - z| < \epsilon$. Then, the fact that $M_{f_n}(\cdot)$ is a martingale and the definition of A^u implies that

$$E_{\mathcal{F}_t} f_n(x(\tau \wedge (t + \Delta))) = \lambda E_{\mathcal{F}_t} \int_t^{\tau \wedge (t+\Delta)} ds \int_\Gamma [f_n(x(s) + q(x(s), \gamma))$$

$$- f_n(x(s))] \Pi(d\gamma). \quad (8.6)$$

The right side of (8.6) is $\leq \lambda \Delta$. Letting $n \to \infty$, (8.6) implies that

$$P_{\mathcal{F}_t} \left(\sup_{0 \leq s \leq \Delta} |x(s + t) - x(t)| > 6\epsilon \right) \leq \lambda \Delta. \quad (8.7)$$

Since ϵ is arbitrary, this shows that the jump rate is $\leq \lambda$. Equation (8.4) follows from this. Using (8.4), (8.6) implies that (8.3) holds. Finally, by using a sequence $f_n(\cdot)$ which approximate the indicator functions of sets H, we can get (8.5).

Let $\{\tau_n\}$ denote the jump times of $x(\cdot)$. The last sentence of (A7.1) can be used to show that there are unique $\{g_n, n < \infty\}$ such that $x(\tau_n) - x(\tau_n^-) = q(x(\tau_n^-), g_n)$. Then (8.3)–(8.5) can be used to show that $\{\tau_{n+1} - \tau_n, n < \infty\}$ and $\{g_n, n < \infty\}$ are mutually independent sequences of independent random variables, and

$$P_{\mathcal{F}_{\tau_n^-}}(g_n \in H) = \Pi(H)$$

$$P_{\mathcal{F}_{\tau_n}}(\tau_{n+1} - \tau_n \geq t) = e^{-\lambda t}.$$

The Poisson measure can be constructed from $\{\tau_{n+1} - \tau_n, g_n, n < \infty\}$.

We now show how to get the driving Wiener process $w(\cdot)$. The basic idea is to remove the jumps from $x(\cdot)$ and then use the method of Section 6. By (8.3)–(8.5), the process defined by

$$M_f'(t) = \sum_{s \leq t}[f(x(s)) - f(x(s^-))] - \lambda \int_0^t ds \int_\Gamma [f(x(s) + q(x(s), \gamma))$$

$$- f(x(s))]\Pi(d\gamma) \tag{8.8}$$

is an $\mathcal{F}_t$-martingale. The process defined by the difference $M_f''(t) \equiv M_f(t) - M_f'(t)$ is also an $\mathcal{F}_t$-martingale. Since, we have subtracted the jumps, $M_f''(\cdot)$ has continuous paths w.p.1. By following the procedure used in Section 6, the quadratic variation of $M_f''(\cdot)$ can be shown to be (see (6.6))

$$\int_0^t |f_x'(x(s))\sigma(x(s))|^2 ds.$$

Now, $w(\cdot)$ is defined as in Section 6 (below (6.6)) by using the martingales $M_f''(\cdot)$ and the same sequence of functions used in the construction below (6.6) to get the $w_n(\cdot)$ and the limit $w(\cdot)$. Thus, we can write

$$x(t) = x + \int_0^t b(x(s), u(s))ds + \int_0^t \sigma(x(s))dw(s) + J(t), \tag{8.9}$$

where $J(\cdot)$ is a piecewise constant right continuous process satisfying (8.3)–(8.5). The demonstration is complete.

3

Controlled Stochastic Differential Equations

0. Outline of the Chapter

In this chapter, we introduce the stochastic control problem and discuss various classes of controls, approximations to these classes and questions concerning the existence of an optimal control. We work with one particular type of cost functional for simplicity in the development. In Section 1, ordinary admissible controls are introduced and it is shown why they might not be adequate for our needs. Section 2 introduces the notion of "relaxed control" for deterministic problems and shows how to use them to prove the existence of an optimal control. It is also shown that these "generalized" or relaxed controls can be approximated by piecewise constant ordinary controls.

Stochastic relaxed controls are introduced in Section 3 and existence and uniqueness of solutions to the controlled stochastic differential equations are discussed. The martingale problem formulation of Chapters 2.6 and 2.8 is revisited in Section 4, where it is shown how to verify the martingale property of the $M_f(\cdot)$. Then the existence of an optimal stochastic relaxed control is proved. The method of proof involves showing that the limit of a sequence of 'minimizing' processes solves the martingale problem for a particular differential operator associated with a given control. This result is then used to show that the "limit" process satisfies a controlled stochastic differential equation. This type of proof will be used over and over again in this book and it is worthwhile to understand it well.

Definitions. Let $\mathcal{U}$, the set of control values, be a compact set in some Euclidean space, and let $\mathcal{F}_t$ be a given filtration. Let $w(\cdot)$ and $N(\cdot,\cdot)$ be an $\mathcal{F}_t$-(R^k-valued) standard Wiener process and stationary Poisson measure resp. Let $\lambda < \infty$ and $\Pi(\cdot)$ be the jump "rate" and jump distribution of $N(\cdot,\cdot)$, resp., and let Γ be a compact set which is the range space of the jump and does not contain the origin. A $\mathcal{U}$-valued $\mathcal{F}_t$-adapted process is called an *ordinary admissible control* or simply an *admissible control*.

We will use the following assumption throughout the chapter except where otherwise mentioned.

A1.1. $q(\cdot,\cdot)$ *is a bounded and continuous function from* $R^r \times \Gamma$ *to* R^r; $q(x,0) \equiv 0$, $b(\cdot,\cdot)$ *is a continuous function from* $R^r \times \mathcal{U}$ *to* R^r *and* $\sigma(\cdot)$ *is*

a continuous ($r \times k$-*matrix-valued*) *function on* R^r. *Also for some* $K < \infty$
and all $\alpha \in \mathcal{U}$

$$|\sigma(x)| + |b(x, \alpha)| \leq K(1 + |x|).$$

1. Controlled SDE's; Introduction

The basic model of a controlled SDE is that given by (2.7.7) for the
controlled jump diffusion. We often write (2.7.7) symbolically in the form

$$dx = b(x, u)dt + \sigma(x)dw + \int_\Gamma q(x, \gamma)N(dtd\gamma). \tag{1.1}$$

Of course, if the jumps are absent, then we simply drop the last term.
For many purposes of approximation and optimization when we are interested in the limits of sequences of processes which satisfy (1.1), the class
of ordinary admissible controls defined above and used in (1.1) is not large
enough. For example, there might not be an optimal control in that class.
Consider the following problem. Let $\{u_n(\cdot)\}$ be a sequence of deterministic
controls which are defined by $u_n(t) = \text{sign}(\sin nt)$, and let $x_n(\cdot)$ be the
solution to (1.1) when $u_n(\cdot)$ is used. For arbitrary $u(\cdot)$ and corresponding
solution $x(\cdot)$, let the cost functional be

$$V_T(x, u) = E_x^u \int_0^T k(x(s), u(s))ds$$

for some bounded and continuous function $k(\cdot)$. The value of the cost
$V_T(x, u)$ actually depends on the joint distribution of $(u(\cdot), w(\cdot), N(\cdot))$
and not just on $u(\cdot)$, and so does the expectation E_x^u. We omit the $(w(\cdot),
N(\cdot))$ dependence in the notation for the sake of simplicity. By the remark
following Theorem 2.3.3, the sequence of solutions $\{x_n(\cdot), n < \infty\}$ is tight
in $D^r[0, T]$, and hence it has a weakly convergent subsequence. But for
any weakly convergent subsequence $\{x_{n_i}(\cdot), i < \infty\}$ we cannot necessarily find an ordinary admissible control $\bar{u}(\cdot)$ such that $x_{n_i}(\cdot) \Rightarrow x(\cdot)$ and
$V_T(x, u_{n_i}) \to V_T(x, \bar{u})$, where $x(\cdot)$ satisfies

$$dx = b(x, \bar{u})dt + \sigma(x)dw + \int_\Gamma q(x, \gamma)N(dtd\gamma).$$

We will often deal with weakly convergent sequences of controlled stochastic differential equations. We need a definition of admissible control which
is broad enough so that the limit process can be represented as a controlled
process where the control is "admissible" in some appropriately extended
sense. It is also important that the infimum of the value functions over this
class of extended admissible controls equals the infimum of the value functions over the class of ordinary admissible controls, and that each control
in the extended class of controls be well approximated by some ordinary

admissible control. The class of 'relaxed admissible controls' has been introduced to handle just such problems [B3, F1, F2, K11, W1]. In order to introduce the concept in the easiest way, we first consider the simpler deterministic problem.

2. Relaxed Controls: Deterministic Case

Consider the ODE (ordinary differential equation)

$$\dot{x} = b(x, u), \tag{2.1}$$

where $u(\cdot)$ is a $\mathcal{U}$-valued measurable function, $b(\cdot)$ satisfies (A1.1) and the cost functional is

$$V_T(x, u) = \int_0^T k(x(s), u(s))ds,$$

where $k(\cdot)$ is a real valued bounded and continuous function and $x = x(0)$.

Definition. Recall that $\mathcal{B}(\mathcal{U} \times [0, \infty))$ is the σ-algebra of Borel subsets of $\mathcal{U} \times [0, \infty)$. Let $\mathcal{R}(\mathcal{U} \times [0, \infty))$ denote the set of measures $m(\cdot)$ on $\mathcal{B}(\mathcal{U} \times [0, \infty))$ which satisfy: $m(\mathcal{U} \times [0, t]) = t$ for all t. Such measures are called *admissible relaxed controls*. Their relationship to ordinary controls will be seen below. For notational purposes, we suppose that the controls are defined on the time interval $[0, \infty)$, even if the control process is of interest only over a finite interval. For any $B \in \mathcal{B}(\mathcal{U})$, we will write $m(B \times [0, t])$ simply as $m(B, t)$ for notational convenience. Since $m(\mathcal{U}, t) = t$ for all t, and $m(B, \cdot)$ is nondecreasing, it is absolutely continuous. Hence the derivative $\dot{m}(B, t) \equiv m_t(B)$ exists almost everywhere for each B. We can define the $m_t(B)$ such that for each t, $m_t(\cdot)$ is a measure on $\mathcal{B}(\mathcal{U})$ with total mass unity, $m.(B)$ is Borel measurable for all B and also $m_t(d\alpha)dt = m(d\alpha dt)$. We will also refer to the derivatives $m_t(\cdot)$ as an admissible relaxed control, and no confusion should arise from this usage.

For a relaxed control $m(\cdot)$, i.e., for $m(\cdot) \in \mathcal{R}(\mathcal{U} \times [0, \infty))$, and $x(0) = x$, redefine the controlled ODE (2.1) as

$$\dot{x} = \int_{\mathcal{U}} b(x, \alpha)m_t(d\alpha) \equiv \tilde{b}(x, t), \tag{2.2a}$$

or equivalently,

$$x(t) = x + \int_0^t \int_{\mathcal{U}} b(x(s), \alpha)m(d\alpha ds) = x + \int_0^t \int_{\mathcal{U}} b(x(s), \alpha)m_s(d\alpha)ds. \tag{2.2.b}$$

Similarly, we write the associated cost functional in the form

$$V_T(x, m) = \int_0^T \int_{\mathcal{U}} k(x(s), \alpha)m_s(d\alpha)ds = \int_0^T \int_{\mathcal{U}} k(x(s), \alpha)m(d\alpha ds). \tag{2.3}$$

We can represent any ordinary admissible control $u(\cdot)$ as a relaxed control by using the *definition* $m_t(d\alpha) = \delta_{u(t)}(\alpha)d\alpha$, where $\delta_u(\alpha)$ is the *Dirac delta function* concentrated at the point $u = \alpha$. Thus, the measure valued derivative $m_t(\cdot)$ of the relaxed control representation of $u(t)$ is a measure which is concentrated at the point $u(t)$. The class of relaxed controls is not intended to be and obviously cannot be used in an actual application. They are introduced for purely mathematical reasons; in particular, for the purpose of facilitating the study of approximation and convergence. It is most important to know that the infimum of the costs over the relaxed controls equals the infimum over the ordinary controls.

An example of a relaxed control. Let $\alpha_i \in \mathcal{U}$, $i = 1, 2$. Let $m_t(\cdot)$ be the measure which takes value $1/2$ at each point α_i. Then (2.2a) can be written as

$$\dot{x} = \frac{1}{2}[b(x, \alpha_1) + b(x, \alpha_2)].$$

More generally, let α_i, $i \leq q$, be in $\mathcal{U}$ and let $f_i(\cdot)$ be non-negative measurable functions such that $\sum_1^q f_i(t) = 1$. Define

$$m_t(d\alpha) = \sum_1^q f_i(t)\delta_{\alpha_i}(\alpha)d\alpha.$$

Then (2.2a) can be written as

$$\dot{x} = \sum_1^q f_i(b)b(x, \alpha_i).$$

In general, let $\dot{x}(\cdot)$ and $\dot{v}(\cdot)$ be measurable functions which satisfy

$$\begin{pmatrix} \dot{x} \\ \dot{v} \end{pmatrix} \in \overline{co} \left\{ \begin{matrix} b(x, \mathcal{U}) \\ k(x, \mathcal{U}) \end{matrix} \right\},$$

where $\overline{co}$ is the closure of the convex hull. Then, there is a relaxed control $m_t(\cdot)$ which attains these derivatives for (2.2a), (2.3).

Working with relaxed controls is essentially equivalent to working with a new problem where the set of possible velocities and cost rates $(b(x, \mathcal{U}), k(x, \mathcal{U}))$ is replaced by its convex hull.

The topology of the space of relaxed controls. The "compact weak" topology will be used on $\mathcal{R}(\mathcal{U} \times [0, \infty))$. Recall that $m^n(\cdot) \to m(\cdot)$ in this topology if and only if

$$\int_0^\infty \int_{\mathcal{U}} f(\alpha, s)m^n(d\alpha ds) \to \int_0^\infty \int_{\mathcal{U}} f(\alpha, s)m(d\alpha ds)$$

for each $f(\cdot) \in C_0(\mathcal{U} \times [0, \infty))$. The space $\mathcal{R}(\mathcal{U} \times [0, \infty))$ is compact under this topology. Thus any sequence has a convergent subsequence. This is one of the keys to the usefulness of the concept of a relaxed control.

The continuity and growth condition which we have assumed on $b(\cdot)$ is not adequate to assure that (2.1) or (2.2) have unique solutions for each admissible control, relaxed or ordinary. A solution does exist for each such control, however. When optimality is referred to, it is optimality over *all* solutions.

For a first application of the concept of relaxed control, we have the existence result of Theorem 2.1.

Theorem 2.1 [B3, W1]. *Under the assumptions in this section on $b(\cdot)$ and $k(\cdot)$, there exists an optimal control in the class of admissible relaxed controls.*

Proof. Note that $|\int_{\mathcal{U}} b(x, \alpha) m_t(d\alpha))| \leq K(1 + |x|)$. Thus, for the initial condition $x = x(0)$ in any compact set, the set of solutions to (2.2) over all admissible controls is uniformly bounded and equi-continuous on any interval $[0, T]$. Let admissible relaxed controls $m^n(\cdot)$ be such that

$$V_T(x, m^n) \to \inf_m V_T(x, m) \equiv \overline{V}_T(x),$$

where the inf is over all admissible relaxed controls, and let $x^n(\cdot)$ be the solution to (2.2) corresponding to the control $m^n(\cdot)$. Then $\{x^n(\cdot), m^n(\cdot), n < \infty\}$ is in a compact set in $C^r[0, \infty) \times \mathcal{R}(\mathcal{U} \times [0, \infty))$. Let n' index a convergent subsequence with limit denoted by $(x(\cdot), \overline{m}(\cdot))$. Thus, $x^{n'}(\cdot) \to x(\cdot)$ uniformly on $[0, T]$. Because of this uniformity of convergence and the properties of $\{m^{n'}(\cdot), n' < \infty\}$,

$$\left| \int_0^t \int_{\mathcal{U}} b(x^n(s), \alpha) m^{n'}(d\alpha ds) - \int_0^t \int_{\mathcal{U}} b(x(s), \alpha) m^{n'}(d\alpha ds) \right| \to 0$$

uniformly on $[0, T]$. Thus

$$x^{n'}(t) = x + \int_0^t \int_{\mathcal{U}} b(x^n(s), \alpha) m^{n'}(d\alpha ds) \to x(t)$$

$$= x + \int_0^t \int_{\mathcal{U}} b(x(s), \alpha) \overline{m}(d\alpha ds)$$

$$= x + \int_0^t \int_{\mathcal{U}} b(x(s), \alpha) \overline{m}_s(d\alpha) ds. \tag{2.4}$$

Similarly, we have

$$V_T(x, m^{n'}) = \int_0^T \int_{\mathcal{U}} k(x^n(s), \alpha) m^n(d\alpha ds)$$

$$\to \int_0^T \int_{\mathcal{U}} k(x(s), \alpha) \overline{m}(d\alpha ds) = V_T(x, \overline{m}). \tag{2.5}$$

Thus, since $V_T(x, m^{n'}) \to V_T(x)$ and $(x(\cdot), \overline{m}(\cdot))$ satisfies (2.2), $\overline{m}(\cdot)$ is indeed an optimal admissible relaxed control.   Q.E.D.

Remark. Notice the convenience of use of the relaxed controls in getting the limits. Part of the convenience stems from the fact that, in the relaxed control formulation, the dynamics and cost terms are "linear" in the control. To see why relaxed controls are often needed when working with the convergence of a sequence of solutions, consider the following example. Let $u_n(t) = \text{sign}[\sin(nt)]$, and let $m^n(\cdot)$ be the relaxed control representation of $u_n(\cdot)$; i.e., $m^n(\cdot)$ is the relaxed control whose derivative $m_t^n(d\alpha)$ is the measure concentrated at the point $u_n(t)$. Then $\{m^n(\cdot), n < \infty\}$ is compact. Obviously, the only limit point of this set is the measure $\hat{m}(\cdot)$ defined by the derivative $\hat{m}_t(1) = \hat{m}_t(-1) = 1/2$; i.e., the measure concentrated at ± 1 with mass $1/2$ at each point. If $x^n(\cdot)$ is the path of (2.1) associated with $u_n(\cdot)$, then $x^n(\cdot) \to x(\cdot)$, where

$$\dot{x} = \frac{1}{2}[b(x, 1) + b(x, -1)] = \int_{\mathcal{U}} b(x, \alpha)\hat{m}_t(d\alpha).$$

In general, we cannot find an admissible ordinary control $u(\cdot)$ such that $\dot{x}(t) = b(x(t), u(t))$. There might not be an optimal control in the class of ordinary admissible controls unless the set $(b(x, \mathcal{U}), k(x, \mathcal{U}))$ is convex for each x; in this latter case the "relaxed" and "ordinary" controls are essentially equivalent.

Although the relaxed controls are a device with primarily a mathematical use, they can always be approximated by ordinary controls, as shown by Theorem 2.2. The theorem is sometimes referred to as the "Chattering Theorem", since the approximations might "oscillate fast." We sometimes use $x(t, m)$ or $x(t, u)$ to denote the solution to (2.2) under the controls $m(\cdot)$ or $u(\cdot)$.

Theorem 2.2 [B3]. *Assume the conditions of Theorem 2.1, and suppose that (2.2) has a unique solution for each initial condition $x(0) = x$ and admissible relaxed control. Let $T < \infty$ and $\gamma > 0$. There is a finite set $\{\alpha_1^\gamma, \ldots, \alpha_{k_\gamma}^\gamma\} \equiv \mathcal{U}^\gamma \subset \mathcal{U}$ and a $\delta > 0$ such that for any admissible relaxed control $m(\cdot)$, there is a $\mathcal{U}^\gamma$-valued ordinary admissible control $u_m^\gamma(\cdot)$, which is constant on each interval $[i\delta, i\delta + \delta)$, $i\delta + \delta \leq T$, and is such that*

$$\sup_{t \leq T} |x(t, m) - x(t, u_m^\gamma)| \leq \gamma,$$

$$|V_T(x, m) - V_T(x, u_m^\gamma)| \leq \gamma. \tag{2.5}$$

Proof. We follow the classical method. Given $\rho > 0$, let B_i^ρ, $i \leq N_\rho < \infty$, be disjoint measurable sets whose diameter is less than ρ and whose union is $\mathcal{U}$. Fix points $\alpha_i^\rho \in B_i^\rho$ and define $\mathcal{U}_\rho = \{\alpha_i^\rho, i \leq N_\rho\}$. Let $m(\cdot)$ be an admissible relaxed control. For $\Delta > 0$ and $i \geq 0$, define

$$\tau_{ij}^{\rho\Delta} = \int_{i\Delta}^{i\Delta+\Delta} m_s(B_j^\rho)ds = m(B_j^\rho, i\Delta + \Delta) - m(B_j^\rho, i\Delta).$$

$\tau_{ij}^{\rho\Delta}$ is the total integrated time that the relaxed control takes values in the set B_j^ρ on the time interval $[i\Delta, i\Delta + \Delta)$. Clearly, $\sum_{j=1}^{N_\rho} \tau_{ij}^{\rho\Delta} = \Delta$. For $i \geq 0$, divide the interval $[i\Delta + \Delta, i\Delta + 2\Delta)$ into N_ρ successive subintervals of lengths $\tau_{ij}^{\rho\Delta}$, $j = 1, \ldots, N_\rho$.

Define the ordinary admissible control $u_m^{\rho\Delta}(\cdot)$ as follows. On $[0, \Delta)$, it is constant with any value in $\mathcal{U}_\rho$. On the jth subinterval of $[i\Delta + \Delta, i\Delta + 2\Delta)$, let $u_m^{\rho\Delta}(\cdot)$ take the value α_j^ρ. Recall that this subinterval has length $\tau_{ij}^{\rho\Delta}$.

Let $m^{\rho\Delta}(\cdot)$ denote the relaxed control representation of $u_m^{\rho\Delta}(\cdot)$. By construction, $m^{\rho\Delta}(\cdot) \to m(\cdot)$ as $\rho \to 0$ and $\Delta \to 0$. By the "compactness" technique of the proof of Theorem 2.1 and the uniqueness of the solution of (2.2) under $m(\cdot)$, $x(\cdot, m^{\rho\Delta}) \to x(\cdot, m)$ as $\rho \to 0$ and $\Delta \to 0$. Also $V_T(x, m^{\rho\Delta}) \to V_T(x, m)$.

We next show that, as $\rho \to 0$ and $\Delta \to 0$,

$$\sup_{\substack{t \leq T \\ m(\cdot)}} |x(t, m) - x(t, u_m^{\rho\Delta})| \to 0, \tag{2.6}$$

$$\sup_{m(\cdot)} |V_T(x, m) - V_T(x, u_m^{\rho\Delta})| \to 0. \tag{2.7}$$

Suppose that (2.6) is false. Then there are $\delta_0 > 0$, $m_n(\cdot)$ and ρ_n and Δ_n such that $\rho_n \to 0$, $\Delta_n \to 0$ and

$$\sup_{t \leq T} |x(t, m_n) - x(t, u_{m_n}^{\rho_n \Delta_n})| \geq \delta_0, \text{ all } n. \tag{2.8}$$

Let $\hat{m}_n(\cdot)$ denote the relaxed control representation of $u_{m_n}^{\rho_n \Delta_n}(\cdot)$. Let us extract a convergent subsequence of both

$$\{x(\cdot, m_n), m_n(\cdot), n < \infty\}, \quad \{x(\cdot, \hat{m}_n), \hat{m}_n(\cdot), n < \infty\}.$$

Let the limits be denoted by $(\hat{x}(\cdot), \hat{m}(\cdot))$ and $(\tilde{x}(\cdot), \tilde{m}(\cdot))$, resp. Both pairs satisfy (2.2). But $\hat{m}(\cdot) = \tilde{m}(\cdot)$. Hence $\hat{x}(\cdot) = \tilde{x}(\cdot)$ by uniqueness. This yields a contradiction to (2.8). Thus (2.6) holds. The result (2.7) is similarly proved. By using the $u_m^{\rho\Delta}(\cdot)$ and setting $\mathcal{U}^\gamma = \mathcal{U}_\rho$ for small enough ρ and Δ we get the theorem, except for the non-constancy of the lengths of the intervals on which $u_m^{\rho\Delta}(\cdot)$ is constant. Another simple approximation will take care of that and we omit the details.   Q.E.D.

3. Stochastic Relaxed Controls

Definition. The *admissible stochastic relaxed controls* are defined just as the admissible deterministic controls were in the last section, but with the added requirement of "adaptivity." Given a filtration $\mathcal{F}_t$, let $w(\cdot)$ and $N(\cdot)$ be as defined above (A1.1). Let $m(\cdot)$ be a random variable with values in $\mathcal{R}(\mathcal{U} \times [0, \infty))$, such that for each $B \in \mathcal{B}(\mathcal{U})$, the function defined by $m(B, t) \equiv m(B \times [0, t])$ is $\mathcal{F}_t$-adapted. Then we say that $m(\cdot)$ is an *admissible relaxed control* or, alternatively, that $(m(\cdot), w(\cdot), N(\cdot))$ is *admissible* or that $m(\cdot)$ is *admissible with respect to* $(w(\cdot), N(\cdot))$, depending on convenience. If the $(w(\cdot), N(\cdot))$ are obvious, then we usually drop them from the notation. By the definition, each sample path of $m(\cdot)$ satisfies the properties of a deterministic admissible relaxed control.

If $m(\cdot)$ is an admissible relaxed control then, analogously to the situation for the deterministic case, there is a "derivative" process $m_t(\cdot) = \dot{m}(t, \cdot)$ such that $m_t(d\alpha)dt = m(dt d\alpha)$ and $m_{\cdot}(B)$ is $\mathcal{F}_t$-adapted for each $B \in \mathcal{B}(\mathcal{U})$. We will also call the $m_t(\cdot)$ an admissible relaxed control. As below (2.3), any ordinary admissible control can be represented as a relaxed control, via the definition of the derivative $m_t(d\alpha) = \delta_{u(t)}(\alpha)d\alpha$.

The construction of the derivative $m_t(\cdot)$ of $m(\cdot)$ requires a little more care in the stochastic case than in the deterministic case, owing to the $\mathcal{F}_t$-adaptability requirement. The construction is roughly as follows. Let $\overline{\mathcal{U}}^n$ be a finite partition of $\mathcal{U}$, with $\overline{\mathcal{U}}^{n+1} \supset \overline{\mathcal{U}}^n$, and set $\overline{\mathcal{U}} = \bigcup_n \overline{\mathcal{U}}^n$. Let the diameters of the sets of $\overline{\mathcal{U}}^n$ be less than $1/n$. For $B \in \overline{\mathcal{U}}$, define

$$m_t(B) = \lim_\delta \frac{1}{\delta}[m(B, t) - m(B, t - \delta)].$$

The limit is defined for almost all (ω, t) for each B. We can choose the values on the null sets, so that the $m_t(B)$ can be extended to a measure on $\mathcal{B}(\mathcal{U})$ with the desired properties. The procedure is similar to that used to construct regular conditional distributions, as in [B9, Ch. 4.3].

Relaxed feedback control. An ordinary admissible control $u(\cdot)$ is called a *feedback control* if there is a $\mathcal{U}$-valued Borel measurable function $u_0(\cdot)$ such that $u(t) = u_0(x(t))$ for almost all ω, t. There is an analogous definition for the relaxed controls. For each x, let $m_{fb}(x, \cdot)$ be a probability measure on $(\mathcal{U}, \mathcal{B}(\mathcal{U}))$ and suppose that for each $B \in \mathcal{B}(\mathcal{U})$, $m_{fb}(\cdot, B)$ is Borel measurable as a function of x. If for almost all (ω, t), the derivative $m_t(\cdot)$ of a relaxed control $m(\cdot)$ can be written as $m_t(\cdot) = m_{fb}(x(t), \cdot)$, then $m(\cdot)$ is said to be a *relaxed feedback control*. Of course, any feedback control $u(x)$ can be represented as a relaxed feedback control via the form $\delta_{u(x)}(\alpha)d\alpha$.

Controlled stochastic differential equations. We will use (A2.7.1) and (A1.1) throughout this and the next section. Let $m(\cdot)$ be admissible with respect to $(w(\cdot), N(\cdot))$. The following model of a controlled SDE will be used:

$$dx = \int_{\mathcal{U}} b(x,\alpha)m_t(d\alpha)dt + \sigma(x)dw + \int_{\Gamma} q(x,\gamma)N(dsd\gamma). \qquad (3.1)$$

Define $\tilde{b}(\cdot)$ by $\tilde{b}(x,t) = \int b(x,\alpha)m_t(d\alpha)$. By (A1.1), $\tilde{b}(\cdot)$ satisfies the condition (A2.3.1). Hence Theorem 2.7.1 holds, where the $K_i(x,T)$ do not depend on the chosen control, and the convergence in Theorem 2.3.3 is uniform in the control. Hence, by the remark following Theorem 2.3.3, the set of solutions $x(\cdot, m)$ to (3.1) for arbitrary admissible $(m(\cdot), w(\cdot), N(\cdot))$ is tight in $D^r[0, \infty)$. If there is no "jump term", then all processes which are limits of weakly convergent subsequences have continuous paths w.p.1.

Usually, the control $m(\cdot)$ which is used will be obvious, and we write $x(\cdot)$ for the solution to (3.1). Otherwise we will write $x(\cdot, m)$.

Remark. Note that neither the variance nor the "jump term" are controlled. These omissions are made for the sake of simplicity only.

Existence and uniqueness of solutions of (3.1) by the measure transformation method. Under certain conditions, if (3.1) has a unique (weak sense) solution in the absence of control, then a unique (weak sense) solution can be defined for all controls, via the measure transformation method of Chapter 2.4. Let $\hat{b}_1(\cdot, \cdot)$ be a bounded and continuous function of (x, α), and suppose that (A1.1) holds for functions $b(\cdot)$, $\sigma_1(\cdot)$, $\sigma_2(\cdot)$ (functions of x only) and $q(\cdot, \cdot)$. Let $\sigma_1^{-1}(\cdot)$ be bounded and continuous. For a given filtration $\mathcal{F}_t$, let $\hat{w}_1(\cdot)$, $w_2(\cdot)$ and $N(\cdot, \cdot)$ be mutually independent standard vector-valued $\mathcal{F}_t$-Wiener processes and an $\mathcal{F}_t$-Poisson measure, resp. Finally, let $m(\cdot)$ be admissible with respect to $(\hat{w}_1(\cdot), w_2(\cdot), N(\cdot, \cdot))$.

Let us represent the *uncontrolled system* by

$$dx = b(x)dt + \begin{pmatrix} \sigma_1(x)d\hat{w}_1 \\ \sigma_2(x)dw_2 \end{pmatrix} + \int_{\Gamma} q(x,\gamma)N(dtd\gamma). \qquad (3.2)$$

Suppose that (3.2) has a unique weak sense solution on each $[0,T]$, and define $\hat{b}_{1,m}(\cdot)$ by $\hat{b}_{1,m}(x,t) = \int \hat{b}_1(x,\alpha)m_t(d\alpha)$. Following the measure transformation technique discussed in Chapter 2.4, define the measure $\hat{P}$ by

$$\hat{P}(A) = \int_A R(T)P(d\omega),$$

where P denotes the original measure on the probability space and

$$R(T) = \exp\left[\int_0^T (\sigma_1^{-1}(x(s))\hat{b}_{1,m}'(x(s),s))d\hat{w}_1(s)\right.$$

$$\left. - \frac{1}{2}\int_0^T |\sigma_1^{-1}(x(s))\hat{b}_{1,m}(x(s),s))|^2 ds\right]. \tag{3.3}$$

Define the process $w_1(t) = \hat{w}_1(t) - \int_0^t \sigma_1^{-1}(x(s))\hat{b}_{1,m}(x(s),s)ds$. Then, as in Chapter 2.4, under the measure $\hat{P}$ the $(w_1(\cdot), w_2(\cdot))$ are mutually independent standard vector-valued $\mathcal{F}_t$-Wiener processes. Also $N(\cdot,\cdot)$ is an $\mathcal{F}_t$-Poisson measure with the same statistics as under the measure P, and is independent of the Wiener processes.

The equation (3.2) can be rewritten as

$$dx = b(x)dt + \left(\begin{array}{c} \int_U \hat{b}_1(x,\alpha)m_t(d\alpha)dt \\ 0 \end{array}\right) + \left(\begin{array}{c} \sigma_1(x)dw_1 \\ \sigma_2(x)dw_2 \end{array}\right)$$

$$+ \int_\Gamma q(x,\gamma)N(dtd\gamma). \tag{3.4}$$

Equation (3.4) has a unique weak sense solution on each $[0,T]$ and $m(\cdot)$ is admissible with respect to $(w_1(\cdot), w_2(\cdot), N(\cdot,\cdot))$. We have just shown that if the control system is of the form (3.4) and the uncontrolled system has a unique weak sense solution on each $[0,T]$, then so does the controlled system. The above method for constructing solutions of controlled systems appears to have been introduced in [B1] for "path dependent" controls and has been heavily used in stochastic control theory since then.

4. The Martingale Problem Revisited

We will recapitulate the discussion concerning the martingale problem of Chapter 2, Sections 6 and 8 for the model (3.1). Recall the definition of the differential operator A^α given in (2.7.8):

$$A^\alpha f(x) = f_x'(x)b(x,\alpha) + \frac{1}{2}\text{trace } f_{xx}(x)\sigma(x)\sigma'(x)$$

$$+ \lambda\int_\Gamma [f(x + q(x,\gamma)) - f(x)]\Pi(d\gamma). \tag{4.1}$$

The A^α is just the differential operator of (1.1) with the control fixed at the value α.

Definition. Let $m(\cdot)$ be an admissible relaxed control for (3.1). Then, in analogy to the definition of the differential operator A^u, defined below (2.7.8) we define A^m to be the differential operator of (3.1) under $m(\cdot)$. The value of A^m acting on $f(x)$ at time t can be written as

$$(A^m f)(x,t) = \int_U A^\alpha f(x)m_t(d\alpha).$$

Then, as for (2.8.1), the process $M_f(\cdot)$ defined by

$$M_f(t) = f(x(t)) - f(x) - \int_0^t \int_{\mathcal{U}} A^\alpha f(x(s)) m_s(d\alpha) ds$$

$$= f(x(t)) - f(x) - \int_0^t (A^m f)(x(s), s) ds \qquad (4.2)$$

is an $\mathcal{F}_t$-martingale.

Definition. Now, consider the converse problem as in Chapter 2.8. Let $x(\cdot)$ have paths in $D^r[0, \infty)$ and let $m(\cdot)$ be an $\mathcal{R}(\mathcal{U} \times [0, \infty))$-valued random variable. Let $\mathcal{F}_t$ be a filtration such that $\mathcal{F}_t \supset \mathcal{B}(x(s), m(s), s \le t)$, and suppose that $M_f(\cdot)$ is an $\mathcal{F}_t$-martingale for each $f(\cdot) \in C_0^2(R^r)$. Then we say that $(x(\cdot), m(\cdot))$ *solves the martingale problem for the operator* A^α. Theorem 2.8.1 also holds for the case of relaxed controls.

On proving that $M_f(\cdot)$ is a martingale. In the sequel, we will often need to show that processes of the form $M_f(\cdot)$ are martingales with respect to an appropriate filtration. We now describe the usual way in which this is done. The method will be used frequently. Let p, q, be arbitrary integers. Let $\varphi(\cdot)$ and $\varphi_j(\cdot)$, $j \le p$, be real valued, bounded and continuous functions of (α, s) with compact support. Define the function $(\varphi, m)_t = \int_0^t \int_{\mathcal{U}} \varphi(\alpha, s) m(d\alpha ds)$. Let T be the (at most countable) set of points where $P(x(t) \ne x(t^-)) > 0$. It is the complement of the set T_p introduced in Chapter 1.5. Let $h(\cdot)$ be an arbitrary bounded and continuous function of its arguments. Let $t_i \le t \le \tau + t$, $i \le q$, where t_i, t and $t + \tau$ are not in T. Suppose that we have shown that

$$Eh((\varphi_j, m)_{t_i}, x(t_i), j \le p, i \le q)[M_f(t + \tau) - M_f(t)] = 0. \qquad (4.3)$$

The relationship (4.3) will usually be obtained by showing that it holds for the limit of some weakly convergent sequence of processes. See Theorem 5.1 below. Since $h(\cdot)$ is arbitrary, (4.3) implies that (w.p.1)

$$E[M_f(t + \tau) - M_f(t) \,|\, (\varphi_j, m)_{t_i}, x(t_i), j \le p, i \le q] = 0. \qquad (4.4)$$

Since p, q and $\varphi_j(\cdot)$ are arbitrary and $\{t_i\}$ ranges over all times $\le t$ (excluding a countable set), we can conclude that (w.p.1)

$$E[M_f(t + \tau) - M_f(t) \,|\, m(s), x(s), s \le t] = 0. \qquad (4.5)$$

Since $M_f(\cdot)$ has paths in $D[0, \infty)$ and t and $t + \tau$ are arbitrary (excluding values in T), (4.5) implies that $M_f(\cdot)$ is a martingale on the filtration $\mathcal{F}_t = \mathcal{B}(m(s), x(s), s \le t)$.

     Thus $(x(\cdot), m(\cdot))$ solves the martingale problem for the operator A^α and filtration $\mathcal{F}_t$. The "driving" processes $w(\cdot)$ and $N(\cdot)$ are constructed as in Chapter 2.8. If $\sigma^{-1}(x)$ exists for all x, then we can construct $w(\cdot)$ to be an $\mathcal{F}_t$-Wiener process. Otherwise, we need to augment the probability space and the $\mathcal{F}_t$ by the addition of an "independent" Wiener process, as discussed in Chapter 2.

An extension: the martingale problem for the triple $(x(\cdot),\ w(\cdot),$ $m(\cdot))$. Sometimes, we need to show that a given triple $(x(\cdot),\ w(\cdot),\ m(\cdot))$ solves an SDE of the type (3.1), for some $N(\cdot)$, where $w(\cdot)$ *is* the driving Wiener process. A convenient way to do this is to show that the triple solves a certain martingale problem.

First, let us look at (3.1) itself. Let $m(\cdot)$ be admissible with respect to $(w(\cdot), N(\cdot))$, and let $x(\cdot)$ solve (3.1). Define the differential operator A_1^α for the augmented system as follows. Since $w(t)$ takes values in R^k, we need to use $f(\cdot) \in C^2(R^r \times R^k)$. Define A_1^α by

$$A_1^\alpha f(x,w) = f_x'(x,w)b(x,\alpha)+\frac{1}{2}\text{trace } f_{xx}(x,w)\sigma(x)\sigma'(x)+\frac{1}{2}\sum_i f_{w_iw_i}(x,w)$$

$$+ \sum_{i,j} f_{x_iw_j}(x,w)\sigma_{ij}(x) + \lambda \int_\Gamma [f(x + q(x,\gamma)) - f(x)]\Pi(d\gamma). \qquad (4.6)$$

The operator A_1^α is just the differential operator of the pair $(x(\cdot),\ w(\cdot))$, satisfying (3.1), when the control value is fixed at α. Define

$$(A_1^m f)(x,w,t) = \int_{\mathcal{U}} A_1^\alpha f(x,w)m_t(d\alpha).$$

For $f(\cdot) \in C_0^2(R^{r+k})$, define the process $M_f(\cdot)$ by

$$M_f(t) = f(x(t),w(t)) - f(x,0) - \int_0^t \int_{\mathcal{U}} A_1^\alpha f(x(s),w(s))m_s(d\alpha)ds. \qquad (4.7)$$

If $(x(\cdot),m(\cdot),w(\cdot))$ satisfies (3.1), then $M_f(\cdot)$ is a martingale.

Now, consider the converse problem. Let a triple $(x(\cdot),\ w(\cdot),\ m(\cdot))$ be given and define $\mathcal{F}_t = \mathcal{B}(x(s),w(s),m(s),\ s \le t)$. The $(x(\cdot),w(\cdot))$ have paths in $D^{r+k}[0,\infty)$ and $m(\cdot)$ takes values in $\mathcal{R}(\mathcal{U} \times [0,\infty))$. We are not told whether the triple satisfies (3.1) for some $N(\cdot)$. But, we wish to check whether that is the case. To do this, we use the idea of the martingale problem of Chapter 2.8, or earlier in this section. Suppose that $M_f(\cdot)$ can be shown to be an $\mathcal{F}_t$-martingale for all $f(\cdot) \in C_0^2(R^{r+k})$. We then say that the triple solves the martingale problem for the operator A_1^α. By the method of Chapter 2.8, it can then be shown that the given triple satisfies (3.1) for some Poisson measure $N(\cdot)$ and that $m(\cdot)$ is admissible.

The need to work with the augmented system arises in certain technical arguments, where the triple is given as a weak limit of a certain sequence, and we wish to check whether or not the "$w(\cdot)$" component of the limit can be used as the driving Wiener process in (3.1).

5. Approximations, Weak Convergence and Optimality

In this section, we do three things which provide powerful techniques which will be used throughout the book. (1) In Theorem 5.1, we let $\{m^n(\cdot)\}$ be a sequence of admissible controls for (3.1) (with $x^n(\cdot)$ denoting the associated solution) and show that the weak limits satisfy (3.1). The proof illustrates the use of the *martingale method* and, in particular, shows the usefulness of the "martingale problem" formulation discussed in Chapters 2.6, 2.8 and in the last section. (2) In Theorem 5.2, we discuss the approximation of an admissible relaxed control by an admissible ordinary control. (3) In Theorem 5.3, the existence of an optimal admissible control is established.

1. **Theorem 5.1.** *Assume* (A2.7.1), (A1.1) *and let* $m^n(\cdot)$ *be admissible for each* n. *Let* $x^n(\cdot)$ *denote the corresponding solution and* $w^n(\cdot)$ *the "driving" Wiener process. Then* $\{x^n(\cdot), m^n(\cdot), w^n(\cdot), n < \infty\}$ *is tight. There is a filtration* $\mathcal{F}_t$ *such that the limit* $(x(\cdot), m(\cdot), w(\cdot))$ *of any weakly convergent subsequence satisfies* (3.1) *for some* $\mathcal{F}_t$-*Poisson measure* $N(\cdot)$, *where the limit* $w(\cdot)$ *is a standard vector valued* $\mathcal{F}_t$-*Wiener process and* $m(\cdot)$ *is admissible. Similarly, the limit* $(x(\cdot), m(\cdot))$ *of any weakly convergent subsequence of* $\{x^n(\cdot), m^n(\cdot), n < \infty\}$ *solves the martingale problem for the operator* A^α *and there is a standard vector-valued Wiener process* $w(\cdot)$ *and Poisson measure* $N(\cdot)$ *such that* $(x(\cdot), m(\cdot), w(\cdot), N(\cdot))$ *solve* (3.1) *and* $m(\cdot)$ *is admissible.*

Proof. The sequence $\{x^n(\cdot), n < \infty\}$ is tight in $D^r[0, \infty)$ by Theorem 2.3.3. $\{w^n(\cdot), n < \infty\}$ is obviously tight. The sequence $\{m^n(\cdot), n < \infty\}$ is tight since $\mathcal{R}(\mathcal{U} \times [0, \infty))$ is compact. With a slight abuse of notation, let n also index a weakly convergent subsequence of $\{x^n(\cdot), m^n(\cdot), w^n(\cdot), n < \infty\}$ with limit denoted by $(x(\cdot), m(\cdot), w(\cdot))$. We will use the "martingale problem" method of Section 4 to characterize the limit triple. Recall the definition of A_1^α in (4.6), as the differential operator of the pair $(x(\cdot), w(\cdot))$ in (3.1), with the control fixed at α.

Let $f(\cdot) \in C_0^2(R^{r+k})$. By the definition of A_1^α and A_1^m, the process

$$M_f^n(t) = f(x^n(t), w^n(t)) - f(x, 0) - \int_0^t \int_{\mathcal{U}} A_1^\alpha f(x^n(s), w^n(s)) m^n(d\alpha ds)$$

is a martingale with respect to the filtration $\mathcal{F}_t^n = \mathcal{B}(m^n(s), w^n(s), x^n(s), s \leq t)$. Hence $E_{\mathcal{F}_t}[M_f^n(t+\tau) - M_f^n(t)] = 0$.

Thus, for $h(\cdot)$, $\varphi_j(\cdot)$, t_i, t, τ, p, and q, defined as above (4.3), we have

$$Eh(w^n(t_i), x^n(t_i), (\varphi_j, m^n)_{t_i}, j \leq p, i \leq q)\cdot$$

$$\cdot [f(x^n(t+\tau), w^n(t+\tau)) - f(x^n(t), w^n(t))$$

$$- \int_t^{t+\tau} \int_{\mathcal{U}} A_1^\alpha f(x^n(s), w^n(s)) m^n(d\alpha ds)] = 0. \tag{5.1}$$

Let us now use the Skorohod representation (Theorem 1.3.1), so that we can suppose that $x^n(\cdot) \to x(\cdot)$, $w^n(\cdot) \to w(\cdot)$, $m^n(\cdot) \to m(\cdot)$ w.p.1 in the appropriate topologies. Under this representation, the $m^n(\cdot)$ converges (w.p.1) in the "compact-weak" topology; e.g., for $\varphi(\cdot)$ real valued, continuous and with compact support,

$$\int_0^\infty \int_{\mathcal{U}} \varphi(\alpha, s) m^n(d\alpha ds) \to \int_0^\infty \int_{\mathcal{U}} \varphi(\alpha, s) m(d\alpha ds) \text{ w.p.1}$$

By the weak convergence, the Skorohod representation and the continuity of $A_1^\alpha f(x, w)$ in (x, w, α), we have

$$E \left| \int_t^{t+\tau} \int_{\mathcal{U}} A_1^\alpha f(x^n(s), w^n(s)) m^n(d\alpha ds) \right.$$

$$\left. - \int_t^{t+\tau} \int_{\mathcal{U}} A_1^\alpha f(x(s), w(s)) m^n(d\alpha ds) \right| \to 0.$$

Also,

$$E \left| \int_t^{t+\tau} \int_{\mathcal{U}} A_1^\alpha f(x(s), w(s))(m^n(d\alpha ds) - m(d\alpha ds)) \right| \to 0,$$

all uniformly on each bounded time interval.

Combining these observations and taking limits in (5.1) yields

$$Eh(w(t_i), x(t_i), (\varphi_j, m)_{t_i}, j \leq p, i \leq q)[f(x(t+\tau), w(t+\tau)) - f(x(t), w(t))$$

$$- \int_t^{t+\tau} \int_{\mathcal{U}} A_1^\alpha f(x(s), w(s)) m(d\alpha ds)] = 0. \tag{5.2}$$

This, in turn, yields the first assertion of the theorem, by virtue of the discussion in Section 4. The second assertion is proved in a similar way. Q.E.D.

2. Approximations to relaxed controls. By using the method of characterizing the limit of a weakly convergent subsequence introduced in Theorem 5.1, we can obtain an analog of Theorem 2.2 for the system (3.1).

Remark on Notation. We let E_x^m denote the expectation under control $m(\cdot)$, and with $x(0) = x$. The expectation and the value of the cost function depend on the joint distribution of $(m(\cdot), w(\cdot), N(\cdot))$, but we omit the other processes for notational simplicity.

Theorem 5.2. *Assume* (A2.7.1) *and* (A1.1) *and let the solution of* (3.1) *be unique in the weak sense for the admissible pair* $(m(\cdot), w(\cdot))$. *Define the cost function*

$$V_T(x, m) = E_x^m \int_0^T \int_{\mathcal{U}} k(x(s), \alpha) m_s(d\alpha) ds,$$

where $k(\cdot)$ *is continuous and* $|k(x, \alpha)| \leq K(1 + |x|^p)$, *for some* $p < \infty$ *and* $K_1 < \infty$. *Given* $T < \infty$ *and* $\gamma > 0$, *there is a finite set* $\{\alpha_1^\gamma, \ldots, \alpha_{k_\gamma}^\gamma\} = \mathcal{U}^\gamma \subset \mathcal{U}$, *a* $\delta > 0$, *and a* $\mathcal{U}^\gamma$-*valued ordinary admissible stochastic control* $u_m^\gamma(\cdot)$ *which is constant on each interval* $[i\delta, i\delta + \delta)$ *and is such that for all* m

$$
\begin{aligned}
P_x^m\big(\sup_{t \leq T} |x(t, u_m^\gamma) - x(t, m)| > \gamma\big) &\leq \gamma, \\
|V_T(x, m) - V_T(x, u_m^\gamma)| &\leq \gamma.
\end{aligned}
\tag{5.3}
$$

If the solution to (3.1) *is unique in the weak sense for each admissible* $m(\cdot)$, *then* (5.3) *holds for all* $m(\cdot)$ *simultaneously.*

Remark. In (5.3), we assume that the $x(\cdot, u_m^\gamma)$ and $x(\cdot, m)$ are defined on the same probability space simply for notational convenience. Via a Skorohod representation, we can do this with no loss of generality.

Proof. Given $m(\cdot)$, $w(\cdot)$, define the ordinary admissible control $u_m^{\rho\Delta}(\cdot)$ in the same way that it was done for the deterministic case in Theorem 2.2. Let $m^{\rho\Delta}(\cdot)$ denote the relaxed control representation of $u_m^{\rho\Delta}(\cdot)$. Let $x^{\rho\Delta}(\cdot)$ denote an associated[1] solution to (3.1). (There exists a solution, but it might not be unique.)

Just as in Theorem 5.1, the set $\{x^{\rho\Delta}(\cdot), m^{\rho\Delta}(\cdot), w(\cdot), \rho > 0, \Delta > 0\}$ is tight. Let $(\tilde{x}(\cdot), \tilde{m}(\cdot), \tilde{w}(\cdot))$ denote the limit of a weakly convergent subsequence. By Theorem 5.1, this triple satisfies (3.1) for some Poisson measure $\tilde{N}(\cdot)$, and $\tilde{m}(\cdot)$ is admissible. By the construction of the $u_m^{\rho\Delta}(\cdot)$ and $m^{\rho\Delta}(\cdot)$, the limit $(\tilde{m}(\cdot), \tilde{w}(\cdot), \tilde{N}(\cdot))$ must have the same probability law as $(m(\cdot), w(\cdot), N(\cdot))$ has. This, together with the uniqueness condition, yields the first line of (5.3).

By Theorem 2.3.2, for $t \leq T$ and any $p < \infty$, we have $E_x^m |x(t)|^{2p} \leq K_p(x, T) < \infty$, where K_p does not depend on $m(\cdot)$. This implies the uniform (in the control) integrability of $\int_0^T \int_{\mathcal{U}} k(x(s), \alpha) m_s(d\alpha) ds$. This uniform

---

[1] A solution might not exist on the same probability space that the original $m(\cdot)$, $w(\cdot)$ were defined on. But there exists a space on which a process with the same distribution as $(m^{\rho\Delta}(\cdot), w(\cdot))$ can be defined, as well as an associated solution $x^{\rho\Delta}(\cdot)$. This is enough for our purposes.

integrability, together with the weak convergence, yields the second line of (5.3). The last assertion of the theorem is proved by using an argument of "contradiction," similar to what was done in Theorem 2.2.   Q.E.D.

3. Existence of an optimal relaxed control. The proof of Theorem 5.1 can be reworded to yield the existence of an optimal control in the class of admissible relaxed controls. By $\inf_m$, we mean the inf over all solutions corresponding to admissible controls.

Theorem 5.3. *Assume* (A2.7.1), (A1.1), *and let* $k(\cdot)$ *satisfy the conditions in Theorem 5.2. Let* $m^n(\cdot)$ *be admissible and let* $V_T(x, m^n) \to \overline{V}_T(x) = \inf_m V_T(x, m)$. *Then there are* $(x(\cdot), \overline{m}(\cdot), w(\cdot), N(\cdot))$ *solving* (3.1), *with* $\overline{m}(\cdot)$ *admissible and such that* $V_T(x, \overline{m}) = \overline{V}_T(x)$.

4

Controlled Singularly Perturbed Systems

0. Outline of the Chapter

Many control problems can be modelled by systems of differential equations where the state variable can be divided into two coupled groups. Those in the first group change at a "normal" rate, and those in the second group change at a much "faster" rate. Such systems can be loosely termed "two time scale" systems, and have been the subject of a great deal of work. We refer the reader to [B2, K4, K5], and the references contained therein. For linear systems models, the first group might represent the low frequency part of the system and the second group the high frequency part.

For purposes of discussion and loosely speaking, write the equations in the following form:

$$\dot{x}^\epsilon = f(x^\epsilon, z^\epsilon, u, \text{noise}), \tag{0.1}$$

$$\epsilon \dot{z}^\epsilon = g(x^\epsilon, z^\epsilon, u, \text{noise}). \tag{0.2}$$

The ϵ is a small parameter and is meant to indicate that the "fast" variable is z^ϵ. It simply parametrizes the relative rates of variation of x^ϵ and z^ϵ.

For many problems, the dimension of the combined state vector is high and one wishes to simplify the analysis and design process by working with a simpler system which incorporates the essential features of the actual system. If the z^ϵ truly varies fast, then the form of the dynamical equations suggests that some sort of averaging principle can be used to average out the fast variable in (0.1) and obtain an "almost equivalent" averaged system, which then could be used for the desired purposes. It must be shown that the averaged system really has the essential properties of x^ϵ and that it can be used for the purposes of both qualitative analysis and for purposes of getting good or nearly optimal controls for the actual physical system. This is not an easy task when the systems are stochastic. For the models of this chapter, powerful methods will be presented for accomplishing these goals.

The book [K5] contains numerous examples. While these examples are generally deterministic, there are stochastic versions of nearly all of them. We will briefly discuss only one example, that of case study 7.2 of Chapter 1 of [K5], which models the longitudinal motion of an airplane. There are four state variables: $z = $ (angle of attack, pitch rate), and $x = $ linear combination of z and (normalized incremental velocity, incremental pitch angle). Under typical conditions (e.g., the force of gravity being much smaller that the velocity,...), the z-variable moves at a much faster rate than does the x-variable. The differential equations can be written in the form:

$$\dot{x}^\epsilon = f(x^\epsilon, z^\epsilon, \epsilon x^\epsilon, \epsilon z^\epsilon, u, \text{noise}), \tag{0.3a}$$

$$\epsilon \dot{z}^\epsilon = g(x^\epsilon, z^\epsilon, \epsilon x^\epsilon, \epsilon z^\epsilon, u, \text{noise}), \tag{0.3b}$$

where the effects of the ϵx^ϵ and ϵz^ϵ terms on the right are not important. The control variable is the elevator position.

Let us consider a particular form of (0.3):

$$dx^\epsilon = f(x^\epsilon, z^\epsilon, u)dt + \sigma_1 dw_1,$$

$$\epsilon dz^\epsilon = g(x^\epsilon, z^\epsilon, u)dt + q(\epsilon)\sigma_2 dw_2. \tag{0.4}$$

If $q(\epsilon)$ is bounded away from zero as $\epsilon \to 0$, then z^ϵ "blows" up to white noise as $\epsilon \to 0$. See Chapter 10. Something similar happens if $q(\epsilon) = O(\epsilon^\alpha)$ for $\alpha > 1/2$. If $q(\epsilon) = O(\epsilon^\alpha)$ for $\alpha < 1/2$, then the effects of the noise w_2 disappear as $\epsilon \to 0$. If $q(\epsilon) = O(\sqrt{\epsilon})$, then the "stretched out" process defined by $z_0^\epsilon(t) = z^\epsilon(\epsilon t)$ satisfies an Itô equation of the type

$$dz_0^\epsilon = g(x_0^\epsilon, z_0^\epsilon, u)dt + \sigma_2 dw_0,$$

where w_0 is a Wiener process and $x_0^\epsilon(t) = x^\epsilon(\epsilon t)$. Then z^ϵ doesn't "blow up." It simply moves faster as $\epsilon \to 0$. In this chapter, we will use the value $q(\epsilon) = \sqrt{\epsilon}$.

The analysis of (0.4) can be quite complicated and difficult to analyze for small ϵ, and the possibility of simplification by the use of some averaging method is quite attractive.

In this chapter, we discuss the basic averaging methods for the diffusion and jump-diffusion model. In the first few sections, we work with the diffusion model only, in order to present the basic ideas in a relatively simple way. Section 1 introduces the singularly perturbed diffusion model. A key part in the analysis is played by the "fast process" under a time scale that is "stretched out" so that it evolves as a standard diffusion, and with the "slow variable x" fixed. The averaged system is defined by averaging out the fast variable z from the dynamics of the "slow system" via use of the invariant measure associated with the stationary fixed-x "stretched out" process. The section proves the basic tightness and weak convergence results, in that it shows that the limit processes are legitimate control problems for an "averaged system." The martingale methods and the averaging techniques that are used are quite powerful, and are of use for a much wider class of problems than we treat here. Perturbed test functions are not used or needed. We introduce a technique to prove that certain "limit" occupation measures are actually invariant measures. These measures are then used to do the averaging which yields the correct averaged system. The methods are relatively simple, particularly when compared with alternative approaches to problems that overlap ours.

In Section 2, it is shown that the optimal cost functions for the singularly perturbed system converges to the optimal cost function for the "averaged problem," and that good controls for the averaged problem are also good for the original problem. We work at first with the simplest case, where the problem is of interest over a finite time only. A method is introduced for the approximation (for mathematical purposes only) of optimal controls for the averaged problem by controls that can be adapted to the singularly perturbed problem in order to get the inequalities that are needed to show the convergence. In Section 3, we discuss extensions to both discounted problems and to problems where the stopping time can be chosen to be the random time that some exit criterion is met or target set reached. These are only a few examples of the type of control problems which can be handled by the methods of this chapter.

Section 4 is concerned with the average cost per unit time problem. We work with "pathwise" average costs and show that the limits of the costs are cost values for some controlled averaged problem. This problem involves a double averaging, since we must first average out the fast variable and then we must do the averaging which is associated with the average cost per unit time (over an infinite time interval) nature of the problem. Essentially, the results state that the average cost per unit time problem for the averaged system is a good approximation to that for the singularly perturbed system, and that good controls for the averaged problem are also good for the singularly perturbed problem. The above results are extended to the jump-diffusion process in Section 5.

The chapter covers the essentials of the averaging and approximation methods for singularly perturbed Itô or jump-diffusion models. Another approach to the average cost per unit time problem (over an infinite time interval) appears in Chapter 5, and the reflected diffusion model is treated there also. There are extensions of the methods to cover the problems of impulsive control, the so-called singularly perturbed "singular" control problem, to controlled "delay" equations, and to virtually all the standard forms of the system and cost function. The last section concerns comparisons with alternative approaches.

1. Problem Formulation. Finite Time Interval

The diffusion model. Until Section 5, we work with the singularly perturbed diffusion model

$$dx^\epsilon = dt \int_{\mathcal{U}} G(x^\epsilon, z^\epsilon, \alpha) m_t^\epsilon(d\alpha) + \sigma(x^\epsilon, z^\epsilon) dw_1, \quad x \in R^r \qquad (1.1)$$

$$\epsilon dz^\epsilon = H(x^\epsilon, z^\epsilon) dt + \sqrt{\epsilon} v(x^\epsilon, z^\epsilon) dw_2, \quad z \in R^{r'} \qquad (1.2)$$

$$a(x, z) = \sigma(x, z)\sigma'(x, z).$$

The $w_i(\cdot)$ are standard vector-valued Wiener processes with respect to some filtration $\mathcal{F}_t$ on the probability space $(\Omega, P, \mathcal{F})$, and $m^\epsilon(\cdot)$ is an admissible relaxed control (see Chapter 3.3) and the control parameter α takes values in a compact set $\mathcal{U}$. The model (1.1), (1.2) is the most common one used for stochastic control problems under singular perturbations, and it will allow us to develop the basic ideas in a relatively simple way. The ideas developed for this model will be extended to more complex cases in the following sections and chapters. The model allows a relatively simple and generic description of the basic methods used for averaging and approximation.

In Section 2, we use the cost criterion of the 'finite time' form

$$V_T^\epsilon(x, z, m^\epsilon) = E_{x,z}^{m^\epsilon} \int_0^T \int_{\mathcal{U}} k(x^\epsilon(s), z^\epsilon(s), \alpha) m_s^\epsilon(d\alpha) ds + E_{x,z}^{m^\epsilon} g(x^\epsilon(T))$$

(1.3)

$$V_T^\epsilon(x, z) = \inf_m V_T^\epsilon(x, z, m).$$

We use $E_{x,z}^{m^\epsilon}$ to denote the expectation under control $m^\epsilon(\cdot)$, and with initial condition (x, z). The distribution of $(x^\epsilon(\cdot), z^\epsilon(\cdot))$, (and hence the value of the cost) depends on the *joint distribution* of $m^\epsilon(\cdot)$, $w_1(\cdot)$, $w_2(\cdot)$, and not only on the initial condition and the control. But, for simplicity, we omit the $w_i(\cdot)$ from the notation of the cost and expectation. The $\inf_m$ denotes the inf over all admissible $(m(\cdot), w_1(\cdot), w_2(\cdot))$.

It is hard to deal in any general way with the case where the fast system is also controlled. The main difficulty is due to the fact that the "stationary measures" which are used to average out the fast variable depend on the control which is used in the fast system. This makes it hard to define the "averaged problem." Results for one class appear in Chapter 5, and there is a further discussion at the end of this chapter. Similar problems occur in the deterministic case, and it is commonly dealt with there by supposing that the choice of control for the fast system does not alter the steady state value of that system, for each fixed value of the slow variable, i.e., that the fast system is asymptotically stable and the control is chosen in a class such that the limit point of that fast system does not depend on the control *when x is fixed*. This assumption essentially "decouples" the fast and slow systems. The assumption is reasonable and yields good results. Unfortunately, it does not seem possible to find a stochastic analog of this approach which works in any generality. Except for a few comments on one case in Chapter 6 and some remarks at the end of this chapter, we prefer to leave it as a topic for further research.

The fixed-x rescaled "fast" process. One of the basic techniques used in the book involves the exploitation of the "time scale" differences between $x^\epsilon(\cdot)$ and $z^\epsilon(\cdot)$ in order to approximate (1.1) by a simpler "averaged" system, which will be "essentially" (1.1) with the z averaged out in some appropriate way. To underscore the time scale differences, let us look at (1.1) and (1.2) in a "stretched out" time scale. Define the processes $z_0^\epsilon(t) = z^\epsilon(\epsilon t)$ and $x_0^\epsilon(t) = x^\epsilon(\epsilon t)$. Then

$$dx_0^\epsilon = dt \, \epsilon \int_\mathcal{U} G(x_0^\epsilon, z_0^\epsilon, \alpha) m_t^\epsilon(d\alpha) + \sqrt{\epsilon} \sigma(x_0^\epsilon, z_0^\epsilon) d\tilde{w}_1 \qquad (1.4)$$

$$dz_0^\epsilon = H(x_0^\epsilon, z_0^\epsilon) dt + v(x_0^\epsilon, z_0^\epsilon) d\tilde{w}_2, \qquad (1.5)$$

where the $\tilde{w}_i(\cdot)$ are standard vector-valued Wiener processes. In fact, $\tilde{w}_i(t) = w_i(\epsilon t)/\sqrt{\epsilon}$. In (1.5), the process $x_0^\epsilon(\cdot)$ "varies slowly" (under the conditions listed below). The equation (1.5) will appear frequently in the analysis. The fact that $x_0^\epsilon(\cdot)$ is "nearly constant" over long time periods when ϵ is small suggests that we can treat $z_0^\epsilon(\cdot)$ as though $x_0^\epsilon(\cdot)$ were "fixed" on appropriate large time intervals. It also suggests that if this "fixed-x" process were essentially stationary (modulo the short term effects of the initial condition), then it could be used to average out the $z^\epsilon(\cdot)$ in (1.1) and (1.3) with respect to the "limit" stationary measure. With this in mind, we also define the *fixed-x process* $z_0(\cdot|x)$ (*written simply as* $z_0(\cdot)$ *if the value of x is clear*) by

$$dz_0 = H(x, z_0) dt + v(x, z_0) d\tilde{w}_2. \qquad (1.6)$$

Here x is a *parameter*. This parameter will sometimes be replaced by a *random variable*.

Notation and Assumption. *Let A_x^0 denote the differential operator of the fixed-x process $z_0(\cdot|x)$, A^ϵ that of $(x^\epsilon(\cdot), z^\epsilon(\cdot))$, and $A^{0,\epsilon}$ that of $(x_0^\epsilon(\cdot), z_0^\epsilon(\cdot))$.*

We will use the following assumption.

A1.1. *For each initial condition and each x, (1.6) has a unique weak sense solution. For each x, $z_0(\cdot|x)$ has a unique invariant measure $\mu_x(\cdot)$. There is a continuous matrix valued function $\overline{\sigma}(\cdot)$ such that*

$$\overline{\sigma}(x)\overline{\sigma}'(x) = \int \sigma(x, z)\sigma'(x, z)\mu_x(dz) \equiv \overline{a}(x).$$

The factorization in (A1.1) is a convenience, but is not essential to the development.

The averaged system. Define the averaged system and value function by

$$\overline{G}(x, \alpha) = \int G(x, z, \alpha)\mu_x(dz), \quad \overline{k}(x, \alpha) = \int k(x, z, \alpha)\mu_x(dz),$$

$$dx = dt \int_\mathcal{U} \overline{G}(x, \alpha) m_t(d\alpha) + \overline{\sigma}(x) dw, \qquad (1.7)$$

$$V_T(x,m) = E_x^m \int_0^T \int_{\mathcal{U}} \overline{k}(x(s),\alpha)m_s(d\alpha)ds + E_x^m g(x(T)) \qquad (1.8)$$

$$V_T(x) = \inf_m V_T(x,m).$$

Here, $m(\cdot)$ is an admissible relaxed control with respect to the standard vector-valued Wiener process $w(\cdot)$. The distribution of the solution to (1.7) depends on the initial condition x, as well as on the joint distribution of $(m(\cdot), w(\cdot))$, but again we omit the $w(\cdot)$ from the notation. The particular $w(\cdot)$ associated with $m(\cdot)$ will usually be obvious. The $\inf_m$ denotes the inf over all admissible pairs $(m(\cdot), w(\cdot))$. Except for use on the cost function, we will usually drop the m^ϵ, x and z on $E_{x,z}^{m^\epsilon}$ when they are obvious.

Notation. *Let* $\overline{A}^\alpha$ *denote the differential generator of* (1.7) *with control fixed at* α, *and let* $\overline{A}^m$ *be that associated with the relaxed control* $m(\cdot)$: *I.e.,*

$$(\overline{A}^m f)(x,t) = \int_{\mathcal{U}} \overline{A}^\alpha f(x)m_t(d\alpha).$$

Assumptions.

A1.2.
$$G(x,z,\alpha) = G_0(x,z) + G_1(x,\alpha),$$
$$k(x,z,\alpha) = k_0(x,z) + k_1(x,\alpha).$$

The functions $\sigma(\cdot)$ *and the* $G_i(\cdot)$ *are continuous and have at most a linear growth in* x (*i.e.,* $|\sigma(x)| \leq K(1+|x|)$ *for some* $K < \infty$), *uniformly in* z, α. *The* $k_i(\cdot)$ *and* $g(\cdot)$ *are continuous and have at most a polynomial growth in* x, *uniformly in* z, α.

Define

$$\overline{G}_0(x) = \int G_0(x,z)\mu_x(dz), \quad \overline{k}_0(x) = \int k_0(x,z)\mu_x(dz).$$

A1.3. $H(\cdot)$ *and* $v(\cdot)$ *are continuous and have at most a linear growth in* z, *uniform in* x.

A1.4. *For each admissible relaxed control,* (1.7) *has a unique weak sense solution.*

A1.5. *For each admissible relaxed control,* (1.1), (1.2) *has a unique weak sense solution.*

A1.6. *For the sequence of admissible relaxed controls* $\{m^\epsilon(\cdot)\}$ (*the sequence will depend on the application*), *there is a function* $0 \leq \hat{g}(z) \to \infty$ *as* $|z| \to \infty$, *a* $K_1 < \infty$, *and* $\Delta_\epsilon \to 0$ *as* $\epsilon \to 0$ *such that* $\epsilon/\Delta_\epsilon \to 0$ *and*

$$\sup_{t \leq T} \frac{1}{\Delta_\epsilon} \int_t^{t+\Delta_\epsilon} E\hat{g}(z^\epsilon(s))ds \leq K_1 < \infty.$$

The condition (A1.6) is, of course, satisfied if $\{z^\epsilon(t), t \leq T, \epsilon > 0\}$ is tight (i.e., bounded in the sense of probability). We prefer the form (A1.6), since it is a little weaker and sometimes easier to verify. But, for purposes of simplifying the reading the reader could simply assume the tightness of $\{z^\epsilon(t), t \leq T, \epsilon > 0\}$. In some applications of (A1.6), we require it for only certain cited sequences $\{m^\epsilon(\cdot)\}$. The stability question raised by (A1.6) is discussed in Chapter 9, where criteria are given for the verification of (A1.6).

Remarks on the Conditions. By the uniqueness of $\mu_x(\cdot)$ for each x, it is weakly continuous in x. I.e., for any real valued, continuous and bounded $f(\cdot)$, $\int f(z)\mu_x(dz)$ is continuous in x. Nevertheless, $\overline{G}(x, \alpha)$ and $\overline{\sigma}(x)$ are not necessarily Lipschitz continuous in x even if $G(x, z, \alpha)$ and $\sigma(x)$ are, and we require (A1.4) in order for the solution to (1.7) to be well defined. In any case, whether or not the uniqueness holds, each moment of the set of random variables $\{x^\epsilon(t), \epsilon \geq 0, t \leq T\}$ is bounded uniformly in ϵ, by the growth conditions in (A1.2) and Theorem 2.3.2.

The "separation of variables" assumption (A1.2) seems to be "usually" satisfied in applications. It is used here in order to have a well defined "average" problem, whose behavior represents well the behavior of $x^\epsilon(\cdot)$ and $V_T^\epsilon(x, z, m^\epsilon)$ for small enough ϵ and any sequence of admissible relaxed controls $\{m^\epsilon(\cdot)\}$. In order to illustrate the point, we give some examples using classical controls $u^\epsilon(\cdot)$, rather than relaxed controls. Let $\int z\mu_x(dz) = 0$, $\int z^2\mu_x(dx) > 0$, all x. Let $u_i(\cdot)$, $i = 0, 1$, be smooth functions of x, and let $G(x, z, u) = zu$. If the control $u^\epsilon(\cdot)$ takes values of the form $z^\epsilon(t) \cdot u_0(x^\epsilon(t)) = u^\epsilon(t)$, then the drift term for the averaged process will be $\int z^2\mu_x(dz)u_0(x) \equiv \overline{G}(x)$. If the control $u^\epsilon(\cdot)$ takes values $u^\epsilon(t) = u_1(x^\epsilon(t))$, then the drift term of the limit process will be $\int z\mu_x(dz)u_1(x) = \overline{G}(x) = 0$ for any $u_1(x)$. Thus, for different classes of sequences $\{m^\epsilon(\cdot)\}$, the *limit models* are actually different. Such problems arise when $m_t^\epsilon(\cdot)$ is "too wild" for small ϵ. Then the $z^\epsilon(\cdot)$ process is not averaged out in a simple way for all controls, by integrating with respect to $\mu_x(dz)$, and the drift term which appears in the limit will depend on the "interaction" between $z^\epsilon(\cdot)$ and $\dot{m}^\epsilon(\cdot) = m_\bullet^\epsilon(\cdot)$ for small ϵ. If the control action is delayed or if there is a constraint which guarantees that the $m_t^\epsilon(\cdot)$ are (uniformly in ϵ) continuous in t, then the separation part of (A1.2) can be dropped.

Convergence of the costs. We first prove tightness of a sequence of occupation measures, for use below. Then the first averaging result will be proved. Let $I_A(x)$ denote the indicator function defined by $I_A(x) = 1$ if $x \in A$ and is zero otherwise.

Lemma 1.1. *Assume* (A1.6) *for a sequence* $\{m^\epsilon(\cdot)\}$ *of relaxed admissible controls. For each* $\Delta > 0$, *define the occupation measure* $\hat{P}_t^{\epsilon,\Delta}(\cdot)$ *by*

$$\hat{P}_t^{\epsilon,\Delta}(F) = \frac{\epsilon}{\Delta} \int_{t/\epsilon}^{(t+\Delta)/\epsilon} I_F(z_0^\epsilon(s)) ds.$$

Let $\Delta = \Delta_\epsilon$ satisfy (A1.6). Then for any bounded sequence $\{t_\epsilon\}$ the sequence of measure-valued random variables $\{\hat{P}_{t_\epsilon}^{\epsilon,\Delta}(\cdot), \epsilon > 0\}$ is tight.

Proof. For any $N < \infty$, we have

$$E_{x,z}^{m^\epsilon} \hat{P}_{t_\epsilon}^{\epsilon,\Delta}(|z| \geq N) = \frac{\epsilon}{\Delta} \int_{t_\epsilon/\epsilon}^{(t_\epsilon+\Delta)/\epsilon} P_{x,z}^{m^\epsilon}(|z_0^\epsilon(s)| \geq N) ds.$$

For some $K < \infty$,

$$K \geq \frac{\epsilon}{\Delta} \int_{t_\epsilon/\epsilon}^{(t_\epsilon+\Delta)/\epsilon} E_{x,z}^{m^\epsilon} \hat{g}(z_0^\epsilon(s)) ds$$

$$\geq \frac{\epsilon}{\Delta} \inf_{|z| \geq N} \hat{g}(z) \int_{t_\epsilon/\epsilon}^{(t_\epsilon+\Delta)/\epsilon} P_{x,z}^{m^\epsilon}(|z_0^\epsilon(s)| \geq N) ds.$$

The conclusion then follows from Theorem 1.6.1 and (A1.6).   Q.E.D.

Theorem 1.2. *Assume* (A1.1) *to* (A1.5) *and let* $\{m^\epsilon(\cdot)\}$ *be admissible relaxed controls such that* (A1.6). *is satisfied. Then* $\{x^\epsilon(\cdot), m^\epsilon(\cdot)\}$ *is tight. Let* $(x(\cdot), m(\cdot))$ *denote the limit of a weakly convergent subsequence (indexed by* ϵ_n). *Then there is a standard vector-valued Wiener process* $w(\cdot)$ *such that* $m(\cdot)$ *is admissible with respect to* $w(\cdot)$ *and* $(x(\cdot), m(\cdot), w(\cdot))$ *satisfy* (1.7). *Also, uniformly in* x *and* z *in each compact set,*

$$V_T^{\epsilon_n}(x, z, m^{\epsilon_n}) \to V_T(x, m). \tag{1.9}$$

Remark. The theorem states that the limits of $\{x^\epsilon(\cdot), m^\epsilon(\cdot)\}$ represent *well defined control problems*, where the limit model is the averaged system (1.7) with the averaged cost (1.8). It is not a priori obvious that any limit pair $(x(\cdot), m(\cdot))$ is consistent, in the sense that there is a Wiener process $w(\cdot)$ such that the limit $x(\cdot)$ can be represented by an averaged system, where the control limit is the $m(\cdot)$ and $w(\cdot)$ is the driving Wiener process, and $m(\cdot)$ is admissible with respect to $w(\cdot)$. Nor is it a priori obvious that the costs $V_T^\epsilon(x, z, m^\epsilon)$ converge as in (1.9). If (1.9) were not the case, then the limit pair $(x(\cdot), m(\cdot))$ would not be too useful. The "averaging result" of Theorem 1.2 suggests that the "important" behavior of $x^\epsilon(\cdot)$ can be captured by the behavior of some averaged system when ϵ is small. Note that we do not require convergence of the transition probabilities of $z_0(\cdot \mid x)$.

Outline of Proof. The proof uses several generally useful techniques and proceeds roughly as follows: (a) tightness of $\{x^\epsilon(\cdot), m^\epsilon(\cdot)\}$ is proved via standard techniques for Itô equations; (b) under a (temporary) assumption that the operator A^ϵ can be replaced (asymptotically) by $\overline{A}^\alpha$, the operator for the averaged system, the limit $(x(\cdot), m(\cdot))$ of any weakly convergent subsequence of $\{x^\epsilon(\cdot), m^\epsilon(\cdot)\}$ is shown to satisfy (1.7); (c) the quantity to be averaged (to justify the above use of $\overline{A}^\alpha$) is represented in terms of an occupation measure $\hat{P}_{t_\epsilon}^{\epsilon,\Delta}(\cdot)$, $t_\epsilon \to t$; (d) via the use of another "martingale" method, the $\{\hat{P}_{t_\epsilon}^{\epsilon,\Delta}(\cdot)\}$ is shown to be a tight sequence of measure valued random variables and the weak limits are shown to be the invariant measures $\mu_{x(t)}(\cdot)$ of the fixed-x process with $x = x(t)$; (e) the costs are shown to converge to the cost for the limit averaged system.

Proof. We do the proof in detail and break it into several parts in order to expose the salient features, most of which recur throughout the book. The details are similar to those of Theorem 3.5.1.

Part 1. Tightness. $\{m^\epsilon(\cdot)\}$ is tight, since the space of the relaxed control $\mathcal{R}(U \times [0,T])$ is compact, as noted in Theorem 3.5.1. The sequence $\{x^\epsilon(\cdot)\}$ is tight in the Skorohod topology by Theorem 2.3.3 and the growth condition in (A1.2). Since each $x^\epsilon(\cdot)$ is continuous w.p.1, the limit of any weakly convergent subsequence is also continuous w.p.1.

Part 2. We next show how to characterize the limit pair $(x(\cdot), m(\cdot))$, using the martingale method of Theorem 3.5.1. Henceforth, for notational simplicity, let ϵ (and not ϵ_n) index a weakly convergent subsequence.

As in Theorem 3.5.1, let $f(\cdot)$ be a smooth real valued function of x with compact support and let $h(\cdot)$ be a real valued, bounded and continuous function of its arguments. Let the $\varphi(\cdot)$ and $\varphi_j(\cdot)$ below be real valued and continuous functions with compact support. Define the function

$$(m, \varphi)_t = \int_0^t \int_{\mathcal{U}} \varphi(s, \alpha) m(d\alpha ds).$$

By the definition of the differential operator A^ϵ, we have

$$(A^\epsilon f)(x, z, t) = f_x'(x) \int_{\mathcal{U}} G(x, z, \alpha) m_t^\epsilon(d\alpha)$$

$$+ \text{trace} \frac{f_{xx}(x)}{2} \cdot a(x, z). \tag{1.10}$$

Let q, t, t_i, τ, $i \le q$, be given such that $t_i \le t \le t + \tau$, $i \le q$.
By Itô's Formula (2.2.13),

$$E[f(x^\epsilon(t + \tau)) - f(x^\epsilon(t)) - \int_t^{t+\tau} (A^\epsilon f)(x^\epsilon(s), z^\epsilon(s), s) ds$$

$$\mid x^\epsilon(u), m_u^\epsilon(\cdot), u \le t] = 0.$$

Thus, as in Theorem 3.5.1, for any $\varphi_1(\cdot), \ldots, \varphi_p(\cdot)$, we have

$$Eh(x^\epsilon(t_i), (m^\epsilon, \varphi_j)_{t_i}, i \le q, j \le p)\cdot$$

$$\cdot \left[f(x^\epsilon(t+\tau)) - f(x^\epsilon(t)) - \int_t^{t+\tau} (A^\epsilon f)(x^\epsilon(s), z^\epsilon(s), s)ds \right] = 0. \quad (1.11)$$

Henceforth, assume that the Skorohod representation Theorem 1.3.1 is used, so that we can assume that the convergences are all w.p.1 in the appropriate topology. Then, since the limit $x(\cdot)$ is continuous w.p.1, convergence in the Skorohod topology implies that

$$m^\epsilon(\cdot) - m(\cdot) \to 0, \quad \sup_{t \le T} |x^\epsilon(t) - x(t)| \to 0 \quad w.p.1., \quad (1.12)$$

where the convergence of $\{m^\epsilon(\cdot)\}$ is in the weak topology or, equivalently, in the Prohorov metric. Thus, as $\epsilon \to 0$, the $x^\epsilon(t)$ and $(m^\epsilon, \varphi_j)_t$ converge to their limits $x(t)$ and $(m, \varphi_j)_t$, resp.

In the subsequent parts of the proof, it will be shown that

$$\left| \int_t^{t+\tau} (A^\epsilon f)(x^\epsilon(s), z^\epsilon(s), s)ds - \int_t^{t+\tau} \int_\mathcal{U} \overline{A}^\alpha f(x^\epsilon(s)) m_s^\epsilon(d\alpha)ds \right| \xrightarrow{\epsilon} 0$$
$$(1.13)$$

in probability. Since the function $\overline{A}^\alpha f(x)$ is continuous in (x, α), (1.12) and (1.13) imply that (similarly to the situation in Theorem 3.5.1)

$$\int_t^{t+\tau} \int_\mathcal{U} \overline{A}^\alpha f(x^\epsilon(s)) m_s^\epsilon(d\alpha)ds \to \int_t^{t+\tau} \int_\mathcal{U} \overline{A}^\alpha f(x(s)) m_s(d\alpha)ds \quad (1.14)$$

w.p.1. Since $f(\cdot)$ has compact support, we can assume that the integrals in (1.13) and (1.14) are all bounded. Using these facts and (1.11) yields

$$Eh(x(t_i), (m, \varphi_j)_{t_i}, i \le q, j \le p) \left[f(x(t+\tau)) - f(x(t)) \right.$$

$$\left. - \int_t^{t+\tau} \int_\mathcal{U} \overline{A}^\alpha f(x(s)) m_s(d\alpha)ds \right] = 0. \quad (1.15)$$

Define the filtration $\mathcal{F}_t = \mathcal{B}(x(s), m_s(\cdot), s \le t)$. Owing to the arbitrariness of $t_i, \varphi_j(\cdot), q, p$ and $h(\cdot)$, (1.15) implies that the expectation of the bracketed term conditioned on $\mathcal{F}_t$ is zero. Thus

$$f(x(t)) - \int_0^t \int_\mathcal{U} \overline{A}^\alpha f(x(s)) m_s(d\alpha)ds$$

is a $\mathcal{F}_t$-martingale for each chosen $f(\cdot)$, and we can suppose that as in Theorem 3.5.1 (augmenting the probability space by adding an independent Wiener process if needed) there is a standard $\mathcal{F}_t$-Wiener process $w(\cdot)$, such that $(x(\cdot), m(\cdot), w(\cdot))$ satisfy (1.7). Also $m(\cdot)$ must be admissible, since $w(\cdot)$ is a $\mathcal{F}_t$-Wiener process.

Part 3. Proof of (1.13). By the weak convergence and the Skorohod representation (and, in particular, the result (1.12)), we have

$$\int_t^{t+\tau} \int_{\mathcal{U}} G_1(x^\epsilon(s), \alpha) m_s^\epsilon(d\alpha) ds \rightarrow \int_t^{t+\tau} \int_{\mathcal{U}} G_1(x(s), \alpha) m_s(d\alpha) ds,$$

$$\int_0^T \int_{\mathcal{U}} k_1(x^\epsilon(s), \alpha) m_s^\epsilon(d\alpha) ds \rightarrow \int_0^T \int_{\mathcal{U}} k_1(x(s), \alpha) m_s(d\alpha) ds, \qquad (1.16)$$

w.p.1, as $\epsilon \rightarrow 0$.

To show (1.13), we need to show that as $\epsilon \rightarrow 0$,

$$\left| \int_t^{t+\tau} f_x'(x^\epsilon(s)) G_0(x^\epsilon(s), z^\epsilon(s)) ds \right.$$

$$\left. - \int_t^{t+\tau} f_x'(x^\epsilon(s)) \overline{G}_0(x^\epsilon(s)) ds \right| \rightarrow 0, \qquad (1.16')$$

$$\left| \int_t^{t+\tau} \text{trace } f_{xx}(x^\epsilon(s)) a(x^\epsilon(s), z^\epsilon(s)) ds \right.$$

$$\left. - \int_t^{t+\tau} \text{trace } f_{xx}(x^\epsilon(s)) \overline{a}(x^\epsilon(s))) ds \right| \rightarrow 0. \qquad (1.16'')$$

To prove the first line of (1.16') we will show that there is a sequence $\Delta \rightarrow 0$ such that $\epsilon/\Delta \rightarrow 0$ and for all t and sequences $t_\epsilon \rightarrow t$, we have

$$\frac{1}{\Delta} \int_{t_\epsilon}^{t_\epsilon + \Delta} f_x'(x^\epsilon(s)) G_0(x^\epsilon(s), z^\epsilon(s)) ds \xrightarrow{P} f_x'(x(t)) \overline{G}_0(x(t)). \qquad (1.17)$$

By (1.17) and the fact that the integrand of the left side of (1.17) is bounded, (1.16') converges as desired.

By a change to the "stretched out" time scale $s \rightarrow \epsilon s$ and using the tightness of $x^\epsilon(\cdot)$, the left hand side of (1.17) can be written as

$$\frac{\epsilon}{\Delta} \int_{t_\epsilon/\epsilon}^{(t_\epsilon + \Delta)/\epsilon} f_x'(x^\epsilon(t)) G_0(x^\epsilon(t), z_0^\epsilon(s)) ds + \delta^{\epsilon,\Delta}(t), \qquad (1.18)$$

where, as $\Delta \rightarrow 0$ and $\epsilon/\Delta \rightarrow 0$, $\delta^{\epsilon,\Delta}(t) \xrightarrow{P} 0$ uniformly in $t \leq T$ and in the sequence $t_\epsilon \rightarrow t$. Henceforth, suppose that $\Delta \rightarrow 0$ and $\epsilon/\Delta \rightarrow 0$ in the way required by Lemma 1.1. Recall the definition of the sample occupation measure $\hat{P}_t^{\epsilon,\Delta}(\cdot)$ of Lemma 1.1. By Lemma 1.1, $\{\hat{P}_t^{\epsilon,\Delta}(\cdot)\}$ is a tight sequence of measure valued random variables. With an abuse of notation, we let (ϵ, Δ) index a weakly convergent subsequence, with limit denoted by $\hat{P}_t(\cdot)$. It will be shown below that

$$\hat{P}_t(\cdot) = \mu_{x(t)}(\cdot) \qquad (1.19)$$

(w.p.1). Since the main term of (1.18) can be represented as

$$\int f_x'(x^\epsilon(t))G_0(x^\epsilon(t), z)\hat{P}_{t_\epsilon}^{\epsilon,\Delta}(dz),$$

we get (1.17) by using the representation (1.19) of the limit together with
the fact that $x^\epsilon(t) \to x(t)$ w.p.1 (under the Skorohod representation).

Part 4. Proof of (1.19). Let $q(\cdot)$ be a smooth real valued function of
z with compact support. The process

$$M_q^\epsilon(t) = q(z^\epsilon(t)) - \int_0^t (A^\epsilon q)(x^\epsilon(s), z^\epsilon(s))ds$$

$$= q(z_0^\epsilon(t/\epsilon)) - \int_0^{t/\epsilon} (A^{0,\epsilon}q)(x_0^\epsilon(s), z_0^\epsilon(s))ds \qquad (1.20)$$

is a martingale. We now show that this martingale property together with
the boundedness of $q(\cdot)$ and $A^{0,\epsilon}q(\cdot)$ implies that

$$\frac{\epsilon}{\Delta}[M_q^\epsilon(t_\epsilon + \Delta) - M_q^\epsilon(t_\epsilon)] \xrightarrow{P} 0, \qquad (1.21)$$

as $\epsilon/\Delta \to 0$ and $\Delta \to 0$.

Let Δ/δ be an integer with $\delta \geq \epsilon$ and $\delta/\Delta \to 0$ as Δ and ϵ go to zero.
Using the martingale property, the variance of the left side of (1.21) is

$$\frac{\epsilon^2}{\Delta^2}E\left\{\sum_{i\delta<\Delta}[M_q^\epsilon(t_\epsilon + i\delta + \delta) - M_q^\epsilon(t_\epsilon + i\delta)]\right\}^2$$

$$= \frac{\epsilon^2}{\Delta^2}\sum_{i\delta<\Delta}E[M_q^\epsilon(t_\epsilon + i\delta + \delta) - M_q^\epsilon(t_\epsilon + i\delta)]^2$$

$$= \left(\frac{\epsilon^2}{\Delta^2}\right)\left(\frac{\Delta}{\delta}\right)\left[O(1) + O\left(\frac{\delta^2}{\epsilon^2}\right)\right] \to 0$$

as $(\Delta, \epsilon) \to 0$.

The expression $\int_{t_\epsilon/\epsilon}^{(t_\epsilon+\Delta)/\epsilon}(A^{0,\epsilon}q)(x_0^\epsilon(s), z_0^\epsilon(s))ds$ occurs in the martingale
difference in the left side of (1.21). Equation (1.21) still holds, if we replace
the $x_0^\epsilon(s)$ in the integral by $x_0^\epsilon(t_\epsilon/\epsilon) = x^\epsilon(t_\epsilon)$. With this replacement and
the definition of A_x^0, the operator of the fixed-x process, the integrand in
the above expression becomes $(A_{x^\epsilon(t_\epsilon)}^0 q(z_0^\epsilon(s)))$. By the definition of $\hat{P}_t^{\epsilon,\Delta}(\cdot)$,
we can write

$$\frac{\epsilon}{\Delta}\int_{t_\epsilon/\epsilon}^{(t_\epsilon+\Delta)/\epsilon}A_{x^\epsilon(t_\epsilon)}^0 q(z_0^\epsilon(s))ds = \int A_{x^\epsilon(t_\epsilon)}^0 q(z)\hat{P}_{t_\epsilon}^{\epsilon,\Delta}(dz). \qquad (1.22)$$

By (1.21) and the fact that $\epsilon q(z^\epsilon(t))/\Delta \to 0$ uniformly in t as $\epsilon/\Delta \to 0$,
the right side of (1.22) goes to zero in probability as $n \to \infty$. By the

convergence of $x^\epsilon(t_\epsilon)$ to $x(t)$, the continuity in (x, z) of $A_x^0 q(z)$, and the convergence of $\hat{P}_{t_\epsilon}^{\epsilon, \Delta}(\cdot)$ to $\hat{P}_t(\cdot)$, the limit of the right side of (1.22) equals, w.p.1,

$$0 = \int A_{x(t)}^0 q(z) \hat{P}_t(dz). \tag{1.23}$$

(1.23) holds for all twice continuously differentiable functions $q(\cdot)$ with compact support w.p.1. Let ω be the generic variable on the space on which the limit $\hat{P}_t(\cdot)$ and $x(\cdot)$ are defined. Then, for almost all ω, (1.23) implies [E2, Prop. 4.9.19] that the values $\hat{P}_t(\omega, \cdot)$ of the random variable $\hat{P}_t(\cdot)$ are invariant measures for the process $z_0(\cdot \mid x(\omega, t))$, the fixed-$x$ process with parameter $x = x(\omega, t)$ (an alternative proof of this fact appears in Chapter 5). By the uniqueness of the invariant measure $\mu_x(\cdot)$ for each x, the chosen subsequence (ϵ, Δ) is irrelevant and (1.19) holds for almost all ω. Thus (1.17) holds. A similar argument is used to get (1.16'').

Part 5. By the methods of Part 4,

$$\int_0^T \int_{\mathcal{U}} k(x^\epsilon(s), z^\epsilon(s), \alpha) m_s^\epsilon(d\alpha) ds \Rightarrow \int_0^T \int_{\mathcal{U}} \bar{k}(x(s), \alpha) m_s(d\alpha) ds, \tag{1.24}$$

$$g(x^\epsilon(T)) \Rightarrow g(x(T)).$$

By Theorem 2.3.2 and the condition (A1.2) on the growth in x of $\sigma(\cdot)$ and $G_i(\cdot)$, each moment of $x^\epsilon(t)$ is bounded uniformly in ϵ and $t \le T$. Since $g(\cdot)$ and $k(\cdot)$ have at most a polynomial growth in x (uniformly in (z, α)), the left hand terms in (1.24) are uniformly (in ϵ) integrable and (1.9) holds. Q.E.D.

Terminal cost depending on z. If the terminal cost depends on z, then we need slightly stronger conditions.

A1.7. $g(\cdot, \cdot)$ *is bounded and continuous.*

A1.8. *Let $P_0(z, t, \cdot \mid x)$ denote the transition function of the fixed-x process $z_0(\cdot \mid x)$. Then $P_0(z, t, \cdot \mid x) \to \mu_x(\cdot)$ weakly as $t \to \infty$, uniformly in each compact (x, z) set.*

The proof of the following result is close to that of Theorem 1.2 and is omitted.

Theorem 1.3. *Assume* (A1.7) *and* (A1.8) *and the conditions of Theorem 1.2. Let $E_{x,z}^m g(x, z)$ replace $E_{x,z}^m g(x)$ and $E_x^m \bar{g}(x)$ replace $E_{x,z}^m g(x)$, where $\bar{g}(x) = \int g(x, z) \mu_x(dz)$. Then Theorem 1.2 continues to hold.*

2. Approximation of the Optimal Controls and Value Functions

Theorem 2.1. *Assume* (A1.1)–(A1.5), *and let* (A1.6) *hold for* $m^\epsilon(\cdot)$ *being the optimal control for* (1.1)–(1.3). *Then*

$$V_T^\epsilon(x, z) \to V_T(x).$$

Remark. The proof implies that if there is a δ-optimal feedback control $u^\delta(x, t)$ for the averaged problem and $u^\delta(\cdot, t)$ were continuous for each t, then that control is 2δ optimal for (1.1), (1.2) for small ϵ. In fact, such a control can be constructed under a Lipschitz condition $\overline{G}(\cdot)$, $\overline{\sigma}(\cdot)$. See [K11].

Proof. Let $m^\epsilon(\cdot)$ be admissible for (1.1), (1.2), with $x^\epsilon(\cdot)$ denoting the associated solution of (1.1). By Theorem 1.2, if $\{x^\epsilon(\cdot), m^\epsilon(\cdot)\}$ converges weakly to $(x(\cdot), m(\cdot))$, then there is a standard vector-valued Wiener process $w(\cdot)$ such that $(m(\cdot), w(\cdot))$ is admissible and yields the solution $x(\cdot)$ to (1.7). Also, $V_T^\epsilon(x, z, m^\epsilon) \to V_T(x, m)$. This implies that for any admissible set $\{m^\epsilon(\cdot), w_1(\cdot), w_2(\cdot)\}$,

$$\underline{\lim}_\epsilon V_T^\epsilon(x, z, m^\epsilon) \geq \inf_m V_T(x, m) = V_T(x).$$

Thus, we need only show that for each $\gamma > 0$, there is an admissible sequence $\{m^\epsilon(\cdot), w_1(\cdot), w_2(\cdot)\}$ such that

$$\lim_\epsilon V_T^\epsilon(x, z, m^\epsilon) \leq V_T(x) + \gamma. \tag{2.1}$$

The sequence which we construct to get (2.1) is for mathematical use only. It is not intended to be used as a practical control. Let $(\overline{m}(\cdot), w(\cdot))$ be an optimal admissible pair for (1.7), (1.8). Such a pair exists by Theorem 3.5.3. Given $\delta > 0$, we first approximate $\overline{m}(\cdot)$ by a simpler admissible (with respect to $w(\cdot)$) control. We will eventually want to use such an approximation on (1.1), (1.2), in order to compare the value of the costs. Essentially, we will approximate the "simpler" control by one which depends on only sampled values of $w(\cdot)$ in such a way that it can be applied to (1.1), (1.2). Recall that by Theorem 3.2.2, we have the following: Given $\gamma > 0$, there is a finite set $\mathcal{U}^\gamma = \{c_1^\gamma, \ldots, c_q^\gamma\} \subset \mathcal{U}$, a $\delta > 0$ and an admissible (with respect to $w(\cdot)$) ordinary control $\tilde{u}^\gamma(\cdot)$ such that $\tilde{u}^\gamma(\cdot)$ has values in $\mathcal{U}^\gamma$ and is constant on the intervals $[i\delta, i\delta + \delta)$, such that

$$V_T(x, \tilde{u}^\gamma) \leq V_T(x, \overline{m}) + \gamma/3.$$

Let $\rho > 0$. Let $\alpha_j \in \mathcal{U}^\gamma$. Since, for each i, the σ-algebra $\mathcal{B}(w(k\rho), k\rho \leq i\delta)$ converges to $\mathcal{B}(w(s), s \leq i\gamma)$ as $\rho \to 0$, the martingale convergence theorem [B9] implies that (the expression defines the function $p_i(\cdot)$)

$$P(\tilde{u}^\gamma(i\delta) = \alpha_i \,|\, \tilde{u}^\gamma(j\delta) = \alpha_j, j < i; w(k\rho), k\rho \le i\delta)$$

$$\equiv p_i(\alpha_j, j \le i; w(k\rho), k\rho \le i\delta) \tag{2.2}$$

$$\rightarrow P(\tilde{u}^\gamma(i\delta) = \alpha_i \,|\, \tilde{u}^\gamma(j\delta) = \alpha_j, j < i; w(s), s \le i\delta)$$

w.p.1 as $\rho \rightarrow 0$. The convergence in (2.2) implies that we can find another ordinary admissible control $u^\gamma(\cdot)$ which is also constant on each interval $[i\delta, i\delta + \delta)$ and where $u^\gamma(i\delta)$ takes values in $\mathcal{U}^\gamma$, *but depends only on the samples* $\{w(k\rho), k\rho \le i\delta, u^\gamma(j\delta), j < i\}$, and is such that

$$V_T(x, u^\gamma) \le V_T(x, \tilde{u}^\gamma) + \gamma/3. \tag{2.3}$$

We now show how to get $u^\gamma(\cdot)$. Select a control $u^{\gamma\rho}(\cdot)$ by choosing its values "at random" according to the sequence of conditional distributions

$$P(u^{\gamma\rho}(i\delta) = \alpha_i \,|\, u^{\gamma\rho}(j\delta) = \alpha_j, j < i; w(k\rho), k\rho \le i\delta)$$

$$= p_i(\alpha_j, j \le i; w(k\rho), k\rho \le i\delta) \tag{2.4}$$

for small enough $\rho > 0$. As $\rho \rightarrow 0$, $(u^{\gamma\rho}(\cdot), w(\cdot)) \Rightarrow (\tilde{u}^\gamma(\cdot), w(\cdot))$. Let $\tilde{x}^\gamma(\cdot)$ and $x^{\gamma\rho}(\cdot)$ denote the solutions to (1.7), under (control, Wiener process) pairs $(\tilde{u}^\gamma(\cdot), w(\cdot))$ and $(u^{\gamma\rho}(\cdot), w(\cdot))$, resp. Now, by the uniqueness condition (A1.4), $(x^{\gamma\rho}(\cdot), u^{\gamma\rho}(\cdot), w(\cdot)) \Rightarrow (\tilde{x}^\gamma(\cdot), \tilde{u}^\gamma(\cdot), w(\cdot))$, as $\rho \rightarrow 0$. By an argument of the type used in Theorem 3.5.3 (or in Theorem 1.2 without the averaging) this implies that $V_T(x, u^{\gamma\rho}) \rightarrow V_T(x, \tilde{u}^\gamma)$. Thus, for small enough ρ, the constructed $u^{\gamma\rho}(\cdot)$ satisfies the conditions required of $u^\gamma(\cdot)$.

We can also assume, w.l.o.g., that the function $p_i(\alpha_j, \ j \le i; \ w(k\rho), k\rho \le i\delta)$ is continuous in the $w(k\rho)$ variables for each value of i and the control variables. If it is not continuous, we simply approximate it by a continuous function as follows: For each small $\rho > 0$, there is a sequence $P^n(\cdot)$ of conditional distributions

$$P^n(u(i\delta) = \alpha_i \,|\, u(j\delta) = \alpha_j, j < i; w(k\rho), k\rho \le i\delta) \tag{2.5}$$

which are continuous in the $w(k\rho)$ variables for each set of values $\{\alpha_j, j \le i\}$, and which converges to the left side of (2.2) w.p.1 for all $\{\alpha_j, j \le i\}$. Then, a weak convergence argument of the sort used in the last paragraph yields that, for large n, we can use the law (2.5) to get a control satisfying (2.3).

By the last paragraph, we can suppose that there are functions $p_i(\cdot)$ which are continuous in the $w(k\rho)$ variables for each value of the other variables and are such that for the control $u^\gamma(\cdot)$ defined by the conditional distributions

$$P(u^\gamma(i\delta) = \alpha_i \,|\, u^\gamma(j\delta) = \alpha_j, j < i; w(k\rho), k\rho \le i\delta)$$
$$= P(u^\gamma(i\delta) = \alpha_i \,|\, u^\gamma(j\delta) = \alpha_j, j < i; w(s), s \le i\delta), \qquad (2.6)$$
$$= p_i(u^\gamma(j\delta) = \alpha_j, j \le i; w(k\rho), k\rho \le i\delta).$$

We have
$$V_T(x, u^\gamma) \le V_T(x, \tilde{u}^\gamma) + \gamma/3 \le V_T(x) + 2\gamma/3.$$

By the first equation of (2.6), we mean that the conditional distribution of $u^\gamma(i\gamma)$ given the past of the control and samples of the past of $w(\cdot)$ equals (w.p.1) the conditional distribution given the past values of the control and the entire past of $w(\cdot)$.

We now adapt the control $u^\gamma(\cdot)$ for use on the original system (1.1), (1.2). We do this only for the case where $\bar{\sigma}^{-1}(x)$ exists and is bounded uniformly in x. (The general degenerate case requires more care, as well as the augmentation of the probability space by the addition of an "independent" Wiener process. See Chapter 2.5 and 2.6 for a discussion of the augmentation.)

Define the process $w^\epsilon(\cdot)$ by

$$w^\epsilon(t) = \int_0^t \bar{\sigma}^{-1}(x^\epsilon(s))\sigma(x^\epsilon(s), z^\epsilon(s))dw_1(s).$$

The process $w^\epsilon(\cdot)$ is a martingale, since it is a stochastic integral. By Chapter 2.5, the quadratic variation of $w^\epsilon(\cdot)$ is

$$\langle w^\epsilon \rangle(t) = \int_0^t \bar{\sigma}^{-1}(x^\epsilon(s))a(x^\epsilon(s), z^\epsilon(s))\bar{\sigma}^{-1'}(x^\epsilon(s))ds. \qquad (2.7)$$

By (2.2.2), for any stopping time $\tau \le T$

$$E_{F_\tau}|w^\epsilon(\tau + \delta) - w^\epsilon(\tau)|^2 = O(\delta),$$

uniformly in ϵ and τ. Thus, Theorem 1.5.1 implies that $\{w^\epsilon(\cdot)\}$ is tight. Let $m^\epsilon(\cdot)$ be an admissible relaxed control (with respect to $w_1(\cdot)$, $w_2(\cdot)$ for (1.1), (1.2)) and let ϵ index a weakly convergent subsequence of $\{x^\epsilon(\cdot), m^\epsilon(\cdot), w^\epsilon(\cdot)\}$, with limit denoted by $(x(\cdot), m(\cdot), \hat{w}(\cdot))$. By Theorem 1.2, there is some standard vector valued Wiener process $w(\cdot)$ such that $(x(\cdot), m(\cdot), w(\cdot))$ satisfies (1.7) and $m(\cdot)$ is admissible with respect to $w(\cdot)$. It can be shown that $\hat{w}(\cdot) = w(\cdot)$. The proof of this latter fact is just the "martingale method" of Theorem 1.2, but applied to the triple $(x^\epsilon(\cdot), m^\epsilon(\cdot), w^\epsilon(\cdot))$ instead of only to $(x^\epsilon(\cdot), m^\epsilon(\cdot))$. We omit the details and only note that by the averaging method used in Part 4 of the proof of Theorem 1.2, the $a(x^\epsilon(s), z^\epsilon(s))$ in (2.7) "averages out" to $\bar{a}(x^\epsilon(s))$. Thus, $\langle w^\epsilon \rangle(\cdot)$ converges weakly to the function with values tI, which is just the quadratic variation of a vector-valued Wiener process.

We let $u^\epsilon(\cdot)$ denote the adaptation of $u^\gamma(\cdot)$ to (1.1) and (1.2) and let $m^\epsilon(\cdot)$ denote its relaxed control representation. The $u^\epsilon(\cdot)$ will be defined as follows: It is the ordinary $\mathcal{U}^\gamma$-valued and piecewise constant control which takes the value $u^\epsilon(i\delta)$ on $[i\delta, i\delta + \delta)$ and satisfies

$$p_i(\alpha_j, j \leq i; w^\epsilon(k\rho), k\rho \leq i\gamma) = P(u^\epsilon(i\delta)$$

$$= \alpha_i \,|\, u^\epsilon(j\delta) = \alpha_j, j < i, w^\epsilon(k\rho), k\rho \leq i\delta)$$

$$= P(u^\epsilon(i\delta) = \alpha_i \,|\, u^\epsilon(j\delta) = \alpha_j, j < i, x^\epsilon(s), w_1(s), w_2(s), w^\epsilon(s), s \leq i\delta).$$

I.e., choose the values of $u^\epsilon(i\delta)$ at random according to the above conditional distributions. Let $m^\gamma(\cdot)$ denote the relaxed control representation of $u^\gamma(\cdot)$. Then, by the continuity of the $p_i(\cdot)$ in the w-arguments for each value of the control value arguments, we have $(m^\epsilon(\cdot), w^\epsilon(\cdot)) \Rightarrow (m^\gamma(\cdot), w(\cdot))$. Hence $(x^\epsilon(\cdot), m^\epsilon(\cdot), w^\epsilon(\cdot)) \Rightarrow (x(\cdot), m^\gamma(\cdot), w(\cdot))$, where the limit satisfies (1.7). Also, for each z and x

$$V_T^\epsilon(x, z, m^\epsilon) \to V_T(x, m^\gamma).$$

(The convergence can be shown to be uniform in (x, z) in each compact set, but we omit the details.) Thus (2.1) holds.   Q.E.D.

Nearly optimal feedback control for (1.1)–(1.3). For the averaged system (1.7), (1.8) to be useful for the actual control problem (1.1)–(1.3), we need to know whether good controls for (1.7), (1.8) are also good for (1.1)–(1.3), uniformly in (small) ϵ. The optimality or "approximate" optimality problem for the averaged system is usually much easier than that for the original system. Such an approximation result is given by the next theorem.

Theorem 2.2. *Assume* (A1.1)–(A1.5). *Let* $u(\cdot)$ *be an admissible ordinary feedback control with values* $u(x, t)$, *and being measurable in* (x, t) *and continuous in* x, *uniformly in* t. *Let* (A1.6) *hold for the sequence of processes* $\{z^\epsilon(\cdot)\}$ *associated with the control* $u(\cdot)$. *Let* (1.7) *have a unique weak sense solution under* $u(\cdot)$. *Then, uniformly in* (x, z) *in each compact set,*

$$V_T^\epsilon(x, z, u) \to V_T(x, u).$$

Remark. The theorem implies that if $u(\cdot)$ is "nearly optimal" for the averaged system, then it is nearly optimal for the system (1.1)–(1.3) for small $\epsilon > 0$. The proof is just a consequence of that of Theorem 2.1.

3. Discounted Cost and Optimal Stopping Problems

The averaging methods of Sections 1 and 2 can be used with all the standard cost functions. In this section, we describe the changes that are required when control stops on hitting a boundary. We work first with a discounted problem. Limitation of space prevents us from treating too many problem types.

The discounted problem. We will use the conditions

A3.1. $\beta > 0$ *and* $k(\cdot)$ *is bounded and continuous.*

A3.2. G *is a set in* R^r *which is the closure of its interior and has a piecewise differentiable boundary.* $g(\cdot)$ *is a bounded and continuous real-valued function.*

We will use the model (1.1), (1.2), but with the cost

$$V_\beta^\epsilon(x, z, m^\epsilon) = E_{x,z}^{m^\epsilon} \int_0^{\tau^\epsilon} \int_{\mathcal{U}} e^{-\beta s} k(x^\epsilon(s), z^\epsilon(s), \alpha) m_s^\epsilon(d\alpha) ds$$

$$+ E_{x,z}^{m^\epsilon} e^{-\beta \tau^\epsilon} g(x^\epsilon(\tau^\epsilon)), \qquad (3.1)$$

where $\tau^\epsilon = \min\{t : x^\epsilon(t) \in \partial G\}$, and ∂G denotes the boundary of G. We use the one-point compactification $\overline{R} = [0, \infty]$ of $[0, \infty)$ for the range space of the stopping time. Hence, any sequence of stopping times is tight.

For the averaged system (1.7), define the stopping time $\tau = \min\{t : x(t) \in \partial G\}$, and the cost

$$V_\beta(x, m) = E_x^m \int_0^\tau \int_{\mathcal{U}} e^{-\beta s} \overline{k}(x(s), \alpha) m_s(d\alpha) ds + E_x^m e^{-\beta \tau} g(x(\tau)). \quad (3.2)$$

If $G = R^r$, then $\tau = \infty$ w.p.1, and the proof that

$$\inf_{m^\epsilon} V_\beta^\epsilon(x, z, m^\epsilon) \equiv V_\beta^\epsilon(x, z) \to V_\beta(x) \equiv \inf_m V_\beta(x, m) \qquad (3.3)$$

is essentially that of Theorem 2.1, under (A1.1)–(A1.6) and (A3.1), (A3.2). Again $\inf_m$ implies the inf over all admissible pairs $(m(\cdot), w(\cdot))$.

If $G \neq R^r$, then we need to treat the limits of $\{\tau^\epsilon\}$. Even if $\{x^\epsilon(\cdot), m^\epsilon(\cdot)\}$ converges weakly to a pair $(x(\cdot), m(\cdot))$, which satisfies (1.7) for some appropriate $w(\cdot)$, it is still not necessarily true that $\tau^\epsilon \to \tau$. For $\phi(\cdot) \in C^r[0, \infty)$, define the function $\tau(\phi) = \inf\{t : \phi(t) \in \partial G\}$. The main problem is that $\tau(\cdot)$ is not continuous at each point $\phi(\cdot)$. If it were, then $\tau^\epsilon = \tau(x^\epsilon(\cdot)) \to \tau(x(\cdot)) = \tau$ and we would have (3.3). In general, we need to assume the following condition.

A3.3. *The function* $\tau(\cdot)$ *is continuous w.p.1 with respect to the measure induced by the limit processes* $x(\cdot)$ *on* $C^r[0, \infty)$, *under each admissible control.*

Conditions such as (A3.3) arise in all problems where one needs to worry about weak convergence together with escape times. Theorem 2.2 and the remark below Theorem 2.1 holds here also.

Discussion of (A3.3). First, let $\overline{a}(x)$ be uniformly positive definite in G. Suppose that there is an open cone C and an $\delta > 0$ such that for each

$y \in \partial G$, we have $\{x : x - y \in C, |x - y| < \delta\} \cap G^0$ = empty set, where G^0 is the interior of G. Then we say that ∂G satisfies the open cone condition, and (via [D6, Theorem 13.8]) (A3.3) holds.

The exit time problem for the case of non-degenerate $\bar{a}(\cdot)$ is relatively easy since the behavior of $x(\cdot)$ on or near the boundary is essentially that of the diffusion term $\int_0^t \bar{\sigma}(x(s))dw(s)$. For a smooth boundary, the non-degeneracy and the properties of the Wiener process imply that if $x(\cdot)$ hits the boundary, then it crosses it infinitely often in an arbitrary small interval after the hitting time. This implies that the function $\tau(\cdot)$ is continuous w.p.1 with respect to the measure induced by $x(\cdot)$.

Verifying (A3.3) for the degenerate case is more difficult, and one usually checks it in each case. Further discussion appears in [K13, p. 64–66]. Frequently, the boundary ∂G can be broken into several pieces, and each considered separately; e.g., (a) a section where the orientation of the "noise" terms is such that the "non-degenerate" considerations discussed above imply (A3.3); (b) a section where the dynamics either guarantee (A3.3) or where escape is impossible due to the direction of the "mean velocity" and orientation of the noise terms; and (c) the remaining section. In many cases, the "remaining section" consists of a finite set of isolated points. We need to show that these points are not accessible from the given initial conditions. Consider, for example, the case depicted in Figure 4.1 where $dx_1 = x_2 dt$, $dx_2 = u\, dt + dw$. The only questionable points are (α) and (β), and these can often be shown to be not accessible from interior points by means of tests such as [S7, Theorem 6.1] (see also [K13, p. 66]).

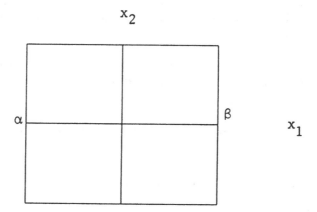

FIGURE 4.1. Questionable points on the boundary.

If the set G is not precisely given or can be altered slightly without losing the meaning of the problem, then a slight alteration to the stopping rule allows us to drop (A3.3). Such alterations are usually allowed when G is chosen largely for numerical reasons; i.e., to guarantee a bounded state space and a "finite" numerical algorithm. The alternative stopping rule is the *randomized stopping* defined as follows.

Randomized Stopping. Let $p(x) \geq 0$ denote a continuous function on G^0 which is nonzero only on $N_\delta(\partial G) \cap G^0$, where $N_\delta(\partial G)$ is the δ-neighborhood of the boundary ∂G, and which goes to infinity as $x \to \partial G$. Then, the rule is to stop $x^\epsilon(\cdot)$ or $x(\cdot)$ at time t with "stopping rate" $p(x^\epsilon(t))dt$ (resp., $p(x(t))dt$). The cost for stopping at x is $g(x)$. Then Theorem 3.1 holds without (A3.3).

Theorem 3.1. *Under* (A1.1)–(A1.6) *and* (A3.1)–(A3.3), (3.3) *holds.*

Stopping Time Problems. Define the costs

$$V^\epsilon(x, z, \tau^\epsilon, m^\epsilon) = E_{x,z}^{m^\epsilon} \int_0^{\tau^\epsilon} \int_{\mathcal{U}} k(x^\epsilon(s), z^\epsilon(s), \alpha) m_s^\epsilon(d\alpha) ds + E_{x,z}^{m^\epsilon} g(x^\epsilon(\tau^\epsilon))$$
(3.4)

$$V(x, \tau, m) = E_x^m \int_0^\tau \int_{\mathcal{U}} \overline{k}(x(s), \alpha) m_s(d\alpha) ds + E_{x,z}^m g(x(\tau)),$$ (3.5)

where τ^ϵ and τ are stopping times for the processes $(x^\epsilon(\cdot), z^\epsilon(\cdot))$ and $x(\cdot)$, resp. Here the "total" control is the choice of both $m(\cdot)$ and the stopping time τ. We also have the following theorem, whose proof is similar to that of Theorem 2.1, and is omitted.

Theorem 3.2. *Assume* (A1.1)–(A1.6) *where* (A1.6) *holds for each* $T < \infty$, *and let* $g(\cdot)$ *and* $k(\cdot)$ *be bounded with* $k(x, z, \alpha) \geq \delta_0 > 0$, *for all* x, z, α. *Then, uniformly in each compact* (x, z) *set,*

$$\inf_{\tau^\epsilon, m^\epsilon} V^\epsilon(x, z, \tau^\epsilon, m^\epsilon) \xrightarrow{\epsilon} \inf_{\tau, m} V(x, \tau, m).$$ (3.6)

4. Average Cost Per Unit Time

In this section we extend the results of Sections 1 to 3 to control on the infinite time interval $[0, \infty)$. Another, and rather powerful, approach to this problem appears in Chapter 5. The method of Chapter 5 can be extended to a fairly broad class of approximation and limit problems on an infinite time interval. For admissible $(m^\epsilon(\cdot), w_1(\cdot), w_2(\cdot))$ for (1.1) and (1.2) and admissible $(m(\cdot), w(\cdot))$ for (1.7), define the "pathwise" average and mean costs

$$\Lambda_T^\epsilon(x, z, m^\epsilon) = \frac{1}{T} \int_0^T \int_{\mathcal{U}} k(x^\epsilon(s), z^\epsilon(s), \alpha) m_s^\epsilon(d\alpha) ds,$$

$$\Lambda^\epsilon(x, z, m^\epsilon) = \overline{\lim}_T E_{x,z}^{m^\epsilon} \gamma_T^\epsilon(x, z, m^\epsilon),$$

$$\Lambda_0^\epsilon(x, z) = \inf_{m^\epsilon} \Lambda^\epsilon(x, z, m^\epsilon). \tag{4.1}$$

$$\Lambda_T(x, m) = \frac{1}{T} \int_0^T \int_{\mathcal{U}} \overline{k}(x(s), \alpha) m_s(d\alpha) ds,$$

$$\Lambda(x, m) = \overline{\lim}_T E_x^m \Lambda_T(x, m),$$

$$\Lambda_0(x) = \inf_m \Lambda(x, m). \tag{4.2}$$

Note that *neither* $\Lambda_T^\epsilon(\cdot)$ *nor* $\Lambda_T(\cdot)$ *involve expectations. They are simply pathwise averages.* Loosely speaking, we will show that as $\epsilon \to 0$ and $T \to \infty$, the pathwise average costs $\Lambda_T^\epsilon(x, z, m^\epsilon)$ will converge to the mean cost per unit time for an "averaged" problem, where the system is (1.7), and (1.8) is replaced by the appropriate "ergodic" cost $\Lambda(x, m)$. The way that $\epsilon \to 0$ and $T \to \infty$ will not be important and this fact is critical in the applications. If the rate of convergence of $\Lambda_T^\epsilon(x, z, m^\epsilon)$ as $T \to \infty$ depended on ϵ (for small ϵ) then the behavior or costs of the averaged system would not provide a good approximation to the behavior or costs of the $x^\epsilon(\cdot)$ over large time intervals.

If $\Lambda(x, m)$ does not depend on x, we write it simply as $\Lambda(m)$. Quite often, under the optimal control for the averaged system, the cost does not depend on x. Then we write $\Lambda_0(x)$ as Λ_0. This occurs when the optimal control is an ordinary feedback control and the corresponding averaged $x(\cdot)$ process is "asymptotically ergodic." See [B2, B8, C2, K12] for more detail on this point. Also under broad conditions, given $\delta > 0$ there will be some sequence $\{m^\epsilon(\cdot)\}$ such that for each x and z, $\Lambda_T^\epsilon(x, z, m^\epsilon)$ converges in probability to within δ of Λ_0 as $\epsilon \to 0$ and $T \to \infty$, and there is no admissible $\{m^\epsilon(\cdot)\}$ for which the limit is less than Λ_0. Furthermore, a good control for the averaged system will also be a good control for the singularly perturbed system. See [K12] for criteria under which there is a smooth δ-optimal control for the averaged system for each $\delta > 0$. Such results help to justify the use of the averaged problem even on an infinite time interval. We will use the following definitions and conditions.

A4.1. *For each $\delta > 0$ and small $\epsilon > 0$, there is a δ-optimal (for the cost functional $\Lambda^\epsilon(x, z, m^\epsilon)$) admissible $m^\epsilon(\cdot)$ such that the corresponding sequence $\{x^\epsilon(n), n = 0, 1, \ldots,$ small $\epsilon > 0\}$ is tight.*

The condition (A4.1) implies that there are δ-optimal policies for which the system $x^\epsilon(t)$ doesn't "explode" as $\epsilon \to 0$ and $t \to \infty$. If the state space is not compact, then such conditions can often be verified by a Liapunov function argument. See Chapter 9 for sufficient conditions. Recall the definition of a relaxed feedback control given in Chapter 3.3.

A4.2. *For each $\delta > 0$, (A1.6) holds for $T = \infty$ and the controls of* (A4.1).

A4.3. *For each admissible feedback relaxed control and each initial condition, (1.7) has a unique weak sense solution. For each admissible feedback control $m_{fb}(\cdot)$, (1.7) has a unique invariant measure $v(m_{fb}, \cdot)$.*

Remark on (A4.3). Let $m_{fb}(\cdot)$ denote a relaxed feedback control. Then (1.7) can be written in the form

$$dx = dt \int \overline{G}(x, \alpha) m_{fb}(x, d\alpha) + \overline{\sigma}(x) dw, \qquad (4.3)$$

The term $\int \overline{G}(x, \alpha) m_{fb}(x, d\alpha)$ might not have regularity properties other than Borel measurability and linear growth in x as $|x| \to \infty$. Often, the measure transformation method of Chapter 3.3 can be used to define a solution and guarantee weak sense uniqueness. Refer to (3.3.4) and just replace $\int \hat{b}(x, \alpha) m_t(d\alpha) dt$ by $\int \hat{b}(x, \alpha) m_{fb}(x, d\alpha) dt$.

Because of the (weak sense) uniqueness of the solution to (4.3) for each initial condition, under (A4.3), the solution is a Feller process; i.e., for each bounded, continuous and real valued function $f(\cdot)$, $E_x f(x(t))$ is continuous in x for each $t \geq 0$.

Relevant information concerning the existence and uniqueness of invariant measures are dealt with in [B2, K23]. For example, we have the following result. Suppose that $\overline{G}(x, \alpha) = G(x) + \hat{G}(x, \alpha)$, where $\hat{G}(x, \alpha)$ is bounded. Then, under a nondegeneracy and stability condition on the system $dx = G(x) dt + \overline{\sigma}(x) dw$, (4.3) can be shown to have a unique invariant measure for any relaxed feedback control. (See [K23], where the results also cover the case of relaxed feedback controls.)

A remark on a criterion for a measure to be an invariant measure for (4.3). Let A be the differential operator of a diffusion process, and suppose that for some measure $v(\cdot)$ and all $g(\cdot) \in C_0^2(R^r)$, we have $(Ag, v) = \int Ag(x) v(dx) = 0$. Then, under quite broad conditions on A [E2, Chapter 4.9] $v(\cdot)$ is an invariant measure for the process $x(\cdot)$ with differential operator A. If the coefficients in A are not continuous, then it is hard to use the results in [E2]. There are alternative approaches; e.g., the one in Chapter 5. But in order to avoid a long discussion at this point, we simply assume the essentially equivalent condition (A4.4) below. Let $A^{m_{fb}}$ denote the differential operator of (4.3). The differential equation in (A4.4) is just the weak form of the Fokker–Planck equation. Any constant solution is an invariant measure.

A4.4. *For each feedback relaxed control $m_{fb}(\cdot)$ for (4.3), the set of functions $\{A^{m_{fb}}g(\cdot),\ g(\cdot) \in C_0^2(R^r)\}$ is measure determining. Equivalently, the equation*

$$(\mu(t), g) = (\mu(0), g) + \int_0^t (A^{m_{fb}}g, \mu(s))ds,\ g(\cdot) \in C_0^2(R^r)$$

has a unique weakly continuous probability measure valued solution $\mu(\cdot)$ for each initial (probability measure) condition $\mu(0)$.

For use below for a relaxed control $m(\cdot)$, define the sample occupation measures $\hat{P}_T^\epsilon(\cdot)$ by

$$\hat{P}_T^\epsilon(B \times C) = \frac{1}{T} \int_0^T I_B(x^\epsilon(s))m_s^\epsilon(C)ds,$$

for $B \in \mathcal{B}(R^r)$, $C \in \mathcal{B}(\mathcal{U})$. The sample values of $\hat{P}_T^\epsilon(\cdot)$ are measures on $(R^r \times \mathcal{U}, \mathcal{B}(R^r \times \mathcal{U}))$. Such occupation measures were used first in [B8] for a problem concerning existence of an optimal (pathwise) ergodic control for a diffusion process and were later used in [K11, K23] to obtain approximation results for "ergodic" cost problems for wide bandwidth noise driven systems.

Keep in mind that in the theorem below the ω is only the canonical variable of the space on which the limit $\hat{P}(\cdot)$ is defined. It is not the canonical variable on the space on which (1.7) is defined. By the notation of the theorem, for each ω not in some null set, there is a relaxed feedback control $m_{fb}^\omega(\cdot)$ and some other probability space on which a stationary solution to (1.7) is defined under $m_{fb}^\omega(\cdot)$.

Theorem 4.1. *Assume (A1.1)–(A1.5), (A4.3), (A4.4), and let $k(\cdot)$ bounded. Let $\{m^\epsilon(\cdot), w_1(\cdot), w_2(\cdot)\}$ denote an admissible sequence such that (A4.1) and (A4.2) hold. Then $\{\hat{P}_T^\epsilon(\cdot),\ \epsilon > 0,\ T < \infty\}$ is a tight sequence of (probability) measure valued random variables. Let $\{\epsilon_n, T_n\}$ index a weakly convergent subsequence of $\{\hat{P}_T^\epsilon(\cdot),\ \epsilon > 0,\ T < \infty\}$, with limit denoted by $\hat{P}(\cdot)$. Let ω be the canonical variable on the sample space on which $\hat{P}(\cdot)$ is defined, and let $\hat{P}(\omega, \cdot)$ denote the sample (measure valued) values. We have*

$$\Lambda_n^{\epsilon_n}(x, z, m^{\epsilon_n}) \Rightarrow \tilde{\Lambda} = \int \int_{\mathcal{U}} \overline{k}(x, \alpha)\hat{P}(dxd\alpha),$$

where the values $\tilde{\Lambda}(\omega)$ have the following representation: For each ω not in some null set there is a relaxed feedback control $m_{fb}^\omega(\cdot)$ for some process satisfying (1.7) with associated invariant measure $v(m_{fb}^\omega, \cdot)$, such that

$$\hat{P}(\omega, dxd\alpha) = m_{fb}^\omega(x, d\alpha)v(m_{fb}^\omega, dx)$$

and

$$\tilde{\Lambda}(\omega) = \Lambda(m_{fb}^{\omega}) = \int \int_{\mathcal{U}} \overline{k}(x, \alpha) m_{fb}^{\omega}(x, d\alpha) v(m_{fb}^{\omega}, dx). \qquad (4.4)$$

Remark. The theorem claims that the sample value of the limit of the pathwise average costs (as $T_n \to \infty$ and $\epsilon_n \to 0$) is the ergodic cost associated with a stationary problem associated with some feedback relaxed control. It does not imply that the $x^{\epsilon}(\cdot)$ converges weakly to such a process in any sense. In Chapter 5, it is shown that a certain "average" of $x^{\epsilon}(\cdot)$ does converge to the limit. For different values of ω, there might be different relaxed controls and associated invariant measures. Since the cost in (4.4) is associated with a stationary problem, we omit the initial condition in the notation, and write $\Lambda(m_{fb}^{\omega})$ for the cost value. The theorem simply characterizes the sample value of the pathwise limit. It states that (for almost all ω) the sample value of the pathwise limit can be represented as the mean ergodic cost associated with some relaxed feedback control. This will be of use in Theorem 4.2 to show that a "nearly" optimal continuous control for the averaged system (1.7) with cost $\Lambda(m)$ is also nearly optimal for (1.1), (1.2) and the "pathwise average" cost function.

We emphasize that the relaxed feedback controls which appear in the proof are used only for mathematical purposes. They simply allow a useful characterization of a pathwise limit, and then an "almost optimality" result as stated at the end of the last paragraph.

Outline of Proof. First, tightness of $\{x^{\epsilon}(\cdot), m^{\epsilon}(\cdot)\}$ is proved, where $\{m^{\epsilon}(\cdot)\}$ is a sequence of δ-optimal controls, for any $\delta > 0$. Then, as usual, the limit $(x(\cdot), m(\cdot))$ of some weakly convergent subsequence is characterized to be an "averaged system." The pathwise cost is written in terms of the occupation measure $\hat{P}_T^{\epsilon}(\cdot)$, and the $z^{\epsilon}(\cdot)$ component of the cost is averaged out. By a type of "martingale argument," the weak limits of the $\{\hat{P}_T^{\epsilon}(\cdot)\}$ are then characterized as a product of a relaxed feedback control and an invariant measure for a stationary process driven by that relaxed feedback control. Finally, the limit of the costs is characterized in terms of the above mentioned feedback control and invariant measure.

Proof. Given $\delta > 0$, let $\{m^{\epsilon}(\cdot)\}$ be δ-optimal controls as in (A4.1). Then condition (A1.2), (A4.1) and Theorem 2.3.2 imply that the set of random variables

$$\{x^{\epsilon}(t), t < \infty, \text{ small } \epsilon > 0\} \qquad (4.5)$$

is tight. This, in turn, implies the tightness of $\{\hat{P}_T^{\epsilon}(\cdot), \text{ small } \epsilon > 0, T < \infty\}$ by Chapter 1.6. The tightness of (4.5), the boundedness and continuity of $k(\cdot)$ and Theorem 2.3.3 imply that

$$\sum_{i=0}^{T/\Delta} \frac{1}{T} E \int_{i\Delta}^{i\Delta+\Delta} \int_{\mathcal{U}} |k(x^{\epsilon}(i\Delta), z^{\epsilon}(s), \alpha) - k(x^{\epsilon}(s), z^{\epsilon}(s), \alpha)| m_s^{\epsilon}(d\alpha) ds \to 0$$

as $\Delta \to 0, \epsilon/\Delta \to 0, T \to \infty$. Thus, the limits (in probability) of $\gamma_T^\epsilon(x, z, m^\epsilon)$ and those of (as $\Delta \to 0, \epsilon/\Delta \to 0, T \to \infty$)

$$\sum_0^{T/\Delta} \frac{1}{T} \int_{i\Delta}^{i\Delta+\Delta} \int_{\mathcal{U}} k(x^\epsilon(i\Delta), z^\epsilon(s), \alpha)m_s^\epsilon(d\alpha)ds \tag{4.6}$$

are the same.

Applying the arguments which were used to average out $z^\epsilon(\cdot)$ in Theorem 1.2 to the terms in (4.6) but using condition (A4.2) and the tightness of the set (4.5), we get that the limits of (4.6) and those of

$$\frac{1}{T} \int_0^T \int_{\mathcal{U}} \overline{k}(x^\epsilon(s), \alpha)m_s^\epsilon(d\alpha)ds = \int_{\mathcal{U}} \overline{k}(x, \alpha)\hat{P}_T^\epsilon(dxd\alpha) \tag{4.7}$$

are the same in the sense that if a subsequence indexed by $(\epsilon_n, \Delta_n, T_n)$ of (4.6) converges to a limit, then (for the subsequence indexed by (ϵ_n, T_n)) (4.7) converges to the same limit (equality w.p.1), and conversely.

We now characterize the limits of $\{\hat{P}_T^\epsilon(\cdot), \epsilon > 0, T < \infty\}$ by a method similar to that used in Part 4 of the proof of Theorem 1.2. Let $q(\cdot)$ be a smooth function of x with compact support. The process

$$M_q^\epsilon(t) = q(x^\epsilon(t)) - q(x(0)) - \int_0^t (A^\epsilon q)(x^\epsilon(s), z^\epsilon(s), s)ds$$

is a zero mean martingale. Since $\sup_{\epsilon,n} E[M_q^\epsilon(n+1) - M_q^\epsilon(n)]^2 < \infty$, we have $M_q^\epsilon(t)/t \xrightarrow{P} 0$ as $t \to \infty$ for all small ϵ. Thus, w.p.1.,

$$0 = \lim_{\epsilon,T} \frac{1}{T} \int_0^T (A^\epsilon q)(x^\epsilon(s), z^\epsilon(s), s)ds$$

$$= \lim_{\epsilon,T} \frac{1}{T} \int_0^T ds \left\{ q_x'(x^\epsilon(s)) \int_{\mathcal{U}} G_1(x^\epsilon(s), \alpha)m_s^\epsilon(d\alpha) \tag{4.8} \right.$$

$$+ \left. q_x'(x^\epsilon(s))G_0(x^\epsilon(s), z^\epsilon(s)) + \frac{1}{2}\text{trace } q_{xx}(x^\epsilon(s)) \cdot a(x^\epsilon(s), z^\epsilon(s)) \right\} ds.$$

By (4.8), we mean that the limit of any weakly convergent subsequence is zero w.p.1 and that the sequence converges to zero in probability. Analogously to what was done in the first part of the proof, the right hand side of (4.8) can be shown to have the same limits (as $\epsilon \to 0, T \to \infty$) as has

$$\frac{1}{T} \int_0^T \left\{ q_x'(x^\epsilon(s)) \left[\int_{\mathcal{U}} G_1(x^\epsilon(s), \alpha)m_s^\epsilon(d\alpha) + \overline{G}_0(x^\epsilon(s)) \right] \right.$$

$$+ \left. \frac{1}{2}\text{trace } q_{xx}(x^\epsilon(s)) \cdot \overline{a}(x^\epsilon(s)) \right\} ds \tag{4.9}$$

$$= \frac{1}{T} \int_0^T \int_{\mathcal{U}} \overline{A}^\alpha q(x^\epsilon(s)) m_s^\epsilon(d\alpha) ds = \int \overline{A}^\alpha q(x) \hat{P}_T^\epsilon(dx d\alpha).$$

The last line of (4.9) follows from the definitions of $\overline{A}^\alpha$ and $\hat{P}_T^\epsilon(\cdot)$. Let $\epsilon_n \to 0$, $T_n \to \infty$ index a weakly convergent subsequence of $\{\hat{P}_T^\epsilon(\cdot), \epsilon > 0, T < \infty\}$, with limit denoted by $\hat{P}(\cdot)$. By (4.8) and the equality of the limits of (4.8) and (4.9), we have (w.p.1)

$$0 = \int \int_{\mathcal{U}} \overline{A}^\alpha q(x) \hat{P}(dx d\alpha). \tag{4.10}$$

With probability one, (4.10) holds for all $q(\cdot) \in C_0^2(R^r)$.

The $\hat{P}(\cdot)$ is a measure valued random variable. Let it be defined on a sample space with canonical variable ω, and let $\hat{P}(\omega, \cdot)$ denote the value of $\hat{P}(\cdot)$ at ω. For each ω, factor $\hat{P}(\omega, \cdot)$ in the product form $\hat{P}(\omega, dx d\alpha) = \beta^\omega(x, d\alpha) v^\omega(dx)$. We can choose $\beta^\omega(x, \cdot)$ such that it is an admissible relaxed feedback control; i.e., we can choose $\beta^\omega(\cdot)$ such that $\beta^\omega(\cdot, B)$ is Borel measurable in x for each $B \in \mathcal{B}(\mathcal{U})$. Write this version as $m_{fb}^\omega(\cdot)$. Then, for almost all ω and for all smooth $q(\cdot)$ with compact support (4.10) implies that

$$0 = \int \int_{\mathcal{U}} \overline{A}^\alpha q(x) m_{fb}^\omega(x, d\alpha) v^\omega(dx). \tag{4.11}$$

Now (A4.4) implies that $v^\omega(\cdot)$ is an invariant measure of (1.7) under the relaxed feedback control $m_{fb}^\omega(\cdot)$. By (A4.3), the invariant measure $v(m_{fb}^\omega, \cdot)$ is unique. Hence $v^\omega(\cdot) = v(m_{fb}^\omega, \cdot)$.

Finally, taking limits in (4.7) (with ϵ_n, Δ_n replacing ϵ, Δ, resp.) yields the limit (in probability)

$$\int \int_{\mathcal{U}} \overline{k}(x, \alpha) \hat{P}(\omega, dx d\alpha) = \int \int_{\mathcal{U}} k(x, z, \alpha) \mu_x(dz) m_{fb}^\omega(x, d\alpha) v(m_{fb}^\omega, dx)$$

$$= \Lambda(m_{fb}^\omega),$$

as desired.   Q.E.D.

An "approximate optimality" result. The results of Theorem 4.1 can be used to obtain an analog of Theorem 2.1. We require the following definition and conditions.

Definition. Let $\delta > 0$. Let $m_{fb}(\cdot)$ be a feedback relaxed control and let $x(\cdot)$ be any associated *stationary* solution to (1.7), with $\Lambda(m_{fb})$ the cost for the stationary solution. Let $u(\cdot)$ be an ordinary admissible feedback control such that the cost $\Lambda(x, u) = \Lambda(u)$ does not depend on x. If $\Lambda(u) \leq \Lambda(m_{fb}) + \delta$ for all such feedback relaxed controls and associated stationary solutions, then $u(\cdot)$ is *said to be δ-optimal.*

A4.5. *For each $\delta > 0$, there is a continuous feedback ordinary admissible control $u^\delta(\cdot)$ such that under $u^\delta(\cdot)$, (1.7) has a unique weak sense solution* *for each initial condition and a unique invariant measure $v(u^\delta, \cdot)$, and $u^\delta(\cdot)$ is δ-optimal for* (1.7).

A4.6. *Under $u^\delta(\cdot)$ applied to* (1.1), (1.2),

$$\{x^\epsilon(n), \ n = 1, 2, \ldots, \text{ small } \epsilon > 0\}$$

is tight.

Remark. The existence of the $u^\delta(\cdot)$ of (A4.5) can be proved under quite broad conditions, if the $\bar{a}(\cdot)$ is non-degenerate. A proof is in [K12]. In that reference, the existence of an optimal measurable feedback control is shown under a stability condition. It is then proved that the cost is a continuous function of the feedback control (with the L_1-topology used on the space of feedback controls). This allows us to "smooth" the optimal control to get a δ-optimal continuous control. The details of [K12] also hold if the system is of the partitioned form where $x = (x_1, x_2)$ and

$$dx_1 = b_1(x)dt + \sigma_1(x)dw_1$$
$$dx_2 = b_2(x, u)dt + \sigma_2(x)dw_2,$$

where $\sigma_2(x)$ is smooth and uniformly non-degenerate and $w_1(\cdot)$ and $w_2(\cdot)$ are mutually independent, and the system is stable for all controls.

Theorem 4.2. *Assume* (A1.1)–(A1.5), (A4.3)–(A4.6), *with $k(\cdot)$ bounded. Assume the tightness and bounds in* (A4.1)–(A4.2) *for both the $m^\epsilon(\cdot)$ of* (A4.1) *and with $u^\delta(\cdot)$ being used for* (1.1), (1.2). *Then for each $\delta > 0$,*

$$\Lambda_T^\epsilon(x, z, u^\delta) \xrightarrow{P} \Lambda(u^\delta) = \int \bar{k}(x, u^\delta(x))v(u^\delta, dx) \qquad (4.13)$$

as $\epsilon \to 0$ and $T \to \infty$, and

$$\lim_{\epsilon \to 0, \, T \to \infty} P\big(\Lambda_T^\epsilon(x, z, m^\epsilon) \leq \Lambda_T^\epsilon(x, z, u^\delta) - 2\delta\big) = 0. \qquad (4.14)$$

The convergence in (4.13) *and* (4.14) *is uniform for the initial conditions in any compact set if the tightness bounds in* (A4.1) *and* (A4.2) *are.*

The proof follows from Theorem 4.1, assumptions (A4.4), (A4.5), and the fact that the tightness and bounds in (A4.1) and (A4.2) hold when either $m^\epsilon(\cdot)$ or $u^\delta(\cdot)$ is used on (1.1), (1.2). It will be seen in Chapter 5 that (A4.3) is not needed.

5. Jump-Diffusion Processes

All of the results of Sections 1 to 4 can be extended to the jump-diffusion model (5.1), (5.2).

$$dx^\epsilon = dt \int G(x^\epsilon, z^\epsilon, \alpha) m_t^\epsilon(d\alpha) + \sigma(x^\epsilon, z^\epsilon) dw_1$$

$$+ \int q_1(x^\epsilon, z^\epsilon, \gamma) N_1(dt d\gamma) \tag{5.1}$$

$$\epsilon dz^\epsilon = H(x^\epsilon, z^\epsilon) dt + \sqrt{\epsilon} v(x^\epsilon, z^\epsilon) dw_2$$

$$+ \int q_2(x^\epsilon, z^\epsilon, \gamma) N_2^\epsilon(dt d\gamma). \tag{5.2}$$

We assume the following condition.

A5.1. $N_1(\cdot)$ and $N_2^\epsilon(\cdot)$ are mutually independent Poisson measures and independent of the $w_i(\cdot)$. $N_1(\cdot)$ has jump rate λ_1 and jump distribution $\Pi_1(\cdot)$. $N_2^\epsilon(\cdot)$ has jump rate λ_2/ϵ and jump distribution $\Pi_2(\cdot)$. The $q_i(\cdot)$ are bounded and are continuous in (x, z) for each value of γ, and $q_i(x, z, 0) = 0$.

The averaged process. Using the notation of Section 1, we define the differential operator $\overline{A}^\alpha$ of the averaged process under control value α to be, for $f(\cdot) \in C_0^2(R^r)$,

$$\overline{A}^\alpha f(x) = f_x'(x) \overline{G}(x, \alpha) + \frac{1}{2} \text{trace } f_{xx}(x) \cdot \overline{a}(x)$$

$$+ \lambda_1 \int [f(x + q_1(x, z, \gamma)) - f(x)] \Pi_1(d\gamma) \mu_x(dz). \tag{5.3}$$

$\overline{A}^\alpha$ is the operator for a controlled jump-diffusion process whose jumps can be constructed as follows. The jump rate is λ_1. Let $\{\tau_n, n < \infty\}$ denote the jump times. Then $x(\tau_n) - x(\tau_n^-) = q(x(\tau_n^-), z_n, g_n)$, where z_n and g_n are mutually independent, g_n has distribution $\Pi_1(\cdot)$ and z_n has distribution $\mu_{x(\tau_n^-)}(\cdot)$. In fact, (g_n, z_n) is conditionally independent of $\{x(s), m_s(\cdot), s < \tau_n\}$, given $x(\tau_n^-)$.

The fixed x-process. The analog of the fixed-x process (1.6) is (5.4), where $N_2(\cdot)$ is a Poisson measure with jump rate λ_2 and jump distribution $\Pi_2(\cdot)$.

$$dz_0 = H(x, z_0) dt + v(x, z_0) d\tilde{w}_2 + \int q_2(x, z, \gamma) N_2(dt d\gamma). \tag{5.4}$$

Theorems 5.1 and 5.2 are the analogs of Theorems 1.2 and 2.1. There are also analogs of the results in Sections 3 and 4, and we leave their formulation to the reader.

Theorem 5.1. *Assume* (A1.1)–(A1.5), *(except for* (A1.4)) *and* (A5.1) *applied to* (5.1), (5.2), (5.4). *Let* $\{m^\epsilon(\cdot)\}$ *be admissible relaxed controls for* (5.1), (5.2), *such that* (A1.6) *is satisfied. Then* $\{x^\epsilon(\cdot), m^\epsilon(\cdot)\}$ *is tight. Let* $(x(\cdot), m(\cdot))$ *denote the limit of a weakly convergent subsequence. Then the differential operator of* $x(\cdot)$ *is the* $\overline{A}^\alpha$ *of* (5.3), *with* $\alpha = m$. *There is a standard vector-valued Wiener process* $w(\cdot)$ *such that* $m(\cdot)$ *is non-anticipative with respect to* $w(\cdot)$ *and*

$$dx = dt \int_{\mathcal{U}} \overline{G}(x,\alpha)m_t(d\alpha) + \overline{\sigma}(x)dw + d\overline{J}, \qquad (5.5)$$

$$\overline{J}(t) = \sum_{s \le t}[x(s) - x(s^-)],$$

where

$$P(x(\cdot) \text{ jumps into } B \text{ on } (t, t+\Delta] \,|\, x(s), m_s(\cdot), w(s), s \le t)$$

$$= \lambda\Delta \int I_B(q(x(t), z, \gamma))\mu_{x(t)}(dz)\Pi_1(d\gamma) + o(\Delta), \qquad (5.6)$$

$$P(> \text{ one jump of } x(\cdot) \text{ on } (t, t+\Delta] \,|\, x(s), m_s(\cdot), w(s), s \le t) = o(\Delta). \quad (5.7)$$

Proof. The tightness of $\{x^\epsilon(\cdot)\}$ follows from the arguments of Theorem 2.7.1 and the criterion of Theorem 1.5.1. To characterize the differential operator associated with the limit $(x(\cdot), m(\cdot))$ of any weakly convergent subsequence of $\{x^\epsilon(\cdot), m^\epsilon(\cdot)\}$, the martingale and averaging method used in the proof of Theorem 1.2 yields that $\overline{A}^m$ is the differential operator of $x(\cdot)$, under $m(\cdot)$. The rest of the proof follows from Theorem 2.8.1, suitably adapted to the problem with a control.   Q.E.D.

The construction of the driving Poisson measure for (5.5). When $q_1(\cdot)$ depends on z and there is a control, it is difficult to deal with questions of uniqueness and approximation for (5.5). *So, for the rest of this section, we suppose that* $q_1(\cdot)$ *does not depend on* z.

Theorem 5.2. *Assume the conditions of Theorem 5.1. Let* $q_1(\cdot)$ *not depend on* z *and suppose that for each value of* x, *the value of* $q_1(x, \gamma)$ *determines* γ. *Then there is a Poisson measure* $N(\cdot)$ *(jump rate* λ_1 *and jump distribution* $\Pi_1(\cdot)$) *such that* $m(\cdot)$ *is admissible with respect to* $(w(\cdot), N(\cdot))$ *and* $(w(\cdot), N(\cdot))$ *are mutually independent. Also*

$$dx = dt \int \overline{G}(x,\alpha)m_t(d\alpha) + \overline{\sigma}(x)dw + \int q(x,\gamma)N(dtd\gamma). \qquad (5.8)$$

The proof is a consequence of Theorem 5.1 and the fact that we can reconstruct the jump times τ_n and magnitudes g_n which yield $N(\cdot)$ directly from $x(\cdot)$. See also Theorem 2.8.1.

Theorem 5.3. *Assume the conditions of Theorem 5.2 and let (5.8) have a unique weak sense solution for each initial condition and admissible $(m(\cdot)$, $w(\cdot)$, $N(\cdot))$. For the cost functions (1.3), (1.8), we have*

$$V_T^\epsilon(x, z) \to V_T(x).$$

Under the additional conditions (A1.7), (A1.8), we can let $g(\cdot)$ depend on z.

Remark on the Proof. The proof is essentially the same as that in Theorem 2.1. The only (slight) difference concerns the construction of the comparison controls $u^\gamma(\cdot)$ in Theorem 2.1, due to the presence of the Poisson measure. Given $\gamma > 0$, we can get a γ-optimal control (for the averaged problem (5.8), (1.8)) which is analogous to that in Theorem 2.1, but where the conditional probabilities for the control values (2.6) depend also on samples of the past jump values and times. The method is essentially that of Theorem 2.1, and we omit the details. The results of Section 4 can also be extended to the jump-diffusion case.

6. Other Approaches

Bensoussan and Blankenship [B2], [K4] take a quite different approach to the problem of singular perturbations. They work with the system

$$\begin{aligned}
dx^\epsilon &= G(x^\epsilon, z^\epsilon, u)dt + dw_1 \\
\epsilon dz^\epsilon &= H(x^\epsilon, z^\epsilon, u)dt + \sqrt{\epsilon}dw_2,
\end{aligned} \tag{6.1}$$

where the $w_i(\cdot)$ are mutually independent Wiener processes. For $\beta > 0$, the cost is

$$V^\epsilon(x, z, u) = E_{x,z}^u \int_0^\tau e^{-\beta s} k(x^\epsilon(s), z^\epsilon(s), u(s))ds, \tag{6.2}$$

$$V^\epsilon(x, z) = \inf_u V^\epsilon(x, z, u),$$

where τ is the first exit time from a bounded open set. The dynamical terms are either periodic in z, or else the $z^\epsilon(\cdot)$ process is replaced by a "fast" process which is a reflected diffusion in a bounded set. They also work with a bilinear system (in unbounded space) that is perturbed by a nonlinear term.

The methods of [B2, K4] are those of the theory of partial differential equations rather than of weak convergence and considerable detail is required in obtaining the necessary estimates. A typical result is of the following type. Let $q^v(z \mid x)$ denote the density of the invariant measure (which exists under the conditions in [B2, K4]) of the fixed-x process

$$dz_0 = H(x, z_0, v(x, z_0))dt + dw_2$$

where $v(x, \cdot)$ is a feedback control parametrized by x. Define the function

$$\rho(x, p) = \inf_v \int q^v(z \mid x)[k(x, z, v(z)) + p'G(x, z, v(z))]dz. \qquad (6.3)$$

Define $\Phi(\cdot)$ by

$$-\Delta\Phi + \beta\Phi = \rho(x, \Phi_x). \qquad (6.4)$$

Then it is proved that

$$\sup_v |\Phi(x) - V^\epsilon(x, z)| \le C\epsilon \qquad (6.5)$$

for some constant C. Such $O(\epsilon)$ estimates have not been obtained via the weak convergence method.

In general, $\Phi(\cdot)$ is not the value function for an averaged system. But it can be represented as such a value function if our additional conditions are imposed; e.g., $G(x, z, u) = G_0(x, z) + G_1(x, u)$, $k(x, z, u) = k_0(x, z) + k_1(x, u)$ and where $H(\cdot)$ does not depend on u. Because of this, the method does not produce a design procedure when the fast system depends on the control. In fact, at present there is no really satisfactory method for such a case. Because PDE methods are used, it seems quite hard to cover wideband noise driven or degenerate systems.

Consider the system (1.1), (1.2), but drop the "separation" requirements on $G(\cdot)$ and $k(\cdot)$, and let $H(\cdot)$ depend on the control. By using an argument analogous to that of Theorem 1.2, it can be shown that $(x^\epsilon(\cdot), m^\epsilon(\cdot))$ converges to a system of the type

$$x(t) = x(0) + \int G(x, z, \alpha)\hat{m}_t(x, dzd\alpha)dt + \hat{\sigma}(x, t)dv, \qquad (6.6)$$

where

$$\hat{\sigma}(x, t)\hat{\sigma}'(x, t) = \int \sigma(x, z)\sigma'(x, z)\hat{m}_t(x, dzd\alpha).$$

Here $\hat{m}(\cdot)$ is the limit of the sequence which is analogous to the $\{\hat{P}_{t_\epsilon}^{\epsilon, \Delta}(\cdot)\}$ of Theorem 1.2 (and is the occupation measure for (the control, fast state)). The system (6.6) will not be an averaged system in general. The $\hat{m}_t(\cdot)$ can be factored into a product of a relaxed feedback control for the fixed-x fast system and the associated invariant measure.

Another interesting approach is in [B4, B5]. These works model the system as a perturbed strongly continuous semigroup, and make heavy use of the analytic semigroup theory and the associated conditions on the operators and their domains. The optimal stopping and impulsive control problems are treated. Thus, there are at most a finite number of control actions. Then, it is shown that the averaged problem yields a good approximation for small ϵ. An ergodic problem for a continuously acting control is in [B4], where it is assumed that the transition measure of the fixed-x process converges "exponentially" to the invariant measure and the $z^\epsilon(\cdot)$ process does not depend on either the control or on $x^\epsilon(\cdot)$.

5

Functional Occupation Measures and Average Cost per Unit Time Problems

0. Outline of the Chapter

In Chapter 4.4, we dealt with average cost per unit time problems for singularly perturbed systems. A basic question concerned the characterization of the limit (as $\epsilon \to 0$, $T \to \infty$) pathwise cost as the cost associated with an average cost per unit time problem for the averaged system. The concept of feedback relaxed control was introduced and it was shown, under given conditions, that the limit costs were the average costs per unit time for some averaged system with such a feedback relaxed control. The feedback relaxed control was introduced for mathematical reasons, since it allows a characterization of the limit, which then allowed us to prove the "approximate optimality" theorems.

In the course of the analysis, certain occupation measures were introduced. In this chapter, we introduce a new and powerful method using "functional occupation measures." The general ideas have wide applications to problems involving systems of many types, where the functionals of interest are of the pathwise average cost per unit time (on the infinite interval) type. The conditions are weaker than those in Chapter 4.4, and feedback relaxed controls need not be used. Essentially, we construct directly a stationary limit process whose "costs" are the required limits. The cost functions can be much more general than those of Chapter 4.4. The power of the method will be well illustrated by the way hard problems can be treated with reasonable ease.

In Section 1, some probabilistic background is discussed and the functional occupation measure is defined in a simple case. In Section 2 we apply the technique to a simple ergodic cost problem for an uncontrolled diffusion and see how tightness is proved and how the limit measures and the processes that they define are characterized. A control is added in Section 3, and Section 4 is concerned with the singularly perturbed controlled problem. Some comments concerning control of the fast system are in Section 5. In Section 6, the results of Section 4 are extended to the problem of where the "fast" process is reflected on hitting the boundary of a compact set.

The treatment of the (pathwise) discounted problem as the discount factor tends to zero is similar to that of the average cost per unit time problem, and the functional occupation measure point of view can be used. This problem is dealt with in Section 7. It is shown that the pathwise costs (for the optimal policy or not) converge to those for an ergodic cost problem for an averaged system, exactly as in Section 4. Also, "good" controls for the averaged problem with the ergodic cost are also "good" for the pathwise discounted cost problem when the discount factor is small.

1. Measure-Valued Random Variables

Let S_0 be a complete and separable metric space with metric $d(\cdot)$. Then $\mathcal{P}(S_0) \equiv S_1$ is a complete and separable metric space, where we recall that $\mathcal{P}(S_0)$ is the space of probability measures on $(S_0, \mathcal{B}(S_0))$ with the Prohorov metric. Refer to Chapter 1.6 for the rest of the terminology. Let Q_n and Q be S_1-valued random variables. Their probability distributions are, of course, elements of $\mathcal{P}(S_1)$. Suppose that $\{Q_n\}$ is tight (in the sense of a sequence of random variables) and let $Q_n \Rightarrow Q$. If Q is a measure-valued random variable, then we write the sample value either as Q^ω or as $Q^\omega(\cdot)$. Suppose now that the Skorohod representation (Chapter 1.3) is used so we can assume that $\{Q_n(\cdot), n < \infty, Q(\cdot)\}$ are defined on the same sample space. Let the generic variable of that sample space be ω. By the assumed weak convergence and the Skorohod representation $Q_n^\omega(\cdot) \to Q^\omega(\cdot)$ for almost all ω, where the convergence is in the topology of S_1. I.e.,

$$\pi(Q_n^\omega, Q^\omega) \to 0 \tag{1.1}$$

for almost all ω, where $\pi(\cdot)$ denotes the Prohorov metric. For each ω, the value $Q_n^\omega(\cdot)$ is a measure on $(S_0, \mathcal{B}(S_0))$. Let ω_0 denote the canonical point of S_0. Then, for each ω, the probability measure $Q_n^\omega(\cdot)$ induces an S_0-valued random variable which we denote by X_n^ω and whose values are denoted by $X_n^\omega(\omega_0)$. Note that the ω and n index the *measure* and, hence, the *random variable*. With the measure $Q_n^\omega(\cdot)$ given, ω_0 gives the *sample value* of the *random variable associated with that measure*.

Now, Theorem 1.1.3 and the almost everywhere convergence of the measures which is implied by (1.1) implies that for almost all ω

$$X_n^\omega \Rightarrow X^\omega. \tag{1.2}$$

By (1.2), we can use the Skorohod representation for the set of *random variables* $(\{X_n^\omega\}, X^\omega)$ for each *fixed* ω not in some null set N_0. Doing this, let us suppose that for each $\omega \notin N_0$ these random variables are defined on some probability space with generic variable ω_0. Then for each $\omega \notin N_0$ there is a null set N^ω in that probability space such that for $\omega_0 \notin N^\omega$,

$$d(X_n^\omega(\omega_0), X^\omega(\omega_0)) \to 0. \tag{1.3}$$

For our purposes it will usually be sufficient to work with each ω, and then the relation (1.3) will be quite useful. We will use the method of Theorem

1.6.1 for proving that a sequence of $\mathcal{P}(S_0)$-valued random variables $\{Q_n\}$ is tight.

In Chapter 4.4, we dealt with occupation measures of the type which will be described next. For simplicity in the initial exposition, we let $x(\cdot)$ be a weak solution to the uncontrolled equation

$$dx = b(x)dt + \sigma(x)dw, \quad x \in R^r. \tag{1.4}$$

Let A denote the differential operator of (1.4). Define the occupation measure $P^t(\cdot)$ by $P^t(B) = I_B(x(t))$ for $B \in \mathcal{B}(R^r)$. Define the normalized occupation measure $P_T(\cdot)$ by $P_T(B) = \int_0^T P^t(B)dt/T$. Note that $P_T(\cdot)$ is a measure valued random variable with values in $(R^r, \mathcal{B}(R^r))$. Assume the following:

A1.1. $b(\cdot)$ *and* $\sigma(\cdot)$ *are continuous with a growth at most linear as* $|x| \to \infty$.

A1.2. $\{x(t), t < \infty\}$ *is tight.*

A1.3. (1.4) *has a unique invariant measure* $v(\cdot)$.

By (A1.2) and Theorem 1.6.1, the set $\{P_T(\cdot), T \geq 0\}$ is tight. In Theorem 4.1.2, we used the occupancy measure $\hat{P}_t^{\epsilon,\Delta}(\cdot)$, and showed that its weak limits were invariant measures for the fixed-x process $z_0^\epsilon(\cdot \mid x)$ for appropriate values of x. This result was then used to average out the $z^\epsilon(\cdot)$ variable from the "slow" dynamics and the cost function. Similarly, it can be shown that $P_T(\cdot) \Rightarrow \nu(\cdot)$ in probability. The limit is not random owing to the uniqueness in (A1.3).

In this chapter, we take a point of view toward occupancy measures which is more general and which has numerous practical advantages and many new applications. We now introduce the idea in a simple context. The space $D^r[0, \infty) \equiv S_0'$ is complete and separable (under the usual Skorohod topology), and we let $\phi(\cdot)$ denote its generic element. For $\phi(\cdot)$ in $D^r[0, \infty)$ and $t \geq 0$, define the *shifted function* $\phi_t(\cdot) = \phi(t + \cdot)$. Define the *shifted process* $x_t(\cdot) = x(t+\cdot)$, and define the *functional occupation measures* $\tilde{P}^t(\cdot)$ by $\tilde{P}^t(B_0) = I_{B_0}(x_t(\cdot))$ and

$$\tilde{P}_T(B_0) = \frac{1}{T} \int_0^T \tilde{P}^t(B_0)dt, \quad B_0 \in \mathcal{B}(S_0'). \tag{1.5}$$

We call $\tilde{P}_T(\cdot)$ a functional occupation measure since it is an occupation measure for the *process* $x(\cdot)$ and not simply for the random variables $x(t)$.

Under (A1.1) and (A1.2), $\{\tilde{P}_T(\cdot), T < \infty\}$ is a tight sequence of measure-valued random variables (see the next section for the proof). Each sample value of $\tilde{P}_T(\cdot)$ is a measure on $(S_0', \mathcal{B}(S_0'))$. Hence each sample value induces a stochastic process with paths in $D^r[0, \infty)$. Suppose that $\{\tilde{P}_T(\cdot), T < \infty\}$

is tight (considered as a sequence of measure-valued random variables with the Prohorov topology). Let $\tilde{P}(\cdot)$ denote a measure-valued random variable which is a weak limit of a weakly convergent subsequence of $\{\tilde{P}_T(\cdot),$ $T < \infty\}$. Using the Skorohod representation, suppose that the $\tilde{P}_T(\cdot)$ and the $\tilde{P}(\cdot)$ are defined on the same sample space, and let ω denote the generic value of that space. Let $\tilde{P}^\omega(\cdot)$ denote the sample value of $\tilde{P}(\cdot)$. Then, for almost all ω, the measure $\tilde{P}^\omega(\cdot)$ also induces a stochastic process on $D^r[0,\infty)$. It turns out (see the next section) that this process, which we write as $x^\omega(\cdot)$, is a stationary diffusion of the form (1.4). Since ω indexes the *sample value of the measure* $\tilde{P}(\cdot)$, it indexes the entire process $x^\omega(\cdot)$, and *not* the sample values of that process. We start with the simple case (1.4), since it allows a relatively unencumbered treatment of the basic ideas of the functional occupation measure method. The use of functional occupation measures for purposes of weak convergence analysis seems to be new here. They have been used in the theory of large derivations [V1], but both the purpose and methods are totally different here.

2. Limits of Functional Occupation Measures for Diffusions

We work with the special case of Section 1, where the system is (1.4), and $\tilde{P}_T(\cdot)$ is defined by (1.5). Theorem 2.1 gives a criterion for the tightness of the normalized occupation measures.

Theorem 2.1. *Assume (A1.1) and (A1.2). Then $\tilde{P}_T(\cdot)$ is a $\mathcal{P}(S_0')$-valued random variable, and $\{\tilde{P}_T(\cdot),\ T < \infty\}$ is tight.*

Proof. For each $B_0 \in \mathcal{B}(S_0')$, the measurability of the process $x(\cdot)$ implies the (ω, t)-measurability of the function with values $\tilde{P}^t(\omega, B_0)$ at (ω, t). Thus $\tilde{P}_T(\cdot)$ is a random variable for each T. By (A1.1) and (A1.2), the sequence of shifted processes $\{x_t(\cdot)\}$ is tight. Thus for each $\delta > 0$, there is a compact set K_δ in S_0' such that $P(x_t(\cdot) \in S_0' - K_\delta) \leq \delta$, all t. This implies that

$$E\tilde{P}_T(S_0' - K_\delta) = \frac{1}{T} \int_0^T E\tilde{P}^t(S_0' - K_\delta)dt \leq \delta,$$

and the tightness of $\{\tilde{P}_T(\cdot), T < \infty\}$ follows from Theorem 1.6.1.   Q.E.D.

A remark on notation. Let $\{\tilde{P}_{T_n}(\cdot), n < \infty\}$ be a weakly convergent subsequence with limit denoted by $\tilde{P}(\cdot)$. Let ω_0 be the generic variable on the sample space Ω_0 on which the process $x(\cdot)$ is defined, and let $x(\omega_0, \cdot)$ denote the sample path. Then on this sample space, for each ω_0 we have the representation

$$\tilde{P}_T^{\omega_0}(B_0) = \frac{1}{T} \int_0^T I_{B_0}(x_t(\omega_0, \cdot))dt, \qquad (2.1a)$$

$$\tilde{P}_T(B_0) = \frac{1}{T} \int_0^T I_{B_0}(x_t(\cdot))dt, \ B_0 \in \mathcal{B}(S_0'). \tag{2.1b}$$

Now, let us use the Skorohod representation for $\{\tilde{P}_{T_n}(\cdot), n < \infty, \tilde{P}(\cdot)\}$, and let the generic variable of the common sample space Ω be ω. As usual, we *will use the same notation* $\{\tilde{P}_{T_n}(\cdot), n < \infty, \tilde{P}(\cdot)\}$ for the random variables constructed via the Skorohod representation. The distribution of the $\tilde{P}_{T_n}(\cdot)$ constructed on Ω via the Skorohod representation is that of the right side of (2.1b) for $T = T_n$. Let $\overset{D}{=}$ denote equality in the sense of distribution.

Theorem 2.2. *Assume* (A1.1), (A1.2) *and let* $T_n \to \infty$ *index a weakly convergent subsequence of* $\{\tilde{P}_T(\cdot), T < \infty\}$ *with limit denoted by* $\tilde{P}(\cdot)$.

Assume that the Skorohod representation is used and that ω *denotes the generic variable of the sample space on which* $\{\tilde{P}_{T_n}(\cdot), n < \infty, \tilde{P}(\cdot)\}$ *are defined. Then, for almost all* ω, *the sample value* $\tilde{P}^\omega(\cdot)$ *of* $\tilde{P}(\cdot)$ *is the measure of a stationary diffusion process* $\tilde{x}^\omega(\cdot)$ *with paths in* S_0' *and with differential operator A. Let* $\tilde{\mu}^\omega(\cdot)$ *denote the measure of the initial condition of the stationary process* $\tilde{x}^\omega(\cdot)$. *Let* $K(\cdot)$ *denote a bounded and continuous real valued function on* S_0'. *Then*

$$\frac{1}{T_n} \int_0^{T_n} K(x_t(\cdot))dt \Rightarrow \int K(\phi(\cdot))\tilde{P}(d\phi) \tag{2.2}$$

The sample values of the right hand side of (2.2) *can be written as* $\tilde{E}^\omega K(\tilde{x}^\omega(\cdot))$, *where* $\tilde{E}^\omega$ *denotes the expectation under* $\tilde{P}^\omega(\cdot)$. *For* $k(\cdot) \in C_b(R^r)$,

$$\frac{1}{T_n} \int_0^{T_n} k(x(s))ds \Rightarrow \int k(x)\tilde{\mu}(dx), \tag{2.3}$$

where $\tilde{\mu}(\cdot)$ *is the measure valued (on the Borel sets of* R^r) *random variable with values* $\tilde{\mu}^\omega(\cdot)$. *Under* (A1.3), *for almost all* ω

$$\tilde{\mu}^\omega(\cdot) = \mu(\cdot),$$

and the subsequence is irrelevant.

Outline of the Proof. First, the stationarity of the process $\tilde{x}^\omega(\cdot)$ is proved, for almost all ω. Then, an adaptation of the martingale method is used to characterize the $\tilde{x}^\omega(\cdot)$ as solutions to (1.4) for some driving Wiener process $\tilde{w}^\omega(\cdot)$. Finally, the cost function is expressed in terms of the occupation measure and limits taken. The proof differs from previous "martingale method" proofs in that we must work with the occpuation measures, and can take expectations only with respect to them.

The convenience and power of the method should be apparent from the proof and those of the following theorems. Little needs to be known about (1.4). The theorem says essentially that each sample value of the average pathwise cost per unit time is the ergodic cost for a stationary diffusion of the type (1.4). The costs can be *very general*. The cost function used in Chapter 4.4 is only a very special case.

Proof. *Stationarity.* As noted above, we use the Skorohod representation for $\{\tilde{P}_{T_n}(\cdot),\ n < \infty,\ \tilde{P}(\cdot)\}$ all through the proof. Let $G \in S_0'$, and define its left shift $G_c = \{\phi(\cdot):\ \phi_c(\cdot) \in G\}$. Then

$$\tilde{P}_T(G_c) \stackrel{D}{=} \frac{1}{T}\int_0^T I_{G_c}(x_t(\cdot))dt = \frac{1}{T}\int_0^T I_G(x_{t+c}(\cdot))dt$$

$$\tilde{P}_T(G_c) - \tilde{P}_T(G) \stackrel{D}{=} \frac{1}{T}\int_T^{T+c} I_G(x_t(\cdot))dt - \frac{1}{T}\int_0^c I_G(x_t(\cdot))dt.$$

Thus $\tilde{P}_T^\omega(G) - \tilde{P}_T^\omega(G_c) \to 0$ as $T \to 0$ for all ω and G. This and the weak convergence implies that $\tilde{P}^\omega(G) = \tilde{P}^\omega(G_c)$ for all ω and all sets G with $\tilde{P}^\omega(\partial G) = 0$. This, in turn, implies the stationarity.

In the rest of the proof, we will only characterize the process $\tilde{x}^\omega(\cdot)$ induced by the values $\tilde{P}^\omega(\cdot)$ of the limit $\tilde{P}(\cdot)$. By the definition of the sample occupancy measure $\tilde{P}_T^\omega(\cdot)$, for each real-valued, bounded and measurable function $F_0(\cdot)$ on S_0', we have

$$\int F_0(\phi)\tilde{P}_T(d\phi) \stackrel{D}{=} \frac{1}{T}\int_0^T F_0(x_t(\cdot))dt, \tag{2.4}$$

In fact, (2.4) completely determines $\tilde{P}_T(\cdot)$.

We use a form of the martingale method of Chapter 2.6 or Theorem 3.5.1 or Theorem 4.1.2 to characterize the weak limits of $\{\tilde{P}_T(\cdot),\ T < \infty\}$. The method differs slightly from that used in the past chapters since we cannot take expectations except with respect to the occupation measures.

Let $h(\cdot)$ be a bounded and continuous real valued function of its arguments, let q be an arbitrary integer, and let $t_i,\ \tau,\ \tau + s$ be such that $t_i \le \tau \le \tau + s,\ i \le q$. Let $f(\cdot) \in C_0^2(R^r)$. Define the function $F(\cdot)$ by

$$F(\phi(\cdot)) = h(\phi(t_i), i \le q)[f(\phi(\tau+s)) - f(\phi(\tau)) - \int_\tau^{\tau+s} Af(\phi(u))du]. \tag{2.5}$$

The function $F(\cdot)$ is measurable but it is not continuous at all points $\phi(\cdot)$. [This lack of continuity is a consequence of the nature of the Skorohod topology.] It is continuous at each $\phi(\cdot)$ which is continuous [B6, p. 121].

Since the processes induced on S_0' by the $\tilde{P}_T^\omega(\cdot)$ are continuous for each ω and T, the sample values $\tilde{P}^\omega(\cdot)$ of the limit measures also have their support on the set of continuous functions in S_0'. Hence, even though the function $F(\cdot)$ is not continuous at every point in S_0', it is continuous w.p.1 relative to $\tilde{P}^\omega(\cdot)$ for almost all ω, since it is continuous at each point $\phi(\cdot)$ which is a continuous function. Due to this fact and the Skorohod representation, Theorem 1.1.4 implies that

$$\int F(\phi)\tilde{P}^\omega_{T_n}(d\phi) \to \int F(\phi)\tilde{P}^\omega(d\phi), \tag{2.6}$$

for almost all ω. Let $\tilde{E}^\omega$ and $\tilde{x}^\omega(\cdot)$ denote the expectation with respect to $\tilde{P}^\omega(\cdot)$ and the process which is induced by $\tilde{P}^\omega(\cdot)$, resp. The result (2.2) is just a consequence of the weak convergence. By the definition of $F(\cdot)$, the right hand side of (2.6), evaluated at ω equals

$$\int F(\phi)\tilde{P}^\omega(d\phi) = \tilde{E}^\omega h(\tilde{x}^\omega(t_i), i \le q) \cdot$$

$$[f(\tilde{x}^\omega(\tau + s)) - f(\tilde{x}^\omega(\tau)) - \int_\tau^{\tau+s} Af(\tilde{x}^\omega(u))du]. \tag{2.7}$$

It will be shown below that (2.7) equals zero for almost all ω. It follows from this and from the arbitrariness of $h(\cdot)$, τ, $\tau + s$, q, t_i, $f(\cdot)$, that for almost all ω the process $\tilde{x}^\omega(\cdot)$ must solve the martingale problem for the operator A. Hence, for almost all ω, $\tilde{x}^\omega(\cdot)$ solves (1.4) for some standard vector valued Wiener process $\tilde{w}^\omega(\cdot)$.

Let $K(\phi(\cdot)) = k(\phi(0))$. Then, for almost all ω, $K(\cdot)$ is continuous w.p.1 relative to the measure $\tilde{P}^\omega(\cdot)$. Note that

$$\frac{1}{T}\int_0^T k(x(s))ds = \frac{1}{T}\int_0^T K(x_s(\cdot))ds \stackrel{D}{=} \int K(\phi)\tilde{P}_T(d\phi),$$

$$\int K(\phi)\tilde{P}^\omega_{T_n}(d\phi) \to \int K(\phi)\tilde{P}^\omega(d\phi) = \int k(\phi(0))\tilde{P}^\omega(d\phi(0)).$$

But, by the stationarity of $\tilde{x}^\omega(\cdot)$ for almost all ω, this last expression on the right equals (for almost all ω) $\int k(x)\tilde{\mu}^\omega(dx)$. Thus $\int_0^T k(x(s))ds/T \Rightarrow \int k(x)\tilde{\mu}(dx)$. Under the uniqueness condition (A1.3), we must have $\tilde{\mu}^\omega(\cdot) = \mu(\cdot)$ for almost all ω.

Proof that (2.7) equals zero for almost all ω. By Itô's Formula and the representation (2.4), we can write

$$\int F(\phi)\tilde{P}^\omega_T(d\phi) \stackrel{D}{=}$$

$$\frac{1}{T}\int_0^T \left(h(x_t(t_i), i \le q) \left[\int_\tau^{\tau+s} f'_x(x_t(u))\sigma(x_t(u))dw_t(u) \right] \right) dt \tag{2.8}$$

$$= \frac{1}{T}\int_0^T dt \int_{t+\tau}^{t+\tau+s} g(t, u)dw(u)$$

for some bounded and non-anticipative function $g(\cdot)$. This expression goes to zero in probability as $T \to \infty$. This fact together with the weak convergence implies that (2.7) equals zero for almost all ω.   Q.E.D.

Extension to jump-diffusion processes. Theorems 2.1 and 2.2 hold for the jump-diffusion process

$$dx = b(x)dt + \sigma(x)dw + \int q(x,\gamma)N(dtd\gamma), \quad x \in R^r, \qquad (2.9)$$

where we assume (A2.1).

A2.1. $N(\cdot)$ *is a Poisson measure with jump rate* λ *and jump distribution* $\Pi(\cdot)$, *and* $w(\cdot)$ *and* $N(\cdot)$ *are mutually independent.* $q(\cdot)$ *is bounded and continuous,* $q(x,0) = 0$ *and for each value of* x, *the value of* $q(x,\gamma)$ *uniquely determines the value of* γ.

Theorem 2.3. *Assume* (A2.1) *and the conditions of Theorem 2.2. Then the conclusions of Theorem 2.1 and 2.2 hold.*

Remark of the Proof. With the two minor modifications given below, the proof is the same as those of Theorems 2.1 and 2.2 combined.

1. Since the trajectories of (2.9) are not continuous, the argument above (2.6) cannot be used. But, (2.6) follows from the fact that for almost all ω,

$$\tilde{P}^{\omega}(\tilde{x}^{\omega}(\cdot) \text{ is discontinuous at } t_0) = 0, \text{ all } t_0. \qquad (2.10)$$

In turn, (2.10) follows from the weak convergence $\tilde{P}_T(\cdot) \Rightarrow \tilde{P}(\cdot)$ and

$$\tilde{P}_T(\phi(\cdot): \phi(\cdot) \text{ jumps on } [t - \Delta, t + \Delta]) \overset{D}{=}$$

$$\frac{1}{T} \int_0^T I_{\{x(\cdot) \text{ jumps on } [t-\Delta, t+\Delta]\}} dt \to 2\lambda\Delta$$

in probability, as $T \to \infty$. Let ω_0 denote the generic variable on the sample space on which the process $\tilde{x}^{\omega}(\cdot)$ is defined. Then we can also write

$$P(\omega: \tilde{P}^{\omega}(\omega_0: \tilde{x}^{\omega}(\omega_0, \cdot) \text{ is discontinuous on } [t - \Delta, t + \Delta]))$$

$$= \lim_T \frac{1}{T} \int_0^T P(x(\cdot) \text{ is discontinuous on } [t - \Delta, t + \Delta]) dt = 2\lambda\Delta.$$

2. The differential operator A is now defined by (2.7.3), where $\tilde{b}(x,t) = b(x)$. Then Itô's Formula (2.7.5) requires that we add to the right side of (2.8) the term

$$\frac{1}{T} \int_0^T \left(h(x_t(t_i), i \leq q) \int_\tau^{\tau+s} f_x'(x_t(u^-)) + q(x_t(u^-), \gamma) \right) \cdot$$

$$\cdot [N_t(dt d\gamma) - \lambda \Pi(d\gamma) du]. \tag{2.11}$$

Here, $N_t(dud\gamma)$ is the (u, γ)-differential of the t-shifted random measure defined by $N_t([0, u] \times C) = N([0, t + u] \times C) - N([0, t] \times C)$. Since (2.11) goes to zero in probability as $T \to \infty$, we still have that (2.7) equals zero w.p.1.

3. The Control Problem

We now do the controlled form of (1.4). The result is the same for the jump-diffusion case (3.1.1).

For admissible $(m(\cdot), w(\cdot))$ the controlled SDE is written as

$$dx = \int_{\mathcal{U}} b(x, \alpha) m_t(d\alpha) dt + \sigma(x) dw, \tag{3.1}$$

We use the following replacement for (A1.1).

A3.1. $\sigma(\cdot)$ and $b(\cdot)$ are continuous, $\sigma(\cdot)$ and $b(\cdot, \alpha)$ have at most a linear growth as $|x| \to \infty$, uniformly in α.

Recall the definition in Chapter 3.2 of $\mathcal{R}(\mathcal{U} \times [0, \infty))$, the space of measures which are the relaxed controls. Denote this space by S_0''. Under the "compact-weak" topology which we use (see Chapter 3.2) S_0'' is compact, hence it is complete and separable. Define $S_0 = S_0' \times S_0''$.

Recall that A^α denotes the differential operator of (3.1) with the control fixed at $\alpha \in \mathcal{U}$, and that the differential operator of (3.1), under $m(\cdot)$ and acting on $f(\cdot) \in C^2(R^r)$ is defined by $(A^m f)(x, t) = \int A^\alpha f(x) m_t(d\alpha)$. For each fixed $t \geq 0$, define the "shifted" stochastic relaxed control $\Delta_t m(\cdot)$ by its value at time s and on set C: $(\Delta_t m)(s, C) = m(t + s, C) - m(t, C)$, $C \in \mathcal{B}(\mathcal{U})$. The relaxed control $\Delta_t m(\cdot)$ is the one that is actually used on the shifted process $x_t(\cdot)$ with initial condition $x_t(0) = x(t)$, since for fixed t, the derivative with respect to s of $(\Delta_t m)(s, C)$ is $(\Delta_t m)_s(C) = m_{t+s}(C)$ and

$$dx_t(s) = ds \int_{\mathcal{U}} b(x_t(s), \alpha) m_{t+s}(d\alpha) + \sigma(x_t(s)) dw_t(s), \quad s \geq 0.$$

Define the occupation measures $\tilde{P}^t(\cdot)$ and $\tilde{P}_T(\cdot)$ by

$$\tilde{P}^t(B_0 \times C_0) = I_{B_0}(x_t(\cdot)) I_{C_0}(\Delta_t m(\cdot)),$$

$$\tilde{P}_T(\cdot) = \frac{1}{T} \int_0^T \tilde{P}^t(\cdot) dt.$$

Here, $C_0 \in \mathcal{B}(S_0'')$ and $B_0 \in \mathcal{B}(S_0')$. The next theorem is the "controlled" analog of Theorem 2.2.

Theorem 3.1. *Assume* (A1.2) *and* (A3.1). *Then* $\{\tilde{P}_T(\cdot), T < \infty\}$ *is a tight set of* $\mathcal{P}(S_0)$-*valued random variables. Let* $\tilde{P}(\cdot)$ *denote the limit of some weakly convergent subsequence (indexed by* $T_n \to \infty$) *and suppose that the Skorohod representation is used for* $\{\tilde{P}_{T_n}(\cdot), n < \infty, \tilde{P}(\cdot)\}$, *where* ω *is the generic variable of the common probability space. Then, for almost all* ω, *the sample value* $\tilde{P}^\omega(\cdot)$ *induces a process* $(\tilde{x}^\omega(\cdot), \tilde{m}^\omega(\cdot))$ *on* S_0. *The process is stationary in that the distribution of the shifted pair* $(\tilde{x}_t^\omega(\cdot), \Delta_t\tilde{m}^\omega(\cdot))$ *does not depend on* t. *For almost all* ω, *there is a filtration* $\tilde{\mathcal{F}}_t^\omega$ *and an* $\tilde{\mathcal{F}}_t^\omega$-*standard vector-valued Wiener process* $\tilde{w}^\omega(\cdot)$, *such that* $\tilde{m}^\omega(\cdot)$ *is admissible (with respect to* $\tilde{w}^\omega(\cdot)$) *and*

$$d\tilde{x}^\omega(t) = \int_{\mathcal{U}} b(\tilde{x}^\omega(t), \alpha)\tilde{m}_t^\omega(d\alpha)dt + \sigma(\tilde{x}^\omega(t))d\tilde{w}^\omega. \qquad (3.2)$$

Let $\tilde{E}^\omega$ *denote the expectation with respect to* $\tilde{P}^\omega(\cdot)$. *Define the random variable* K *with values*

$$K(\omega) = \tilde{E}^\omega \int_0^1 \int_{\mathcal{U}} k(\tilde{x}^\omega(s), \alpha)\tilde{m}_s^\omega(d\alpha)ds. \qquad (3.3)$$

Then

$$\frac{1}{T_n} \int_0^{T_n} \int_{\mathcal{U}} k(x(t), \alpha)m_t(d\alpha)dt \Rightarrow K. \qquad (3.4)$$

Remark. In the absence of control, (3.4) reduces to (2.3). We use the integral over $[0, 1]$ in (3.3) since the $m_s(\cdot)$ is only defined for almost all s. The expression (3.3) is the average cost per unit time under $\tilde{P}^\omega(\cdot)$, due to the stationarity. I.e., the average cost per unit time for a stationary system is just the cost over the unit interval. Note that, although we start with an arbitrary relaxed control $m(\cdot)$, the limit has a control with stationary increments.

The $\tilde{m}_t^\omega(d\alpha)$ are not limits in any sense of a sequence of functionals of the true control. For each weakly convergent subsequence of $\{\tilde{P}_T(\cdot), T < \infty\}$, we are simply interested in characterizing the pathwise limit cost in terms of a stationary process driven by a control with stationary increments. The cost and the process might depend on ω, the canonical variable of the sample space on which the convergent subsequence and the limit are defined. The stationarity is a consequence of the normalization used in the definition of $\tilde{P}_T(\cdot)$. With the representation given, the theorem can be used to show that the limits of the pathwise costs for any control are no smaller than the infimum of the average (in sense of expectation) cost per unit time. Also, in a specific sense, the best (smallest) limit of the pathwise costs is the same as the best limit when the cost functional contains the mathematical expectation. This point is developed further in Section 4.

Proof. The proof is similar to that of Theorem 2.2. The set of processes $\{\Delta_t m(\cdot), \ t < \infty\}$ is tight and so is the set $\{x_t(\cdot), \ t < \infty\}$. This implies the tightness of $\{\tilde{P}_T(\cdot), \ T < \infty\}$ by Theorem 1.6.1, and we need only characterize the limits of the weakly convergent subsequences and explain (3.4). Let T_n index a weakly convergent subsequence, with limit denoted by $\tilde{P}(\cdot)$. Let $(\tilde{x}^\omega(\cdot), \tilde{m}^\omega(\cdot))$ denote the process induced by $\tilde{P}^\omega(\cdot)$. The asserted stationarity is proved as in Theorem 2.2. Let $\psi(\cdot)$ and $\psi_j(\cdot)$ be in $C_0(\mathcal{U} \times [0, \infty))$ and for $v(\cdot) \in \mathcal{R}(\mathcal{U} \times [0, \infty))$ define

$$(\psi, \nu)_t = \int_{\mathcal{U}} \int_0^t \psi(\alpha, s)\nu(d\alpha ds) = \int_{\mathcal{U}} \int_0^t \psi(\alpha, s)\nu_s(d\alpha)ds.$$

Choose $h(\cdot)$, $f(\cdot)$, q, t_i, τ, $\tau + s$ as in Theorem 2.2 and let p be an arbitrary integer. We now adapt the martingale method of Theorem 2.2 to include the controls. (See Chapter 3.5 also.) Define the function $F_1(\cdot)$ on S_0 by

$$F_1(\phi, \nu) = h(\phi(t_i), (\psi_j, \nu)_{t_i}, i \le q, j \le p)[f(\phi(\tau + s)) - f(\phi(\tau))$$

$$- \int_\tau^{\tau+s} \int_{\mathcal{U}} A^\alpha f(\phi(u))\nu(d\alpha du)].$$

By Itô's Formula and the definition of $\tilde{P}_T(\cdot)$,

$$\int F(\phi, \nu)\tilde{P}_T(d\phi d\nu) \overset{D}{=} \frac{1}{T} \int_0^T F_1(x_t(\cdot), \Delta_t m(\cdot))dt$$

$$= \frac{1}{T} \int_0^T h(x_t(t_i), (\psi_j, \Delta_t m)_{t_i}, i \le q, j \le p)dt \int_\tau^{\tau+s} f'_x(x_t(u))\sigma(x_t(u))dw_t(u).$$

$$(3.5)$$

Let us use the Skorohod representation. By the same argument used in Theorem 2.2, we have, for almost all ω,

$$\int F_1(\phi, \nu)\tilde{P}^\omega(d\phi d\nu) = 0. \qquad (3.6)$$

Equivalently, for almost all ω,

$$\tilde{E}^\omega h(\tilde{x}^\omega(t_i), (\psi_j, \tilde{m}^\omega)_{t_i}, i \le q, j \le p)[f(\tilde{x}^\omega(\tau + s)) - f(\tilde{x}^\omega(\tau))$$

$$- \int_\tau^{\tau+s} \int_{\mathcal{U}} A^\alpha f(\tilde{x}^\omega(u))\tilde{m}^\omega(d\alpha du)] = 0.$$

Analogous to the situation in Theorem 2.2, this implies that the process $(\tilde{x}^\omega(\cdot), \tilde{m}^\omega(\cdot))$ solves the martingale problem with operator A^α and with respect to the filtration $\mathcal{B}(\tilde{x}^\omega(s), \tilde{m}^\omega(\cdot, s), s \le t)$. All the conclusions except the representation on the right hand side of (3.3) follow from this.

Define the function $\overline{K}(\cdot)$ on S_0 by

$$\overline{K}(\phi, \nu) = \int_0^1 \int_{\mathcal{U}} k(\phi(s), \alpha)\nu(d\alpha ds).$$

The representation (3.3) of the right hand side of (3.4) is obtained by noting
that the limit of the left hand side of (3.4) equals the limit of

$$\frac{1}{T_n} \int_0^{T_n} \overline{K}(x_t(\cdot), \Delta_t m(\cdot)) dt \stackrel{D}{=} \int \overline{K}(\phi, \nu) \tilde{P}_{T_n}(d\phi d\nu)$$

$$= \frac{1}{T_n} \int_0^{T_n} dt \int_0^1 ds \int_{\mathcal{U}} k(x(t+s), \alpha) m_{t+s}(d\alpha). \qquad Q.E.D.$$

4. Singularly Perturbed Control Problems

In this section we work with the singularly perturbed diffusion model
(4.1.1), (4.1.2), and the average pathwise cost per unit time (4.4.1). A jump
term can be added with little additional difficulty. The "fast variable" is
averaged out by a method which is different from that used in Theorem
4.1.2.

Let us assume:

A4.1. *For each $\delta > 0$ and $\epsilon > 0$, there is a δ-optimal (for the cost functional $\gamma^\epsilon(x, z, m^\epsilon)$ of (4.4.1)) admissible $m^\epsilon(\cdot)$ such that the corresponding
sequence*

$$\{x^\epsilon(t), z^\epsilon(t), t < \infty, \epsilon > 0\}$$

is tight.

We follow the terminology of Section 3. Define the functional occupation
measures $\tilde{P}^{\epsilon,t}(\cdot)$ and $\tilde{P}^\epsilon_T(\cdot)$ by

$$\tilde{P}^{\epsilon,t}(B_0, C_0) = I_{B_0}(x_t^\epsilon(\cdot)) I_{C_0}(\Delta_t m^\epsilon(\cdot))$$

$$\tilde{P}^\epsilon_T(\cdot) = \frac{1}{T} \int_0^T \tilde{P}^{\epsilon,t}(\cdot) dt.$$

Theorem 4.1. *Assume (A4.1.1) to (A4.1.5) and (A4.1), and use the relaxed controls of (A4.1) for some $\delta > 0$. Then the sequence of $\mathcal{P}(S_0)$-
valued random variables $\{\tilde{P}^\epsilon_T(\cdot), \epsilon > 0, T < \infty\}$ is tight. Let $\tilde{P}(\cdot)$ denote
a weak limit (for some weakly convergent subsequence induced by $\epsilon_n \to 0$,
$T_n \to \infty$) with sample values $\tilde{P}^\omega(\cdot)$. For almost all ω, $\tilde{P}^\omega(\cdot)$ induces a stationary process $(\tilde{x}^\omega(\cdot), \tilde{m}^\omega(\cdot))$ on S_0 in the sense that the distribution of
$(\tilde{x}_t^\omega(\cdot), \Delta_t \tilde{m}^\omega(\cdot))$ does not depend on t. For almost all ω, there is a standard vector-valued Wiener process $\tilde{w}^\omega(\cdot)$ such that $\tilde{m}^\omega(\cdot)$ is admissible with
respect to $\tilde{w}^\omega(\cdot)$ and $(\tilde{x}^\omega(\cdot), \tilde{m}^\omega(\cdot), \tilde{w}^\omega(\cdot))$ satisfy (4.1.7) and for each x
and z*

$$\Lambda_{T_n}^{\epsilon_n}(x, z, m^{\epsilon_n}) \Rightarrow \Lambda(\tilde{m}),$$

where the values of the random variable $\Lambda(\tilde{m})$ *are given by*

$$\Lambda(\tilde{m}^\omega) = \tilde{E}^\omega \int_{\mathcal{U}} \int_0^1 \overline{k}(\tilde{x}^\omega(s), \alpha)\tilde{m}_s^\omega(d\alpha)ds.$$

Remark. In the last equation $\Lambda(\tilde{m}^\omega)$ is the cost for the stationary problem, where the initial condition $\tilde{x}^\omega(0)$ is the random variable with the stationary distribution.

Proof. The $\{x_t^\epsilon(\cdot), \, t < \infty, \, \epsilon > 0\}$ is tight due to (A4.1.2) and (A4.1). The set $\{\Delta_t m^\epsilon(\cdot), \, t < \infty, \, \epsilon > 0\}$ is always tight. These facts imply the asserted tightness of $\{\tilde{P}_T^\epsilon(\cdot), \, \epsilon > 0, \, T < \infty\}$, analogously to the case of Theorem 2.2. Let $\epsilon_n \to 0$, $T_n \to \infty$ index a weakly convergent subsequence with limit $\tilde{P}(\cdot)$. Let us use the Skorohod representation and assume that $\{\tilde{P}_{T_n}^{\epsilon_n}(\cdot), \, \tilde{P}(\cdot), \, n < \infty\}$ are defined on the same sample space with generic variable ω. Let $h(\cdot), \, t_i, \, \psi_j(\cdot), \, p, \, q, \, \tau$ and s all be as defined in the proof of Theorem 3.1, and let $\tilde{E}_T^\omega$ denote the expectation with respect to $\tilde{P}_T^{\epsilon,\omega}(\cdot)$. Define F_1^ϵ as F_1 was defined in Theorem 3.1, but with A^ϵ replacing A^α. Then, analogously to the calculation in Theorem 3.1, by Itô's Formula and the definition of $\tilde{P}_T^\epsilon(\cdot)$ we have

$$\int F_1^\epsilon(\psi, \nu)\tilde{P}_T^\epsilon(d\psi d\nu) \stackrel{D}{=}$$

$$\frac{1}{T} \int_0^T dt h(x_t^\epsilon(t_i), (\Delta_t m^\epsilon, \psi_j)_{t_i}, i \le q, j \le p)[f(x_t^\epsilon(\tau + s))$$

$$- f(x_t^\epsilon(\tau)) - \int_\tau^{\tau+s} du \int_{\mathcal{U}} f_x'(x_t^\epsilon(u))G(x_t^\epsilon(u), z_t^\epsilon(u), \alpha)(\Delta_t m^\epsilon)_u(d\alpha)$$

$$- \frac{1}{2} \int_\tau^{\tau+s} du \, \text{trace} \, f_{xx}(x_t^\epsilon(u)) \cdot a(x_t^\epsilon(u), z_t^\epsilon(u))]\Big)$$

$$= \frac{1}{T} \int_0^T dt \int_\tau^{\tau+s} du h(x_t^\epsilon(t_i), (\Delta_t m^\epsilon, \psi_j)_{t_i}, i \le q, j \le p)$$

$$\cdot f_x'(x_t^\epsilon(u))\sigma(x_t^\epsilon(u)), z_t^\epsilon(u))dw_t(u)). \tag{4.1}$$

It can be shown that, by using the averaging method introduced in Theorem 2.2 we can replace the $G(\cdot)$ and $a(\cdot)$ in (4.1) by $\overline{G}(\cdot)$ and $\overline{a}(\cdot)$ without changing the limits as $\epsilon_n \to 0$ and $T_n \to \infty$. Only a few of the details will be given. The procedure is roughly as follows, and differs from the method used in Theorem 4.1.2. Consider the term (4.2) which is a component of the third line of (4.1):

$$\int_\tau^{\tau+s} f_x'(x_t^\epsilon(u))G_0(x_t^\epsilon(u), z_t^\epsilon(u))du. \tag{4.2}$$

Let $s = n\delta$ for an integer n and $\delta > 0$, and rewrite the integral in (4.2) as

$$\sum_{i=0}^{n-1} \delta \frac{1}{\delta} \int_{\tau+i\delta}^{\tau+i\delta+\delta} f_x'(x_i^\epsilon(u)) G_0(x_i^\epsilon(u), z_i^\epsilon(u)) du. \tag{4.3}$$

For each $t \geq 0$, define the shifted rescaled processes $x_{t0}^\epsilon(\cdot)$ and $z_{t0}^\epsilon(\cdot)$ by $x_{t0}^\epsilon(u) = x^\epsilon(t + \epsilon u)$, $z_{t0}^\epsilon(u) = z^\epsilon(t + \epsilon u)$. Now take one of the summands in (4.3) and change the time scale $s \to \epsilon s$ to get

$$\delta \frac{\epsilon}{\delta} \int_{(\tau+i\delta)/\epsilon}^{(\tau+i\delta+\delta)/\epsilon} f_x'(x_{t0}^\epsilon(u)) G_0(x_{t0}^\epsilon(u), z_{t0}^\epsilon(u)) du. \tag{4.4}$$

We study (4.4) as $\delta \to 0$ and $\epsilon/\delta \to 0$. Note that by the tightness asserted on the first line of the proof, for each $\delta_0 > 0$,

$$\sup_{t<\infty} P\left(\sup_{u\leq\delta} |x_t^\epsilon(u) - x_t^\epsilon(0)| > \delta_0\right) \to 0 \tag{4.5}$$

uniformly in (small) ϵ, as $\delta \to 0$.

Define the occupation measures $\hat{P}_t^{\epsilon,\delta,u}(\cdot)$ and $\hat{P}_{t,T}^{\epsilon,\delta}(\cdot)$ by

$$\hat{P}_t^{\epsilon,\delta,u}(A \times B) = I_A(x_{t+\tau+i\delta}^\epsilon(\epsilon u + \epsilon \cdot)) \cdot I_B(z_{t+\tau+i\delta}^\epsilon(\epsilon u + \epsilon \cdot))$$

$$\hat{P}_{t,T}^{\epsilon,\delta}(\cdot) = \frac{1}{T} \int_0^T \hat{P}_t^{\epsilon,\delta,u}(\cdot) du.$$

Then (4.4) equals

$$\delta \int f_x'(\phi(0)) G_0(\phi(0), \xi(0)) \hat{P}_{t,\delta/\epsilon}^{\epsilon,\delta}(d\phi d\xi).$$

By an "occupation measure argument" similar to that used with Theorem 2.2, and using the fact that $x_{t+\tau+i\delta}^\epsilon(\cdot)$ varies arbitrarily little over the interval $[0, \delta]$ (as implied by (4.5)), we can show that as $\epsilon/\delta \to 0$ and $\epsilon \to 0$, the difference between $\hat{P}_{t,\delta/\epsilon}^{\epsilon,\delta}(\cdot)$ and the measure whose x-component is concentrated on $x_{t+\tau+i\delta}^\epsilon(\cdot)$ and whose z-component is concentrated on the stationary fixed-x $z_0(\cdot)$ process with parameter $x = x_{t+\tau+i\delta}^\epsilon(0)$ converges weakly to zero. This convergence can be shown to be uniform in the variables (t, τ). Thus, as $\epsilon \to 0$ and $\epsilon/\delta \to 0$,

$$E\left| \frac{\epsilon}{\delta} \int_{(\tau+i\delta)/\epsilon}^{(\tau+i\delta+\delta)/\epsilon} f_x'(x_{t0}^\epsilon(u)) G_0(x_{t0}^\epsilon(u), z_{t0}^\epsilon(u)) du \right.$$

$$\left. - \int f_x'(x_t^\epsilon(\tau + i\delta)) G_0(x_t^\epsilon(\tau + i\delta), z) \mu_{x_t^\epsilon(\tau+i\delta)}(dz) \right| \to 0 \tag{4.6}$$

uniformly in τ, t. Thus we can replace the right hand part of the third line
of (4.1) by (their limits are the same)

$$\int_\tau^{\tau+s} \int_{\mathcal{U}} f'_x(x^\epsilon_t(u))\overline{G}(x^\epsilon_t(u),\alpha)(\Delta_t m^\epsilon)_u(d\alpha)du.$$

Similarly for the $a(x,z)$ term in the fourth line of (4.1).

Analogously to the case of Theorem 2.2, the right hand side of (4.1) goes
to zero in probability as $\epsilon \to 0$ and $T \to \infty$. We can also replace the $k(\cdot)$ in
the cost function Λ^ϵ_T defined by (4.4.1) by $\overline{k}(\cdot)$ without changing the limits.
Thus, via an argument as used in the last part of the proof of Theorem 3.1,
it follows that the cost functional has the same limits as has

$$\int \left[\int_0^1 \int_{\mathcal{U}} \overline{k}(\phi(s),\alpha)\nu(d\alpha ds) \right] \tilde{P}^\epsilon_T(d\phi d\nu).$$

The rest of the proof is no different than that used for Theorem 2.2 and
the details are omitted.   Q.E.D.

The approximate optimality theorem. We redefine the δ-optimal feedback control $u^\delta(\cdot)$ of Chapter 4.4, for the cost $\Lambda(x,m)$ and the averaged
system (4.1.7). $u^\delta(\cdot)$ is a stabilizing control; i.e., the associated process is
"asymptotically stationary" in that the limit mean cost doesn't depend on
the initial condition $x = x(0)$, and there is a unique stationary process
under $u^\delta(\cdot)$. Also, if $m(\cdot)$ is any relaxed control which is "stationary" in
the sense that the distribution of $(x_t(\cdot), \Delta_t m(\cdot))$ doesn't depend on the
shift t (where $x(\cdot)$ is the solution under $m(\cdot)$), then $\Lambda(x,u^\delta) = \Lambda(u^\delta) \le$
$\Lambda(x,m) + \delta$. Recall the definition of the minimum cost Λ_0 of Chapter 4.4.
The existence of such δ-optimal controls is discussed in [K12]. We have
the following form of Theorem 4.4.2, without the need for relaxed feedback
controls or (A4.4.3), (A4.4.4).

Theorem 4.2. *Assume (A4.1.1) to (A4.1.5). For each $\delta > 0$, let there be
a continuous δ-optimal feedback control $u^\delta(\cdot)$ for the averaged system and
cost $\Lambda(x,m)$ of (4.4.2). Let $\{x^\epsilon(t), z^\epsilon(t), t < \infty, \epsilon > 0\}$ be tight under
$u^\delta(\cdot)$. Then*

$$\Lambda^\epsilon_T(x,z,u^\delta) \to \Lambda(u^\delta) \le \Lambda_0 + \delta \qquad (4.7)$$

*in probability, uniformly in each compact (x,z)-set, as ϵ and T go to their
limits. For any $\delta_1 > 0$ and sequence of admissible $m^\epsilon(\cdot)$ satisfying the
tightness condition (A4.1)*

$$\limsup_{\epsilon \to 0, T \to \infty} P(\Lambda^\epsilon_T(x,z,m^\epsilon) \le \Lambda_0 - \delta_1) = 0. \qquad (4.8)$$

Proof. (4.7) follows by Theorem 4.1, with the $m^\epsilon(\cdot)$ replaced by $u^\delta(\cdot)$. If
(4.8) is not true, then there is a $\delta_0 > 0$ such that the lim sup is greater

than δ_0. Choose a weakly convergent subsequence of $\{\tilde{P}_T^\epsilon(\cdot)\}$ which attains the limsup in (4.8). Let $\tilde{P}^\omega(\cdot)$ denote the sample value of the limit measure and let $(\tilde{x}^\omega(\cdot), \tilde{m}^\omega(\cdot))$ denote the process induced by $\tilde{P}^\omega(\cdot)$. Then

$$P(\omega: \Lambda(\tilde{m}^\omega) \leq \Lambda_0 - \delta_1) \geq \delta_0,$$

a contradiction to the δ-optimality of $u^\delta(\cdot)$ for small enough δ.   Q.E.D.

5. Control of the Fast System

For the model (4.1.1) and (4.1.2) with cost (4.1.3), we were able to average out the fast variable and approximate the original control problem by an averaged problem. We now discuss a model where the control problem essentially decouples into an average cost per unit time problem for the fast system, and a control problem over $[0, T]$ for the slow system.

The system and cost function of this section will be

$$dx^\epsilon = dt \int_{\mathcal{U}_1} G(x^\epsilon, \alpha_1) m_{1t}^\epsilon(d\alpha_1) + \sigma(x^\epsilon) dw_1 \tag{5.1}$$

$$\epsilon dz^\epsilon = dt \int_{\mathcal{U}_2} H(x^\epsilon, z^\epsilon, \alpha_2) m_{2t}^\epsilon(d\alpha_2) + \sqrt{\epsilon} v(x^\epsilon, z^\epsilon) dw_2 \tag{5.2}$$

$$V_T^\epsilon(x, z, m_1^\epsilon, m_2^\epsilon) = E_{x,z}^{m^\epsilon} \int_0^T \int_{\mathcal{U}} k(x^\epsilon(t), z^\epsilon(t), \alpha_1, \alpha_2) m_{1t}^\epsilon(d\alpha_1) m_{2t}^\epsilon(d\alpha_2) dt. \tag{5.3}$$

We write (5.3) using $m_{1t}^\epsilon(d\alpha_1) m_{2t}^\epsilon(d\alpha_2)$ in lieu of $m_t^\epsilon(d\alpha_1 d\alpha_2)$ since it will be apparent that the controls can be decoupled as $\epsilon \to 0$.

We will use the following assumptions.

A5.1. $\mathcal{U}_i$, $i = 1, 2$, *are compact sets with* $\alpha_1 \in \mathcal{U}_1$ *and* $\alpha_2 \in \mathcal{U}_2$, $\mathcal{U} = \mathcal{U}_1 \times \mathcal{U}_2$.

A5.2. $H(\cdot)$ *is a continuous function of* (x, z, α_2) *and* $v(\cdot)$ *of* (x, z), *satisfying*

$$|H(x, z, \alpha_2)| + |v(x, z)| \leq K(1 + |z|).$$

$G(\cdot)$ *is a continuous function of* (x, α_1) *and* $\sigma(\cdot)$ *of* x *and*

$$|G(x, \alpha_1)| + |\sigma(x)| \leq K(1 + |x|).$$

(5.1), (5.2) *have a unique weak sense solution for each relaxed control.*

A5.3. $k(\cdot)$ *is bounded, continuous and*

$$k(x, z, \alpha_1, \alpha_2) = k_0(x, z) + k_1(x, \alpha_1) + k_2(z, \alpha_2).$$

Define the fixed-x system

$$dz_0 = dt \int_{\mathcal{U}_2} H(x, z_0, \alpha_2) m_{2t}(d\alpha_2) + v(x, z_0) d\tilde{w}_2, \qquad (5.4)$$

where $\tilde{w}_2(\cdot)$ is a standard vector-valued Wiener process.

Define the average cost per unit time problem for $z_0(\cdot)$, with x and α_1 fixed by

$$\Lambda(x, z, \alpha_1, m_2) = \overline{\lim}_{T_1} \frac{1}{T_1} E_{x,z}^{\alpha_1, m_2} \int_0^{T_1} \int_{\mathcal{U}_2} k(x, z_0(s), \alpha_1, \alpha_2) m_{2s}(d\alpha_2) dt. \qquad (5.5)$$

Let $\overline{\Lambda}(x, z, \alpha_1) = \inf_{m_2} \Lambda(x, z, \alpha_1, m_2)$. Define

$$V_T(x, z, m_1) = E_{x,z}^{m_1} \int_0^T \int_{\mathcal{U}_1} \overline{\Lambda}(x(t), z, \alpha_1) m_{1t}(d\alpha_1)_d \, dt, \qquad (5.6)$$

$$V_T(x, z) = \inf_{m_1} V_T(x, z, m_1).$$

Assume

A5.4. $\{z^\epsilon(t), t < T, \epsilon > 0, \text{ all controls}\}$ *is tight.*

A5.5. *For each $\rho > 0$, there is a continuous function $u_2^\rho(\cdot)$ of (x, z) such that $u_2^\rho(x, \cdot)$ is a ρ-optimal feedback control for (5.4) and cost (5.5). Under $u_2^\rho(x, \cdot)$, (5.4) has a unique weak sense solution for each initial condition, and a unique invariant measure $\mu(x, u^\rho, \cdot)$.*

Theorem 5.1. *Assume (A5.1) to (A5.5). Then $\overline{\Lambda}(x, z, \alpha_1)$ does not depend on z, and is continuous on (x, α_1). $V_T(x, z) = V_T(x)$, $V_T(x, z, m_1) = V_T(x, m_1)$ do not depend on z. As $\epsilon \to 0$*

$$\inf_{m_1^\epsilon, m_2^\epsilon} V_T^\epsilon(x, z, m_1^\epsilon, m_2^\epsilon) \to V_T(x).$$

Let $u_1^\rho(\cdot)$ be a continuous ρ-optimal feedback control for (5.1), (5.6), such that (5.1) has a unique weak sense solution under $u_1^\rho(\cdot)$. Then $(u_1^\rho(\cdot), u_2^\rho(\cdot))$ is $(3\rho + \rho T)$-optimal for (5.1)–(5.3) for small ϵ.

Outline of Proof. We only make a few rough remarks, since the method follows the lines of the previous proofs. Without loss of generality (and little change in the cost function) we suppose that the control for (5.1) is ordinary and is constant on the intervals $[j\delta_0, j\delta_0 + \delta_0)$, for small $\delta_0 > 0$. Denote the value of that control at t by $\alpha_1^\epsilon(t)$, and let $m_1^\epsilon(\cdot)$ denote the relaxed control representation. Let $\delta \to 0$ and $\epsilon/\delta \to 0$ and let δ_0 and T be integral multiples of δ. Rewrite (5.3) as (5.7), where we put $x^\epsilon(t)$ constant on the intervals $[i\delta, i\delta + \delta)$ without changing the value of the expression appreciably.

$$\sum_{i=1}^{T/\delta-1} \delta E_{x,z}^{m^\epsilon} \frac{\epsilon}{\delta} \int_0^{\delta/\epsilon} ds \int_{\mathcal{U}_2} k(x^\epsilon(i\delta), z^\epsilon(i\delta + \epsilon s), \alpha_1^\epsilon(i\delta), \alpha_2) m_{2,i\delta+\epsilon s}^\epsilon(d\alpha_2)$$

$$\sim \delta \sum_{i=0}^{T/\delta-1} E_{x,z}^{m^\epsilon} \frac{\epsilon}{\delta} \int_0^{\delta/\epsilon} ds \int_s^{s+1} du \int_{\mathcal{U}_2} \left[k(x^\epsilon(i\delta), \right. \tag{5.7}$$

$$\left. z^\epsilon(i\delta + \epsilon u), \alpha_1^\epsilon(i\delta), \alpha_2) m_{2,i\delta+\epsilon u}^\epsilon(d\alpha_2) \right].$$

The difference between the two sides of (5.7) goes to zero as $\delta \to 0$ and $\epsilon/\delta \to 0$.

Next, let $x^\epsilon(\cdot)$ and $m_1^\epsilon(\cdot)$ converge weakly to $(x(\cdot), m_1(\cdot))$, where $m_1(\cdot)$ is a relaxed control representation of an ordinary control, whose values $\alpha_1(\cdot)$ we denote by $\alpha_1(j\delta_0)$ on $[j\delta_0, j\delta_0 + \delta_0)$. By use of a functional occupation measure argument similar to that used in Sections 3 and 4, we obtain the following. Let $i_\delta \delta \to t \notin [j\delta_0, j = 1, 2, \ldots]$. Then the limits of the i_δth summands on the right side of (5.7) can be characterized as

$$\lim_{T_0} \frac{1}{T_0} E_{x,z}^m \int_0^{T_0} ds \int_{\mathcal{U}_2} k(x(t), z_0(s \mid x(t)), \alpha_1(t), \alpha_2) m_{2s}(d\alpha_2), \tag{5.8}$$

where the $z_0(\cdot \mid x(t))$ and its control $m_2(\cdot)$ are stationary in the sense that the distribution of the s-shifts $(z_0(s + \cdot \mid x(t)), m_2(s + \cdot) - m_2(s, \cdot))$ doesn't depend on s. The occupation measure argument is simpler than that used in the previous section since, due to the expectation $E_{x,z}^{m^\epsilon}$, we use the *expectation* of the sequence of sample occupation measures. The right side of (5.8) can be "nearly" minimized by using the $u^\rho(\cdot)$ of (A5.5) for small ρ. The rest of the details are left to the reader.

6. Reflected Diffusions

We will do the analog of the problems of Sections 2 and 4 for a reflected diffusion. The "Skorohod problem" model of the reflected diffusion will be used [L1]. The following assumption on the boundary will be used. First, we discuss the "reflected" analog of Section 2, and then state the result for the singularity perturbed and controlled problem.

The uncontrolled problem

A6.1. Γ *is the closure of a bounded open set in* R^r *with a twice continuously differentiable boundary* $\partial\Gamma$. *Let* $n(x)$ *denote the outward normal to* $\partial\Gamma$ *at* x, *and let* $\beta(x)$ *denote the reflection direction. Suppose that* $\beta(x)$ *is the restriction to* $\partial\Gamma$ *of a function which is twice continuously differentiable in* *a neighborhood of $\partial\Gamma$ and let there be $\alpha_0 > 0$ such that $-\beta'(x)n(x) \geq \alpha_0$, all $x \in \partial\Gamma$.*

The Skorohod problem. Let $w(\cdot)$ be a standard vector-valued $\mathcal{F}_t$ Wiener process. We say that $x(\cdot)$ solves the *Skorohod problem* if it is $\mathcal{F}_t$-adapted, continuous, takes values in Γ and there is an $\mathcal{F}_t$-adapted function $Y(\cdot)$ such that for $x \in \Gamma$,

$$x(t) = x + \int_0^t b(x(s))ds + \int_0^t \sigma(x(s))dw(s) + Y(t),$$

$$(\text{var } Y) = |Y|(t) = \int_0^t I_{\partial\Gamma}(x(s))d|Y|(s), \tag{6.1}$$

$$Y(t) = \int_0^t \beta(x(s))d|Y|(s).$$

Define the shifted function $\Delta_t|Y|(\cdot) = |Y|(t+\cdot) - |Y|(t)$. Define the occupation measure $\tilde{P}^t(\cdot)$ for the pair of processes $(x_t(\cdot), \Delta_t|Y|(\cdot))$, and then define $\tilde{P}_T(\cdot)$ as in (1.5). In order to get the needed tightness in the theorem below we will need the following result [L1, Theorem 4.1].

Theorem 6.1. *Assume (A6.1) and, for each $T < \infty$ consider the Skorohod problem*

$$x(t) = f(t) + q(t), \quad t \leq T_0, \tag{6.2}$$

where $f(\cdot)$ and $q(\cdot)$ are in $C^r[0,T]$ and

$$q(t) = \int_0^t \beta(x(s))d|q|(s)$$

$$|q|(t) = \int_0^t I_{\partial\Gamma}(x(s))d|q|(s).$$

If $f(\cdot)$ is in a compact set in $C^r[0,T]$, then $(x(\cdot), q(\cdot), |q|(\cdot))$ is in a compact set in $C^{2r+1}[0,T]$.

Theorem 6.2. *Assume (A1.1) and (A6.1) and let $K(\cdot)$ and $k(\cdot)$ be bounded and continuous. Then the set of shifted processes $\{x_t(\cdot), \Delta_t|Y|(\cdot), t < \infty\}$ and $\{\tilde{P}_T(\cdot), T < \infty\}$ are tight. Let $\tilde{P}(\cdot)$ denote the limit of a weakly convergent subsequence of $\{\tilde{P}_T(\cdot), T < \infty\}$. Let ω be the generic variable of the sample space on which $\tilde{P}(\cdot)$ is defined. Then, for almost all ω, the sample value $\tilde{P}^\omega(\cdot)$ induces a stationary process $\tilde{x}^\omega(\cdot)$ satisfying (6.1). Also (2.2) and (2.3) hold.*

Proof. The proof is similar to that of Theorem 2.2. The tightness of $\{x_t(\cdot), \Delta_t|Y|(\cdot), t < \infty\}$ is a consequence of Theorems 6.1 and the tightness of the sequence of shifted processes

$$\left\{\int_0^\bullet b(x_s(u))du, \int_0^\bullet \sigma(x_s(u))dw_s(u), s < \infty\right\}.$$

The tightness of the sequence of measure-valued random variables follows from the tightness of the above set of processes and Theorem 2.1. Let $\tilde{P}(\cdot)$ denote the limit of a weakly convergent subsequence of $\{\tilde{P}_T(\cdot), T < \infty\}$, and let $(\tilde{x}^\omega(\cdot), |\tilde{Y}^\omega|(\cdot))$ denote the process induced by the value $\tilde{P}^\omega(\cdot)$. The $\tilde{x}^\omega(\cdot)$ and $|\tilde{Y}^\omega|(\cdot)$ are continuous processes and $x^\omega(t) \in \Gamma$ for all t. Also $|\tilde{Y}^\omega|(\cdot)$ can increase only when $\tilde{x}^\omega(t) \in \partial\Gamma$. The stationarity of the limit processes is proved as in Theorem 2.2.

We need only characterize the limits $\tilde{x}^\omega(\cdot)$. Let $\phi(\cdot)$ denote the generic path in $D^r[0,\infty)$ associated with the process $\tilde{x}^\omega(\cdot)$, and $y(\cdot)$ the generic path in $D^r[0,\infty)$ associated with the process $|\tilde{Y}^\omega|(\cdot)$. Redefine the function $F(\cdot)$ of (2.5) as follows:

$$F(\phi(\cdot), y(\cdot)) = h(\phi(t_i), y(t_i), i \leq q)[f(\phi(\tau + s)) - f(\phi(\tau))$$

$$- \int_\tau^{\tau+s} Af(\phi(u))du - \int_\tau^{\tau+s} f_x'(\phi(u))\beta(\phi(u))dy(u)].$$

This function is defined for all $\phi(\cdot) \in D^r[0,\infty)$ and for all $y(\cdot)$ in $D^r[0,\infty)$ which are of bounded variation. If $y(\cdot)$ is not of bounded variation, set the value of the function equal to some very large value. Define $\mathcal{F}_t^\omega = \mathcal{B}(\tilde{x}^\omega(s), |\tilde{Y}^\omega|(s), s \leq t)$ and let $f(\cdot) \in C_0^2(R^r)$. Then it can be shown that, for almost all ω the process $\tilde{M}_f(\cdot)$ defined by

$$\tilde{M}_f^\omega(t) = f(\tilde{x}^\omega(t)) - f(\tilde{x}^\omega(0)) - \int_0^t Af(\tilde{x}^\omega(u))du$$

$$- \int_0^t f_x'(\tilde{x}^\omega(u))\beta(\tilde{x}^\omega(u))d|\tilde{Y}^\omega|(u)$$

is an $\mathcal{F}_t$-martingale. This implies that there is an $\tilde{w}^\omega(\cdot)$ such that the triple $(\tilde{x}^\omega(\cdot), \tilde{Y}^\omega(\cdot), \tilde{w}^\omega(\cdot))$ satisfies the Skorohod problem (6.1).   Q.E.D.

The controlled singularly perturbed problem. We use the model (4.1.1), but let the $z^\epsilon(\cdot)$ process satisfy the Skorohod problem:

$$dx^\epsilon = \int_\mathcal{U} G(x^\epsilon, z^\epsilon, \alpha)m_t^\epsilon(d\alpha)dt + \sigma(x^\epsilon, z^\epsilon)dw_1 \qquad (6.3)$$

$$\epsilon dz^\epsilon = H(x^\epsilon, z^\epsilon)dt + \sqrt{\epsilon}v(x^\epsilon, z^\epsilon)dw_2 + Y^\epsilon(t)$$

$$(\text{var } Y^\epsilon) \equiv |Y^\epsilon(t)| = \int_0^t I_{\partial\Gamma}(z^\epsilon(s))d|Y^\epsilon|(s) \qquad (6.4)$$

$$Y^\epsilon(t) = \int_0^t \beta(z^\epsilon(s))d|Y^\epsilon|(s).$$

The fixed-x process is defined by

$$dz_0 = H(x, z_0)dt + v(x, z_0)dw_2 + Y(t)$$

$$|Y(t)| = \int_0^t I_{\partial\Gamma}(z^\epsilon(s))d|Y|(s) \qquad (6.5)$$

$$Y(t) = \int_0^t \beta(z_0(s))d|Y|(s).$$

Theorem 6.3. *Assume (A6.1) and the conditions of Theorems 4.1 or 4.2. Then the conclusions of these theorems continue to hold.* (Of course, in this case the set $\{z^\epsilon(t), \epsilon > 0, t < \infty, \text{ all controls}\}$ is tight since Γ is bounded.)

7. Discounted Cost Problems

Consider the following discounted cost problems, where $\beta > 0$: For Section 2,

$$V_\beta(x) = \beta E_x^m \int_0^\infty e^{-\beta t} k(x(t))dt;$$

for Section 3,

$$V_\beta(x, m) = \beta E_x^m \int_0^\infty e^{-\beta t} dt \int_{\mathcal{U}} k(x(t), \alpha)m_t(d\alpha);$$

for Section 4,

$$V_\beta^\epsilon(x, z, m^\epsilon) = \beta E_{x,z}^{m^\epsilon} \int_0^\infty e^{-\beta t} dt \int_{\mathcal{U}} k(x^\epsilon(t), z^\epsilon(t), \alpha)m_t(d\alpha).$$

As $\beta \to 0$ or $(\beta, \epsilon) \to 0$, we can obtain results analogous to the previous results of this chapter. For example, define the discounted occupation measure $\tilde{P}_\beta^\epsilon(\cdot)$ by:

$$\tilde{P}_\beta^\epsilon(\cdot) = \beta \int_0^\infty e^{-\beta t} \tilde{P}^{\epsilon,t}(\cdot)dt,$$

where $\tilde{P}^{\epsilon,t}(\cdot)$ is defined above Theorem 4.1. By using the discounted occupation measures, all the theorems of this chapter hold if $\beta \to 0$ replaces $T \to \infty$, and the discounted costs are used. The proofs are virtually unchanged.

6

The Nonlinear Filtering Problem

0. Outline of the Chapter

In Chapter 4, we were concerned with the problem of approximations for the singularly perturbed system (4.1.1), (4.1.2), and with the associated problem of control approximations. In this chapter, we will be concerned with approximations for the nonlinear filtering problem. Suppose that we observe the noise corrupted data $y^\epsilon(\cdot)$ defined by $dy^\epsilon = g(x^\epsilon, z^\epsilon)dt + dw_0$, where $w_0(\cdot)$ is a standard vector-valued Wiener process. Owing to the complexity and high dimension of the original system (4.1.1), (4.1.2), the construction of the optimal filter or even the direct construction of an acceptable approximation can be a very hard task. The possibility of using some sort of averaging method, such as in Chapter 4, to get a simpler filter (say, one for an averaged system) is quite appealing. One would use the filter for the averaged system, but the input would be the true physical observations $y^\epsilon(\cdot)$.

It turns out that one can do this, but the averaging is more delicate than that used in Chapter 4. Under the appropriate conditions, the so-called averaged filter provides a good approximation to the true optimal filter for the singularly perturbed system and it can be much simpler. Some standard representation formulas for the nonlinear filter are reviewed in Section 1. These representation formulas will be heavily used in the analysis. The filtering problem and the averaged filter for the singularity perturbed problem is defined in Section 2. In Section 3, we prove that the average filter is "almost optimal" under appropriate conditions. The conditions differ from those used in Chapter 4 mainly in an assumption concerning the convergence of the transition measure for the fixed-x process to the invariant measure of that process. If this condition does not hold, then the difference between the true filter and the averaged filter might not converge to zero, as $\epsilon \to 0$. A counterexample is provided in Section 4. However, even when the cited convergence assumption does not hold, the averaged filter is nearly optimal in a specific and important sense: In particular, it can be shown to be nearly optimal with respect to a large and natural class of alternative filters, as shown in Section 5.

Nonlinear filters are sometimes used for control problems, when the only available data is noise corrupted, and the filter is not too hard to build or approximate. One then considers the conditional distribution of the original system state as the new state variable. It turns out that the averaged filter can also be used for control purposes. One simply assumes that it is the correct optimal filter, and designs the control accordingly. Under broad conditions, this provides a good approximation to the actual optimal control and value function for the original problem. In Section 6, we treat a special case, that of a repair and maintenance problem, where the control is the moment of intervention in the system—to inspect and repair, if necessary. It is shown that the averaged filter can provide a simple and nearly optimal policy. Section 7 is concerned with the robustness of the averaged filter. The robustness is shown to be uniform in the singular perturbations parameter ϵ. In practice, one is more concerned with the robustness of whatever filter is actually constructed than with the actual optimal or even with the averaged filter, which might need to be approximated itself if it is to be used in practice. In Section 8, a useful computational approximation to the averaged filter is constructed, and it is shown to be robust, uniformly in both the (computational) approximation parameter and the singular perturbation parameter. Finally, Section 9 deals with the near-optimality of the averaged filter when used on the infinite time interval, and the cost criterion is the average error per unit time.

1. A Representation of the Nonlinear Filter

In order to prepare ourselves for the analysis of the nonlinear filtering problem for the singularly perturbed system, we first describe a standard representation for the non-linear filter for a signal process that is a jump-diffusion, and where the observation noise is "white Gaussian." Let the signal $x(\cdot)$ be the solution to

$$dx = b(x)dt + \sigma(x)dw + \int q(x,\gamma)N(dtd\gamma), \quad x \in R^r, \qquad (1.1)$$

and let the observation process be $y(\cdot)$, where

$$dy = g(x)dt + dw_0, \quad y \in R^{r''}. \qquad (1.2)$$

We could use Ddw_0 in lieu of dw_0, where D is non-singular, but the resulting problem will be equivalent.

For convenience in the development, we make the following assumption.

A1.1. $b(\cdot), \sigma(\cdot), q(\cdot), g(\cdot)$ *are continuous.* $q(\cdot)$ *and* $g(\cdot)$ *are bounded and* $b(\cdot)$ *and* $\sigma(\cdot)$ *have at most a linear growth as* $|x| \to \infty$. *The standard vector-valued Wiener processes* $w(\cdot)$ *and* $w_0(\cdot)$ *are mutually independent. The Poisson measure* $N(\cdot)$ *has jump rate* $\lambda < \infty$ *and jump distribution* $\Pi(\cdot)$, *and is independent of the Wiener processes.*

For each $t \geq 0$, let E_t denote the expectation conditioned on $\{y(s), s \leq t\}$, and let $\phi(\cdot)$ be a bounded, continuous and real valued function of x. A great deal of effort has gone into the problem of either calculating the $E_t\phi(x(t))$ or getting useful representations for it [E1, K14, F5, L2, Z1]. The so-called "representation theorem" is of fundamental importance in the calculations, since it is the basis for the derivation of the equations for the evolution of the conditional moments or conditional distribution. We now state the representation theorem in the form in which it was initially derived in [K14]. This particular form is equivalent to that derived by use of the measure transformation method [E1, F5, L2, Z1], but it seems to be more convenient to use for the type of weak convergence calculations that the singular perturbation problem requires.

Let $\hat{x}(\cdot)$ denote a process which has the same probability law that $x(\cdot)$ has, but which is independent of $(x(\cdot), y(\cdot), w_0(\cdot))$. If necessary, enlarge the probability space so as to include $\hat{x}(\cdot)$. Define the process $R(\cdot)$ by

$$R(t) = \exp\left[\int_0^t (g(\hat{x}(s)))'dy(s) - \frac{1}{2}\int_0^t |g(\hat{x}(s))|^2 ds\right].$$

It is sometimes helpful to consider E_t as an operator which, when acting on functions of $(y(\cdot), \hat{x}(\cdot))$ simply averages out the $\hat{x}(\cdot)$, with $y(\cdot)$ considered as a parameter in taking the expectation. We can write [K14,E1,L2,K6].

$$E_t\phi(x(t)) = E_t\phi(\hat{x}(t))R(t)/E_t R(t). \tag{1.3}$$

Equation (1.3) defines a probability measure valued process $P(\cdot)$, the conditional distribution, by the relationship

$$(P(t), \phi) = E_t\phi(x(t)), \quad \text{all } \phi(\cdot) \in C_b(R^r). \tag{1.3'}$$

The representation (1.3) was first derived in [K14] for the diffusion process model, and in [S5, W2], for the pure jump process model. The proof for the combined jump-diffusion case is a straightforward extension of the proof for the diffusion case. More recent derivations [E1, L2], use proofs which are more sophisticated and more easily extendible to other cases, but the actual representation formulas which are obtained are identical to (1.3) for all the cases of interest in this chapter. The evolution equations for the moments $E_t\phi(x(t))$ are defined by (1.4) for the case of $\phi(\cdot)$ in $C_0^2(R^r)$ or in $C_b^2(\overline{R}^r)$. Recall that $\overline{R}^r$ is the 'compactified' R^r.

$$dE_t\phi(x(t)) = E_t A\phi(x(t))dt + [E_t\phi(x(t))g(x(t))$$

$$- E_t\phi(x(t))E_t g(x(t))] \cdot [dy - E_t g(x(t))dt]. \tag{1.4}$$

2. The Filtering Problem for the Singularly Perturbed System

We use the signal model (4.5.1), (4.5.2), but without the control terms, and rewrite it as

$$dx^\epsilon = G(x^\epsilon, z^\epsilon)dt + \sigma(x^\epsilon, z^\epsilon)dw_1 + \int q_1(x^\epsilon, z^\epsilon, \gamma)N_1(dtd\gamma), \quad x \in R^r, \quad (2.1)$$

$$\epsilon dz^\epsilon = H(x^\epsilon, z^\epsilon) + \sqrt{\epsilon}v(x^\epsilon, z^\epsilon)dw_2 + \int q_2(x^\epsilon, z^\epsilon, \gamma)N_2^\epsilon(dtd\gamma), \quad z \in R^{r'}.$$

The observational information available at time t is $\{y^\epsilon(s), s \leq t\}$, where

$$dy^\epsilon = g(x^\epsilon, z^\epsilon)dt + dw_0, \quad y \in R^{r''}. \qquad (2.2)$$

Let A^ϵ denote the differential operator of $(x^\epsilon(\cdot), z^\epsilon(\cdot))$.

For notational convenience only we suppose that $x^\epsilon(0) = x(0)$, not depending on ϵ. Let $P(0)$ denote the distribution of $x(0)$.

The following condition will be used.

A2.1. $G(\cdot)$ and $\sigma(\cdot)$ are continuous and for some $K < \infty$ are bounded by $K(1+|x|)$. $H(\cdot)$ and $v(\cdot)$ are continuous and bounded by $K(1+|z|)$. The $g(\cdot)$ and $q_i(\cdot)$ are bounded and continuous. w_0 is independent of $(w_1(\cdot), w_2(\cdot), x^\epsilon(\cdot), z^\epsilon(\cdot))$. The $N_1(\cdot)$ and $N_2^\epsilon(\cdot)$ satisfy (A4.5.1) and are independent of the Wiener processes.

Analogously to the situation for the control problem in Chapter 4, the *averaged* system is (4.5.5) with the control term deleted; i.e.,

$$dx = \overline{G}(x)dt + \overline{\sigma}(x)dw + d\overline{J}(t), \qquad (2.3)$$

where $\overline{J}(\cdot)$ is the jump process described by (4.5.6), (4.5.7) and the initial value is $x(0)$. The observational data for the filter for the averaged system is

$$dy = \overline{g}(x) + dw_0. \qquad (2.4)$$

The $w_0(\cdot)$ here is a standard vector-valued Wiener process which is independent of $(w(\cdot), x(\cdot))$. Also, $\overline{g}(x) = \int g(x, z)\mu_x(dz)$ and $\overline{G}(\cdot)$ and $\overline{\sigma}(\cdot)$ are as defined in Chapter 4.1. $\mu_x(\cdot)$ is the invariant measure of the fixed-x process $z_0(\cdot|x) = z_0(\cdot)$. Let $\overline{A}$ denote the differential operator of $x(\cdot)$. As in Section 1, let E_t denote the expectation, conditioned on $\{y(s), s \leq t\}$.

In this chapter, we are concerned with the relationships between the filtering problem associated with (2.1), (2.2), and that associated with (2.3), (2.4). Usually, the filter for the averaged system is simpler than that for the physical system (2.1), (2.2), since the dimension of x is less than that of (x, z), and the "large" $1/\epsilon$ term which appears in A^ϵ does not appear in $\overline{A}$. With this in mind, the following natural and important question arises:

Suppose that we construct the optimal non-linear filter for the averaged system (2.3), (2.4), but in lieu of the observations $y(\cdot)$ as the input to that filter, we use the physical data $y^\epsilon(\cdot)$ instead. We call the resulting filter the *averaged filter*. When and in what sense is the output process from this averaged filter close to the true optimal filter for (2.1), (2.2)? This question will be answered in the next two sections. The basic result is that the averaged filter approximation to the true optimal filter for the singularly perturbed problem is nearly optimal under broad conditions. Even when it is not nearly optimal in an absolute sense, it is nearly optimal with respect to a large and natural class of alternative filters. We next define the representation for the optimal filter for the singularly perturbed problem as well as for the averaged filter.

The filter for the singularity perturbed system. *Definitions.* The functions $\phi(\cdot)$ used in this section will be in $C_b^2(\overline{R})$ and their first and second order partial derivatives will be bounded. Let E_t^ϵ denote the expectation conditioned on $\{y^\epsilon(s),\ s \leq t\}$. Let $(\hat{x}^\epsilon(\cdot),\ \hat{z}^\epsilon(\cdot))$ be processes with the same probability law that $(x^\epsilon(\cdot),\ z^\epsilon(\cdot))$ has, but which are independent of $(x^\epsilon(\cdot),\ z^\epsilon(\cdot),\ w_0(\cdot))$. Define the process $R^\epsilon(t)$ by

$$R^\epsilon(t) = \exp\left[\int_0^t g(\hat{x}^\epsilon(s), \hat{z}^\epsilon(s))' dy^\epsilon(s) - \frac{1}{2}\int_0^t |g(\hat{x}^\epsilon(s), \hat{z}^\epsilon(s))|^2 ds\right].$$

Equation (1.3) implies the representation

$$E_t^\epsilon \phi(x^\epsilon(t)) = \frac{E_t^\epsilon \phi(\hat{x}^\epsilon(t)) R^\epsilon(t)}{E_t^\epsilon R^\epsilon(t)}. \tag{2.5}$$

Equation (2.5) defines the conditional probability process $P^\epsilon(\cdot)$ by the operation

$$(P^\epsilon(t), \phi) = E_t^\epsilon \phi(x^\epsilon(t)). \tag{2.5'}$$

The filter for the averaged system. Let $\hat{x}(\cdot)$ be a process with the same probability law that $x(s)$ has, but which is independent of all the other processes. Define the process $R(t)$ by

$$R(t) = \exp\left[\int_0^t \overline{g}(\hat{x}(s))' dy(s) - \frac{1}{2}\int_0^t |\overline{g}(\hat{x}(s))|^2 ds\right].$$

Then, by (1.3), the filter for the averaged system can be represented as

$$E_t \phi(x(t)) = \frac{E_t \phi(\hat{x}(t)) R(t)}{E_t R(t)}. \tag{2.6}$$

The equations for the conditional moments of $x(\cdot)$ are given by (1.4) with A replaced by $\overline{A}$. We can define the conditional probability process $P(\cdot)$ by

$$(P(t), \phi) = E_t \phi(x(t)). \tag{2.6'}$$

The filter (2.6) is meaningless in applications since $y^\epsilon(\cdot)$ is the only available data; $y(\cdot)$ is not part of the physical problem data. Hence, any filter must be "driven" by $y^\epsilon(\cdot)$. Our aim is to find a relatively simple and nearly optimal filter.

The averaged filter. Define $\overline{R}^\epsilon(\cdot)$ by

$$\overline{R}^\epsilon(t) = \exp\left[\int_0^t \overline{g}(\hat{x}(s))' dy^\epsilon(s) - \frac{1}{2}\int_0^t |\overline{g}(\hat{x}(s))|^2 ds\right].$$

Then the averaged filter can be represented in the form

$$\overline{E}^\epsilon_t \phi(x^\epsilon(t)) = \frac{E^\epsilon_t \phi(\hat{x}(t))\overline{R}^\epsilon(t)}{E^\epsilon_t \overline{R}^\epsilon(t)}. \tag{2.7}$$

Equation (2.7) defines a probability-measure valued process $\overline{P}^\epsilon(\cdot)$ by the formula

$$(\overline{P}^\epsilon(t), \phi) = \overline{E}^\epsilon_t \phi(x^\epsilon(t)). \tag{2.7'}$$

Sometimes it is useful to view E^ϵ_t as the operator on functions of $(\hat{x}^\epsilon(s), \hat{z}^\epsilon(s), \hat{x}(s), \hat{z}(s), y^\epsilon(\cdot))$ which simply averages out all the functions except the $y^\epsilon(\cdot)$; i.e., where the observation function $y^\epsilon(\cdot)$ is treated as a parameter in taking the expectation. The random variable $\overline{E}^\epsilon_t \phi(x^\epsilon(t))$ (resp. $\overline{P}^\epsilon(\cdot)$) is not usually a conditional expectation (resp., a conditional probability). But, as will be seen in Section 3 below, it is close to the conditional expectation $E^\epsilon_t \phi(x^\epsilon(t))$ (resp., conditional probability $P^\epsilon(t)$) under appropriate conditions. The right hand side of (2.7) is just (2.6), but with $y^\epsilon(\cdot)$ replacing $y(\cdot)$. The expression (2.7) does indeed yield the averaged filter, since it depends on ϵ only via $y^\epsilon(\cdot)$ which replaces $y(\cdot)$ in (2.6). This can also be seen from the moment equations, which are given by

$$d\overline{E}^\epsilon_t \phi(x(t)) = \overline{E}^\epsilon_t (A^\epsilon \phi)(x^\epsilon(t), z^\epsilon(t))dt$$

$$+ [\overline{E}^\epsilon_t \phi(x^\epsilon(t))\overline{g}(x^\epsilon(t)) - \overline{E}^\epsilon_t \phi(x^\epsilon(t))\overline{E}^\epsilon_t \overline{g}(x^\epsilon(t))] \tag{2.8}$$

$$\times (dy^\epsilon - \overline{E}^\epsilon_t \overline{g}(x^\epsilon(t))dt).$$

The term in the last parentheses on the right is *not* generally the differential of an innovations process. The moment equation (2.8) is obtained just as (1.4) is, via use of Itô's Formula on the defining expression (2.7). The fact that the E_t on the left hand side of (1.3) is actually a conditional expectation in (1.3) is not used in the proof of (1.4). The proof uses only the fact that the right hand side of (1.3) is a ratio of conditional expectations.

We now turn to the study of the "error"

$$|\overline{E}^\epsilon_t \phi(x^\epsilon(t)) - E^\epsilon_t \phi(x^\epsilon(t))|.$$

3. The Almost Optimality of the Averaged Filter

In this section, we show that the averaged filter defined by (2.7) or (2.8) is often a good approximation to the optimal filter.
We will use the following conditions.

A3.1. (2.1) *and* (2.3) *have unique weak sense solutions for each initial condition.*

A3.2. *The fixed x process $z_0(\cdot|x)$ has a unique invariant measure $\mu_x(dz)$ for each initial x, and the transition measure $P_0(z, t, \cdot|x)$ of $z_0(\cdot|x)$ satisfies*

$$P_0(z, t, dz|x) \rightarrow \mu_x(dz),$$

weakly as $t \rightarrow \infty$, uniformly in x and z_0 in any compact set. I.e., for $f(\cdot) \in C_b(R^{r'})$ and $z_0(\cdot)$ denoting the fixed-x process with parameter x,

$$E_{z_0} f(z_0(t)) \rightarrow \int f(z)\mu_x(dz)$$

as $t \rightarrow \infty$, uniformly in x and $z(0) = z_0$ in any compact set.

A3.3. $\{z^\epsilon(t), \epsilon > 0, t \le T\}$ *is tight.*

A3.4. $\overline{a}(\cdot)$ *can be factored as in* (A4.1.1).

Theorem 3.1. (a) *Assume* (A2.1) *and* (A3.1) *to* (A3.4). *Then the error process $\rho^\epsilon(\cdot)$ defined by*

$$\overline{E_t^\epsilon}\phi(x^\epsilon(t)) - E_t^\epsilon\phi(x^\epsilon(t)) = \rho^\epsilon(t), \ t \le T, \qquad (3.1)$$

converges weakly to the "zero" process uniformly in $(x, z, P(0))$ in any compact set.
(b) *If the convergence of the measures in* (A3.2) *is dropped, then we still have*

$$(\overline{E_t^\epsilon}\phi(x^\epsilon(t)), x^\epsilon(\cdot), y^\epsilon(\cdot)) \Rightarrow (E_t\phi(x(t)), x(\cdot), y(\cdot)),$$

uniformly in $(x, z, P(0))$ in any compact set.

Remarks on the Proof. 1. Before giving the proof, we explain why a direct averaging method of the type used in Chapter 4 cannot usually be used to "average out" the $z^\epsilon(\cdot)$ process. For notational simplicity, let $x(t)$ and $y(t)$ be scalar valued. Define $\tilde{g}(x, z) = g(x, z) - \overline{g}(x)$. Define the martingale $\tilde{M}^\epsilon(\cdot)$ by

$$\tilde{M}^\epsilon(t) = \int_0^t \tilde{g}(\hat{x}^\epsilon(s), \hat{z}^\epsilon(s))dw_0(s).$$

The main difficulty in the proof of Theorem 3.1 is caused by the fact that the simple averaging out of the $\hat{z}^\epsilon(\cdot)$ in the exponent of the $R^\epsilon(t)$ can yield the incorrect result. Define $\tilde{M}^\epsilon(t) = \int_0^t \tilde{g}(\hat{x}^\epsilon(s), \hat{z}^\epsilon(s))dw_0(s)$. The troublesome term is

$$\exp\left[\tilde{M}^\epsilon(t) - \frac{1}{2}\int_0^t \tilde{g}^2(\hat{x}^\epsilon(s), \hat{z}^\epsilon(s))ds\right].$$

The $\tilde{M}^\epsilon(\cdot)$ is a martingale with quadratic variation $\int_0^t |\tilde{g}(\hat{x}^\epsilon(s), \hat{z}^\epsilon(s))|^2 ds$. The set $\{\tilde{M}^\epsilon(\cdot)\}$ is tight. Suppose that $\{x^\epsilon(\cdot)\}$ is tight and (abusing notation) let ϵ index a weakly convergent subsequence of $\{\tilde{M}^\epsilon(\cdot), x^\epsilon(\cdot), \hat{x}^\epsilon(\cdot), y^\epsilon(\cdot)\}$ with limit denoted by $(\tilde{M}(\cdot), x(\cdot), \hat{x}(\cdot), y(\cdot))$.

The limit $\tilde{M}(\cdot)$ is a martingale with quadratic variation $\int_0^t [g_0(\hat{x}(s))]^2 ds$, where we define

$$[g_0(x)]^2 = \int [\tilde{g}(x, z)]^2 \mu_x(dz).$$

There is a standard Wiener process $\tilde{w}_0(\cdot)$ with respect to which $\hat{x}(\cdot)$ is non-anticipative and such that

$$\tilde{M}(t) = \int_0^t g_0(\hat{x}(s))dw_0(s).$$

It can be shown that $R^\epsilon(t)$ converges weakly to

$$\exp\int_0^t\left[\overline{g}(\hat{x}(s))dy - \frac{1}{2}\int_0^t |\overline{g}(\hat{x}(s))|^2 ds\right] \times \exp\left[\tilde{M}(t) - \frac{1}{2}\int_0^t |g_0(\hat{x}(s))|^2 ds\right],$$

which is not $R(t)$. Thus, some mechanism other than a direct averaging of $R^\epsilon(t)$ and $\phi(x^\epsilon(t))$ must be used to prove the approximation result. We will need to exploit the way that the conditional expectation operator E_t^ϵ is used in (2.5).

2. In the proof, we introduce $Z^\epsilon(t)$, random variables with values in the space of bounded measures on $\overline{R}^r$, the "compactified" R^r: I.e., the values are in $\mathcal{M}(\overline{R}^r)$ (Refer to Chapter 1.6 for the definition of $\mathcal{M}$.) On $\mathcal{M}(\overline{R}^r)$, the weak topology is equivalent to the Prohorov topology (Chapter 1) and the space is a complete and separable metric space under this topology. We could work with $\mathcal{M}(R^r)$ also, but then we would have to add additional details concerning the "essential confinement" of the "mass" to a compact subset.

Outline of Proof. For Part (a), we will show that the difference between the numerators of (2.5) and (2.7) goes to zero as $\epsilon \to 0$ (hence the difference between the denominators goes to zero, since $\phi(x) \equiv 1$). Stochastic differential equations for the numerators are derived and represented in a convenient way by an interchange of integration and conditional expectation. The numerators and solutions to the SDE's are represented in terms of processes $Z^\epsilon(\cdot)$ and $\overline{Z}^\epsilon(\cdot)$, the first being an unnormalized conditional density for the true filter for the singularly perturbed problem, and the second being an unnormalized density of the "Ersatz" conditional measure given by the averaged filter. The coefficients in the SDE's are "averaged" by use of (A3.2). It is then shown that the sequences $\{Z^\epsilon(\cdot)\}$, $\{\overline{Z}^\epsilon(\cdot)\}$ are tight, and that the limits satisfy the same SDE. Since this SDE has a unique solution, the limits are equal. This yields Part (a).

Proof. Part (a) of the theorem will be proved first. For notational simplicity, we do only the case where x, z and y are scalar valued and where there is no jump term in (2.1). The general case is treated in essentially the same way. Let ϵ index a weakly convergent subsequence of $\{x^\epsilon(\cdot), y^\epsilon(\cdot), \hat{x}^\epsilon(\cdot)\}$. We need to show that

$$\frac{E_t^\epsilon \phi(\hat{x}^\epsilon(t)) R^\epsilon(t)}{E_t^\epsilon R^\epsilon(t)} - \frac{E_t^\epsilon \phi(\hat{x}(t)) \overline{R}^\epsilon(t)}{E_t^\epsilon \overline{R}^\epsilon(t)} \Rightarrow \text{zero process} \tag{3.2}$$

for $\phi(\cdot) \in C_b(R^r)$. Since

$$\lim_{\delta \to 0} \inf_{\epsilon > 0} P\left(\inf_{t \le T} E_t^\epsilon R^\epsilon(t) > \delta \right) = 1,$$

$$\lim_{\delta \to 0} \inf_{\epsilon > 0} P\left(\inf_{t \le T} E_t^\epsilon \overline{R}^\epsilon(t) > \delta \right) = 1,$$

we need only show that

$$[E_t^\epsilon \phi(\hat{x}^\epsilon(t)) R^\epsilon(t) - E_t^\epsilon \phi(\hat{x}(t)) \overline{R}^\epsilon(t)] \Rightarrow \text{zero process}. \tag{3.3}$$

Actually, we need only show (3.3) for $\phi(\cdot) \in C_0(R^r)$ or for $\phi(\cdot) \in C_0^2(R^r)$ or in $C_b^2(\overline{R}^r)$ since

$$\lim_{N} \sup_{\epsilon} E\, E_t^\epsilon\, R^\epsilon(t)\, I_{\{|\hat{x}^\epsilon(t)| \ge N\}} = 0,$$

$$\lim_{N} \sup_{\epsilon} E\, E_t^\epsilon\, \overline{R}^\epsilon(t)\, I_{\{|\hat{x}^\epsilon(t)| \ge N\}} = 0.$$

Henceforth, let $\phi(\cdot) \in C_0^2(R^r)$.
 Define

$$\tilde{R}^\epsilon(t) = \exp\left[\int_0^t \overline{g}(\hat{x}^\epsilon(s)) dy^\epsilon(s) - \frac{1}{2} \int_0^t |\overline{g}(\hat{x}^\epsilon(s))|^2 ds \right].$$

By the averaging method of Theorem 4.1.2, the tightness of $\{\hat{x}^\epsilon(\cdot)\}$, and the mutual independence of $(x^\epsilon(\cdot), z^\epsilon(\cdot))$ and $(\hat{x}^\epsilon(\cdot), \hat{x}(\cdot))$, we have

$$\int_0^t \overline{g}(\hat{x}^\epsilon(s))[g(x^\epsilon(s), z^\epsilon(s)) - \overline{g}(x^\epsilon(s))] ds \Rightarrow \text{zero process},$$

$$\int_0^t \overline{g}(\hat{x}(s))[g(x^\epsilon(s), z^\epsilon(s)) - \overline{g}(x^\epsilon(s))]ds \Rightarrow \text{zero process.}$$

From this and the weak convergence $\hat{x}^\epsilon(\cdot) \Rightarrow \hat{x}(\cdot)$, it follows that

$$\sup_{t \leq T} E|E_t^\epsilon \phi(\hat{x}(t))\overline{R}^\epsilon(t) - E_t^\epsilon \phi(\hat{x}^\epsilon(t))\tilde{R}^\epsilon(t)| \xrightarrow{\epsilon} 0.$$

Thus, we need only show that, as $\epsilon \to 0$,

$$E_t^\epsilon \phi(\hat{x}^\epsilon(t))\tilde{R}^\epsilon(t) - E_t^\epsilon \phi(\hat{x}^\epsilon(t))R^\epsilon(t) \Rightarrow \text{zero process.} \qquad (3.4)$$

The process $\hat{x}^\epsilon(\cdot)$ satisfies $d\hat{x}^\epsilon = G(\hat{x}^\epsilon, \hat{z}^\epsilon)dt + \sigma(\hat{x}^\epsilon, \hat{z}^\epsilon)d\hat{w}_1^\epsilon$, where $\hat{w}_1^\epsilon(\cdot)$ is some standard vector-valued Wiener process which is independent of $(y^\epsilon(\cdot)$, $x^\epsilon(\cdot)$, $z^\epsilon(\cdot))$. Recall that $\tilde{g}(x, z) = g(x, z) - \overline{g}(x)$. By Itô's Formula and the independence of $\hat{x}^\epsilon(\cdot)$ and $y^\epsilon(\cdot)$,

$$d(R^\epsilon(t)\phi(\hat{x}^\epsilon(t)) = R^\epsilon(t)d\phi(\hat{x}^\epsilon(t)) + \phi(\hat{x}^\epsilon(t))dR^\epsilon(t)$$

$$dR^\epsilon(t) = R^\epsilon(t)[\overline{g}(\hat{x}^\epsilon(t)) + \tilde{g}(\hat{x}^\epsilon(t), \hat{z}^\epsilon(t))]dy^\epsilon(t) \qquad (3.5)$$

$$d\phi(\hat{x}^\epsilon(t)) = (A^\epsilon \phi)(\hat{x}^\epsilon(t), \hat{z}^\epsilon(t))dt + \phi_x(\hat{x}^\epsilon(t))\sigma(\hat{x}^\epsilon(t), \hat{z}^\epsilon(t))d\hat{w}_1^\epsilon.$$

Also,

$$d\phi(\hat{x}^\epsilon(t))\tilde{R}^\epsilon(t) = \tilde{R}^\epsilon(t)d\phi(\hat{x}^\epsilon(t)) + \phi(\hat{x}^\epsilon(t))d\tilde{R}^\epsilon(t),$$

$$d\tilde{R}^\epsilon(t) = \tilde{R}^\epsilon(t)\overline{g}(\hat{x}^\epsilon(t))dy^\epsilon(t).$$

Integrating (3.5) and taking conditional expectations yields

$$E_t^\epsilon R^\epsilon(t)\phi(\hat{x}^\epsilon(t)) = E_t^\epsilon \phi(\hat{x}^\epsilon(0)) + E_t^\epsilon \int_0^t R^\epsilon(s)(A^\epsilon \phi)(\hat{x}^\epsilon(s), \hat{z}^\epsilon(s))ds$$

$$+ E_t^\epsilon \int_0^t R^\epsilon(s)\phi_x(\hat{x}^\epsilon(s))\sigma(\hat{x}^\epsilon(s), \hat{z}^\epsilon(s))d\hat{w}_1^\epsilon(s) \qquad (3.6)$$

$$+ E_t^\epsilon \int_0^t \phi(\hat{x}^\epsilon(s))R^\epsilon(s)[\overline{g}(\hat{x}^\epsilon(s)) + \tilde{g}(\hat{x}^\epsilon(s), \hat{z}^\epsilon(s))]dy^\epsilon(s).$$

By the independence of $y^\epsilon(\cdot)$ and $(\hat{x}^\epsilon(\cdot)$, $\hat{z}^\epsilon(\cdot)$, $\hat{w}_1^\epsilon(\cdot))$, the conditional expectation of the stochastic integral with respect to $\hat{w}_1^\epsilon(\cdot)$ in (3.6) equals zero, and (3.6) can be written as

$$E_t^\epsilon R^\epsilon(t)\phi(\hat{x}^\epsilon(t)) = E_0^\epsilon \phi(\hat{x}^\epsilon(0)) + \int_0^t [E_s^\epsilon R^\epsilon(s)(A^\epsilon \phi)(\hat{x}^\epsilon(s), \hat{z}^\epsilon(s))]ds$$

$$+ \int_0^t [E_s^\epsilon \phi(\hat{x}^\epsilon(s))R^\epsilon(s)\overline{g}(\hat{x}^\epsilon(s))]dy^\epsilon(s) \qquad (3.7)$$

$$+ \int_0^t [E_s^\epsilon \phi(\hat{x}^\epsilon(s))R^\epsilon(s)\tilde{g}(\hat{x}^\epsilon(s), \hat{z}^\epsilon(s))]dy^\epsilon(s).$$

For use below, define the integrand of the last term on the right of (3.7) to be

$$L^\epsilon(s) = E_s^\epsilon \phi(\hat{x}^\epsilon(s)) R^\epsilon(s) \tilde{g}(\hat{x}^\epsilon(s), \hat{z}^\epsilon(s)).$$

Similarly

$$E_t^\epsilon \tilde{R}^\epsilon(t) \phi(\hat{x}^\epsilon(t)) = E_0^\epsilon \phi(\hat{x}^\epsilon(0)) + \int_0^t [E_s^\epsilon \tilde{R}^\epsilon(s)(A^\epsilon \phi)(\hat{x}^\epsilon(s), \hat{z}^\epsilon(s))] ds$$

$$+ \int_0^t [E_s^\epsilon \phi(\hat{x}^\epsilon(s)) \tilde{R}^\epsilon(s) \overline{g}(\hat{x}^\epsilon(s))] dy^\epsilon(s). \tag{3.8}$$

Define $p(\cdot)$ by

$$(A^\epsilon \phi)(x, z) = \phi_x(x) G(x, z) + \frac{1}{2} \phi_{xx}(x) \sigma^2(x, z) \equiv p(x, z).$$

Define $\overline{p}(x) = \int p(x, z) \mu_x(dz)$. The fact that $\{R^\epsilon(\cdot)\}$ is tight and can be well approximated by the piecewise constant process with values $R^\epsilon(i\Delta)$ on $[i\Delta, i\Delta + \Delta)$ for small Δ together with the method used to average $z^\epsilon(\cdot)$ out in Theorem 4.1.2 can be used to average the $\hat{z}^\epsilon(\cdot)$ out and yield that

$$\int_0^t E_s^\epsilon R^\epsilon(s)[p(\hat{x}^\epsilon(s), \hat{z}^\epsilon(s)) - \overline{p}(\hat{x}^\epsilon(s))] ds \Rightarrow \text{zero process}, \tag{3.9}$$

as $\epsilon \to 0$. The same limit (3.9) holds if $\tilde{R}^\epsilon(\cdot)$ replaces $R^\epsilon(\cdot)$. Thus, the second terms on the right side of (3.7) and (3.8) are asymptotically equivalent to, resp.,

$$\int_0^t E_s^\epsilon R^\epsilon(s) \overline{A} \phi(\hat{x}^\epsilon(s)) ds, \quad \int_0^t E_s^\epsilon \tilde{R}^\epsilon(s) \overline{A} \phi(\hat{x}^\epsilon(s)) ds.$$

We next show that the integrand $L^\epsilon(s)$ in the last integral on the right side of (3.7) goes to zero as $\epsilon \to 0$, for almost all ω, s. We will do this in the following way. Fix $\Delta > 0$. Then we replace the $R^\epsilon(s)$ and $\hat{x}^\epsilon(s)$ in $L^\epsilon(s)$ by $R^\epsilon(i\Delta)$ and $\hat{x}^\epsilon(i\Delta)$, resp., on the interval $[i\Delta, i\Delta + \Delta)$, for each i. The resulting expression is (3.10) below. We then show that the limit as $\epsilon \to 0$ of (3.10) is zero w.p.1. Then we let $\Delta \to 0$ to get the final estimates for the proof of the assertion.

Let $s \in (i\Delta, i\Delta + \Delta)$ and let k be a positive real number in what follows. For $k > 0$ and $s - \epsilon k \geq 0$, we can write

$$E_s^\epsilon \phi(\hat{x}^\epsilon(i\Delta)) R^\epsilon(i\Delta) \tilde{g}(\hat{x}^\epsilon(i\Delta), \hat{z}^\epsilon(s))$$

$$= E_s^\epsilon \{\phi(\hat{x}^\epsilon(i\Delta)) R^\epsilon(i\Delta) E_s^\epsilon [\tilde{g}(\hat{x}^\epsilon(i\Delta), \hat{z}^\epsilon(s)) \mid \hat{x}^\epsilon(u), \hat{z}^\epsilon(u), u \leq s - \epsilon k]\}. \tag{3.10}$$

Note that conditioning on $\{y^\epsilon(u), u \leq s\}$ which is implicit in the definition of E_s^ϵ is irrelevant in the inner conditional expectation, since $\tilde{g}(\hat{x}^\epsilon(i\Delta), \hat{z}^\epsilon(s))$ is independent of $y^\epsilon(\cdot)$.

Let $\hat{z}_0^\epsilon(s - \epsilon k + \epsilon \cdot \mid \hat{x}^\epsilon(s - \epsilon k))$ denote the fixed-x process with parameter $x = \hat{x}^\epsilon(s - \epsilon k)$ and initial condition $\hat{z}^\epsilon(s - \epsilon k)$ at time $s - \epsilon k$. Then, by the tightness of $\{\hat{x}^\epsilon(s),\ \hat{z}^\epsilon(s),\ s \leq T,\ \epsilon > 0\}$ and the growth conditions on the functions $G(\cdot)$ $H(\cdot)$, $\sigma(\cdot)$, $v(\cdot)$, we have that

$$\{\hat{z}^\epsilon(s - \epsilon k + \epsilon \cdot),\ \hat{z}_0^\epsilon(s - \epsilon k + \epsilon \cdot \mid \hat{x}^\epsilon(s - \epsilon k)),\ \epsilon > 0\}$$

is tight and the difference between the two sequences converges weakly to the zero process as $\epsilon \to 0$. This latter fact implies that

$$E \left| E[\tilde{g}(\hat{x}^\epsilon(i\Delta), \hat{z}^\epsilon(s)) \mid \hat{x}^\epsilon(u), \hat{z}^\epsilon(u), u \leq s - \epsilon k] \right.$$

$$\left. - \int \tilde{g}(\hat{x}^\epsilon(i\Delta), z) P_0\left(\hat{z}^\epsilon(s - \epsilon k), k, dz \mid \hat{x}^\epsilon(s - \epsilon k)\right) \right| \xrightarrow{\epsilon} 0, \qquad (3.11)$$

where we recall that $P_0(\cdot \mid x)$ is the transition measure for the fixed-x $z_0(\cdot \mid x)$ process. By (3.11) and (A3.2), as $\epsilon \to 0$ and then $k \to \infty$ we have

$$E \left| \int \tilde{g}(\hat{x}^\epsilon(i\Delta), z)[P_0\left(\hat{z}^\epsilon(s - \epsilon k), k, dz \mid \hat{x}^\epsilon(s - \epsilon k)\right) - \mu_{\hat{x}^\epsilon(s)}(dz)] \right| \to 0.$$

$$(3.12)$$

By (3.11), (3.12), the limits in the mean (as $\epsilon \to 0$ and then $\Delta \to 0$) of the left side of (3.10) are the same as those of (as $\epsilon \to 0$ and then $\Delta \to 0$)

$$E_s^\epsilon \phi(\hat{x}^\epsilon(i\Delta)) R^\epsilon(i\Delta) \int \tilde{g}(\hat{x}^\epsilon(i\Delta), z) \mu_{\hat{x}^\epsilon(s)}(dz). \qquad (3.13)$$

Since $|\hat{x}^\epsilon(i\Delta) - \hat{x}^\epsilon(s)| \xrightarrow{P} 0$ (uniformly in ϵ) as $s - i\Delta \to 0$ and $\int \tilde{g}(x, z) \mu_x(dz) \equiv 0$, we can conclude that (3.13) goes to zero in probability (uniformly in ϵ) as $s - i\Delta \to 0$. Thus, we have proved that $L^\epsilon(\cdot)$ goes to zero in (ω, s)-measure, as $\epsilon \to 0$. This and the fact that

$$\limsup_N E \int_0^t |L^\epsilon(s)|^2 I_{\{|L^\epsilon(s)| \geq N\}} ds < \infty$$

implies that

$$\int_0^t L^\epsilon(s) dy^\epsilon(s) \equiv \hat{p}^\epsilon(t) \Rightarrow \text{zero process,}$$

as asserted.

Define the measure valued processes $Z^\epsilon(\cdot)$ and $\tilde{Z}^\epsilon(\cdot)$, acting on $\phi(\cdot) \in C_b^2(\overline{R}^r)$ as follows:

$$(Z^\epsilon(t), \phi) = E_t^\epsilon R^\epsilon(t) \phi(\hat{x}^\epsilon(t)), \quad (\tilde{Z}^\epsilon(t), \phi) = E_t^\epsilon \tilde{R}^\epsilon(t) \phi(\hat{x}^\epsilon(t)).$$

The $Z^\epsilon(t)$ and $\tilde{Z}^\epsilon(t)$ take values in $\mathcal{M}(\overline{R}^r)$, the space of measures on the compact space $\overline{R}^r$, with the weak (equivalently, Prohorov) topology. (See the second comment before the proof.)

The representations (3.7) and (3.8) imply that the $(Z^\epsilon(\cdot), \phi)$ and $(\tilde{Z}^\epsilon(\cdot), \phi)$ are continuous. Thus, the $Z^\epsilon(\cdot)$ have their paths in $D[\mathcal{M}(\overline{R}'); 0, \infty)$. We now prove tightness of $\{Z^\epsilon(\cdot), \tilde{Z}^\epsilon(\cdot)\}$ by use of the criterion of Theorem 1.5.1 or Theorem 1.6.2.

Given $T < \infty$, let τ be an arbitrary stopping time less than or equal to T. Then it follows from the representation (3.7) that for some $K < \infty$,

$$\lim_\delta \limsup_\epsilon \sup_\tau E|(Z^\epsilon(\tau + \delta) - Z^\epsilon(\tau), \phi)| = 0$$

$$\sup_{\epsilon, t \leq T} E|(Z^\epsilon(t), \phi)| \leq K \sup_x |\phi(x)|.$$

Thus $\{Z^\epsilon(\cdot)\}$ is tight in $D[\mathcal{M}(\overline{R}'); 0, T]$ by Theorem 1.6.2. A similar result holds for $\{\tilde{Z}^\epsilon(\cdot)\}$.

With an abuse of notation, let ϵ index a weakly convergent subsequence of $\{Z^\epsilon(\cdot), \tilde{Z}^\epsilon(\cdot), y^\epsilon(\cdot), \hat{x}^\epsilon(\cdot)\}$, with limit denoted by $(Z(\cdot), \tilde{Z}(\cdot), y(\cdot), \hat{x}(\cdot))$. Then using (3.7) to (3.9) and the approximation results proved above, yields that both $Z(\cdot)$ and $\tilde{Z}(\cdot)$ are measure valued processes which satisfy the equation

$$(U(t), \phi) = (U(0), \phi) + \int_0^t (U(s), \overline{A}\phi)ds + \int_0^t (U(s), \overline{g}\phi)dy(s). \quad (3.14)$$

The initial condition in (3.14) is $(U(0), \phi) = (Z(0), \phi) = (\tilde{Z}(0), \phi) = E_0^\epsilon \phi(x^\epsilon(0))$. Any two solutions of (3.14) with the same initial condition and satisfying

$$\sup_{t \leq T} E|(U(t), \phi)|^2 \leq K \sup_x |\phi(x)|^2, \quad (3.15)$$

for some constant $K < \infty$, must be the same w.p.1 [S10, Theorem IV-1, V-1]. The required inequality (3.15) follows from the weak convergence, Fatou's Lemma, and the fact that there is a $K_1 < \infty$ such that

$$\sup_{\epsilon, t \leq T} E|(\tilde{Z}^\epsilon(t), \phi)|^2 + \sup_{\epsilon, t \leq T} E|(Z^\epsilon(t), \phi)|^2 \leq K_1 \sup_x |\phi(x)|^2.$$

The proof of Part (b) of the theorem is easier than the proof of Part (a), since we can directly average the $g(x^\epsilon(\cdot), z^\epsilon(\cdot))$ in the exponent in (2.7) and the details are omitted.   Q.E.D.

4. A Counterexample to the Averaged Filter

The convergence in (A3.2) is essential to the proof of Theorem 3.1. Mere uniqueness of the invariant measure is not enough, although it was sufficient in Chapter 4. The role of the convergence requirement can be seen from the following simple example. Define $z^{1,\epsilon}(\cdot)$ by $\epsilon \ddot{z}^{1,\epsilon} = -z^{1,\epsilon}$, and define $z^{2,\epsilon}(\cdot)$ by $z^{2,\epsilon} = \dot{z}^{1,\epsilon}$, and let $z = (z^1, z^2)$. Choose the initial condition

$z^\epsilon(0) = z(0)$ such that $z^{1,\epsilon}(t) = \sin(t/\sqrt{\epsilon}+\theta)$, where θ equals either 0 or π, each with probability 1/2. Define the stretched out system $z_0^1(t) = \sin(t+\theta)$, $\dot{z}_0^2(t) = z_0^1(t)$. The invariant measure depends on the initial condition $z_0(0)$. But if $|z^\epsilon(0)| = 1$, then there is a unique invariant measure $\mu(dz)$, and that measure is a uniform distribution on the unit circle. But (A3.2) is violated, since the transition probability does not converge to the invariant measure, as required.

We will work with the special case where the observations are defined by $dy^\epsilon = g(z^\epsilon)dt + dw$, where $g(z) = \text{sign } z^1$, and $w(\cdot)$ is a standard Wiener process and $(\theta, w(\cdot), z^\epsilon(0))$ are mutually independent and $|z^\epsilon(0)| = 1$. Since $\int g(z)\mu(dz) = 0$, the averaged system is

$$P(\theta = 0) = P(\theta = \pi) = \frac{1}{2}, \quad dy = dw_0 \tag{4.1}$$

We will show that the conditional probability $P(\theta = \pi \,|\, y^\epsilon(s),\, s \leq t)$ converges to the filter for the problem of estimating θ with the initial data

$$P(\theta = 0) = P(\theta = \pi) = \frac{1}{2} \tag{4.2a}$$

and observations

$$dy = [I_{\{\theta=0\}} - I_{\{\theta=\pi\}}]dt + dw_0 \tag{4.2b}$$

where $w_0(\cdot)$ is a standard Wiener process which is independent of θ. Loosely speaking, the information concerning θ which is contained in the observations $y^\epsilon(\cdot)$ does not decrease as $\epsilon \to 0$. This would not be true if (A3.2) held. The filters for (4.1) and (4.2) are not the same.

Proof of the Claim. To get the representation of the filter we follow the procedure in Section 2 and use (2.5). Let $\hat{\theta}$ have the same probability law that θ has, but be independent of $(\theta, w(\cdot))$, and let $\hat{z}^\epsilon(\cdot)$ denote the associated process.

Define the function $F(\hat{\theta}) = I_{\{\hat{\theta}=\pi\}}$. Then by (2.5),

$$P(\theta = \pi \,|\, y^\epsilon(s), s \leq t)$$

$$= \frac{\int P(d\hat{\theta})F(\hat{\theta}) \exp\left[\int_0^t g(\hat{z}^\epsilon(s))dy^\epsilon(s) - \frac{1}{2}\int_0^t |g(\hat{z}^\epsilon(s))|^2 ds\right]}{\int P(d\hat{\theta}) \exp\left[\int_0^t g(\hat{z}^\epsilon(s))dy^\epsilon(s) - \frac{1}{2}\int_0^t |g(\hat{z}^\epsilon(s))|^2 ds\right]}. \tag{4.3}$$

Next, we take limits in (4.3) as $\epsilon \to 0$. We have

$$\lim_\epsilon \int_0^t g(\hat{z}^\epsilon(s))g(z^\epsilon(s))ds = \pm t,$$

according to whether

$$\{\theta = 0, \hat{\theta} = 0\}, \quad \{\theta = \pi, \hat{\theta} = \pi\} \quad \text{(limit is } +t) \tag{4.4a}$$

or

$$\{\theta = 0, \hat{\theta} = \pi\}, \quad \{\theta = \pi, \hat{\theta} = 0\} \quad \text{(limit is } -t\text{)}. \tag{4.4b}$$

Also

$$\int_0^t g^2(\hat{z}^\epsilon(s)) ds = t.$$

Define

$$M_0^\epsilon(t) = \int_0^t g(\sin(s/\sqrt{\epsilon})) dw(s),$$

$$M^\epsilon(t) = \int_0^t g(\hat{z}^\epsilon(s)) dw(s) = I_{\{\hat{\theta}=0\}} M_0^\epsilon(t) - I_{\{\hat{\theta}=\pi\}} M_0^\epsilon(t).$$

The sequence $\{M^\epsilon(\cdot), M_0^\epsilon(\cdot), \theta, \hat{\theta}\}$ is tight. Abusing notation, let ϵ index a weakly convergent subsequence with limit denoted by $(M(\cdot), w_0(\cdot), \theta, \hat{\theta})$. We have

$$M(t) = I_{\{\theta=0\}} w_0(t) - I_{\{\hat{\theta}=\pi\}} w_0(t).$$

The process $w_0(\cdot)$ is a standard Wiener process, and is independent of $(\theta, \hat{\theta})$. The fact that $w_0(\cdot)$ is a Wiener process can be seen by noting that it is a continuous martingale whose quadratic variation equals t. (In fact, $M_0^\epsilon(\cdot)$ also is a Wiener process, for each $\epsilon > 0$.) The asserted independence is a consequence of the mutual independence of $(M_0^\epsilon(\cdot), \theta, \hat{\theta})$ and the fact that independence is preserved under weak convergence.

Define $\overline{g}(\cdot)$ by $\overline{g}(0) = 1$, $\overline{g}(\pi) = -1$. Then the weak limit of the process with values $\int_0^t g(\hat{z}^\epsilon(s)) dy^\epsilon(s)$ is the process with values

$$I_{\{\hat{\theta}=0\}}[t\overline{g}(\theta) + w_0(t)] - I_{\{\hat{\theta}=\pi\}}[t\overline{g}(\theta) + w_0(t)].$$

Thus, taking limits in (4.3) yields

$$\frac{\int P(d\hat{\theta}) F(\hat{\theta}) \exp[\overline{g}(\hat{\theta})(t\overline{g}(\hat{\theta}) + w_0(t)) - t/2]}{\int P(d\hat{\theta}) \exp[\overline{g}(\hat{\theta})(t\overline{g}(\hat{\theta}) + w_0(t)) - t/2]}. \tag{4.5}$$

Since (4.5) equals $P(\theta = \pi \mid s\overline{g}(\theta) + w_0(s), s \leq t)$, the claim is proved.

5. The Near Optimality of the Averaged Filter

Under the conditions of Theorem 3.1a, the output of the averaged filter defined by (2.7) or (2.7′) is close to the optimal estimate $E_t^\epsilon \phi(x^\epsilon(t))$ for small $\epsilon > 0$.

If the convergence part of (A3.2) doesn't hold, then the example in the last section shows that the output of the "averaged" filter is not necessarily close to the true conditional expectation, even for small $\epsilon > 0$. Nevertheless, since the estimator $\overline{E}_t^\epsilon \phi(x^\epsilon(t))$ defined by the averaged filter is usually much easier to construct or approximate than $E_t^\epsilon \phi(x^\epsilon(t))$ and does not require the

high gains of the optimal estimator, it is still of considerable interest. We
will show that this averaged filter is "nearly optimal" for small $\epsilon > 0$, in a
specific and very important sense, even if the convergence in (A3.2) doesn't
hold. Let us define an *estimator* of $\phi(x^\epsilon(t))$ to be any bounded measurable
function of $\{y^\epsilon(u),\ u \leq t\}$. The quality of any estimator can be judged
by comparing its properties with those of an appropriate alternative class
of estimators. There is a large and natural class of alternative estimators
with respect to which the estimator $\overline{E}^\epsilon_t \phi(x^\epsilon(t))$ is "nearly optimal" for small
$\epsilon > 0$, even if (A3.2) doesn't hold.

Example. Suppose that we use the averaged filter (2.7) or (2.7'). Then
some typical error criteria which are of interest in applications are functionals of the form

$$\phi(x^\epsilon(t)) - \overline{E}^\epsilon_t \phi(x^\epsilon(t)) \tag{5.1}$$

or of the form

$$\int_0^T |\phi(x^\epsilon(t)) - \overline{E}^\epsilon_t \phi(x^\epsilon(t))|^2 dt. \tag{5.2}$$

Under the conditions of Theorem 3.1b (which does not require the convergence to the invariant measure needed in Theorem 3.1a), the above
functionals converge weakly to, resp.,

$$\phi(x(t)) - E_t \phi(x(t)) \tag{5.1'}$$

or

$$\int_0^T |\phi(x(t)) - E_t \phi(x(t))|^2 dt, \tag{5.2'}$$

where $x(\cdot)$ is the averaged process (2.3), and the conditional expectation
is calculated using the data (2.4). Thus, the distributions of (5.1) or (5.2)
are close to the distributions of (5.1') or (5.2') for small $\epsilon > 0$.

An alternative comparison class of estimates. We now define the
alternative class of estimators which we will compare to the estimator
$\overline{E}^\epsilon_t \phi(x^\epsilon(t))$. Fix $T < \infty$. Let Q denote the set of bounded and measurable functions on $C^{r''}[0, T]$ with values in $C[0, T]$ and which are continu-
ous w.p.1 with respect to Wiener measure on $C^{r''}[0, T]$. Let $\hat{Q} \subset Q$ denote the subset of functions which are non-anticipative in the following
sense: Let $\psi(\cdot)$ be the canonical function in $C^{r''}[0, T]$, let $F(\cdot) \in \hat{Q}$ and let
$F(\psi(\cdot))(t) \equiv F_t(\psi(\cdot))$ denote the value of $F(\psi(\cdot))$ at time t. Then $F_t(\psi(\cdot))$
depends only on the values $\psi(u),\ u \leq t$. For $F(\cdot) \in \hat{Q}$, the functionals
$F(y^\epsilon(\cdot))$ are our alternative comparison class of estimators. Of course, if
only one value of t is of interest, then we can let $F(\cdot)$ be real valued. The
w.p.1 continuity requirement on the elements of $\hat{Q}$ (rather than just
a continuity requirement) is important in applications, as will be seen in
the example in the next section. Such w.p.1 continuity requirements enter

when stopping times are involved in the criteria, e.g., where we wish to approximate the first time that a conditional probability reaches a given value. A similar situation arose in Chapter 4.3 when dealing with a problem where control stopped when some boundary was reached.

Near optimality of the averaged filter. The measure that the process $y(\cdot)$ defined by (2.4) induces on $C^{r''}[0,T]$ is absolutely continuous with respect to Wiener measure. Then, for $F(\cdot) \in \hat{Q}$, the weak convergence, Theorem 1.1.4, and the optimality in the mean square sense of the estimator $E_t \phi(x(t))$ (of $\phi(x(t))$ yields that

$$\phi(x^\epsilon(t)) - F_t(y^\epsilon(\cdot)) \Rightarrow \phi(x(t)) - F_t(y(\cdot)),$$

$$E|\phi(x(t)) - F_t(y(\cdot))|^2 \geq E|\phi(x(t)) - E_t\phi(x(t))|^2. \tag{5.3}$$

Thus, for small $\epsilon > 0$, (5.3) and Theorem 3.1b imply that $\overline{E}_t^\epsilon \phi(x^\epsilon(t))$ is *nearly optimal with respect to any estimator* $F(\cdot) \in \hat{Q}$ in the sense that for $t \leq T$

$$\lim_\epsilon E|\phi(x^\epsilon(t)) - \overline{E}_t^\epsilon \phi(x^\epsilon(t))|^2 = E|\phi(x(t)) - E_t\phi(x(t))|^2$$

$$\leq E|\phi(x(t)) - F_t(y(\cdot))|^2 = \lim_\epsilon E|\phi(x^\epsilon(t)) - F_t(y^\epsilon(\cdot))|^2. \tag{5.4}$$

Similar results hold for the "integral" criterion of (5.2). Consider the following more general case. Let n be an integer and $f(\cdot)$ a bounded real valued and continuous function on $D^r[0,T] \times C^n[0,T]$, and let $\phi_j(\cdot), j \leq n$, be bounded and continuous functions on R^r. For $j \leq n$, let $F_j(\cdot) \in \hat{Q}$ and suppose that

$$Ef(x(u), E_u\phi_j(x(u)), j \leq n, u \leq T)$$

$$\leq Ef(x(u), F_{ju}(y(\cdot)), j \leq n, u \leq T). \tag{5.5a}$$

Then by the weak convergences $x^\epsilon(\cdot) \Rightarrow x(\cdot)$, $y^\epsilon(\cdot) \Rightarrow y(\cdot)$, $\overline{E}_u^\epsilon \phi_j(x^\epsilon(u)) \Rightarrow E_u\phi_j(x(u))$,

$$\lim_\epsilon Ef(x^\epsilon(u), \overline{E}_u^\epsilon \phi_j(x^\epsilon(u)), j \leq n, u \leq T)$$

$$\leq \lim_\epsilon Ef(x^\epsilon(u), F_{ju}(y^\epsilon(\cdot)), j \leq n, u \leq T). \tag{5.5b}$$

The estimators $F(\cdot)$ in $\hat{Q}$ which we discussed above do *not* depend on ϵ. A sequence $F^\epsilon(\cdot) \in \hat{Q}$ can be used provided that the set is uniformly continuous w.p.1 with respect to Wiener measure. If (A3.2) doesn't hold, then loosely speaking, when viewed as a function of $y^\epsilon(\cdot)$ the "smoothness" of the functional $E_t^\epsilon \phi(x^\epsilon(t))$ "decreases" as $\epsilon \to 0$, due to the $1/\epsilon$ term in the differential operator of the $z^\epsilon(\cdot)$ process. The difficulty in working with such high gain systems partly justifies the assumption placed on the estimators of this section that they be (uniformly in ϵ) continuous w.p.1 with respect to Wiener measure. This "increasing" loss of smoothness of the filter can be seen from the example of the last section, where the optimal filter effectively generates a signal $\sin(t/\epsilon)$ and multiplies it by the "observation differential" dy^ϵ.

See [K21] for a discussion of these points in the context of approximately optimal filters for systems where both the observation and driving noise are wideband.

6. A Repair and Maintainance Example

In this section, we treat a simple example which illustrates how both the averaged filter (2.7), (2.7′), and the filter for the averaged system of (2.3), (2.4) can be used to obtain nearly optimal results in a control problem for the singularly perturbed model (2.1), (2.2). The basic approximation idea can also be made when the observation noise is wideband and not white [K21]. See also [K22] for a related problem of approximately optimal filtering and optimal control. The system $(x^\epsilon(\cdot), z^\epsilon(\cdot))$ with which we will work is assumed to satisfy (2.1), but the $x^\epsilon(\cdot)$ is a pure jump process (i.e., there is no diffusion component) with two states, which we denote by G (for Good) and B (for Bad). State B is absorbing and the transition rate from G to B is $p_0(z^\epsilon)$, a bounded and continuous function. Let $z^\epsilon(\cdot)$ satisfy (2.1) but without the jump term (dropped for simplicity in the development only). Let the observation process $y^\epsilon(\cdot)$ given by (2.2). We can suppose that the G and B refer to the "state of operation" or quality of a machine. The rate of "mean wear" of the machine depends on its usage, which is a function of $z^\epsilon(\cdot)$. The assumptions of Theorem 3.1a will be used.

The filtering problem for the averaged system is defined as follows: $x(\cdot)$ is a pure jump Markov process with jump rate from G to B being $\bar{p}_0 = \int p_0(z)\mu_G(dz)$, state B is absorbing and the observational process is

$$dy = \bar{g}(x)dt + dw_0, \quad \bar{g}(x) = \int g(x,z)\mu_x(dz), \quad x = \text{B or G}.$$

Define $P_B^\epsilon(t) = P(x^\epsilon(t) = B|y^\epsilon(s), s \le t)$ and $P_B(t) = P(x(t) = B|y(s), s \le t) = E_t I_B(x(t))$. Define the innovations process $v(\cdot)$ by $v(0) = 0$ and

$$dv = dy - (\bar{g}(B)P_B + \bar{g}(G)(1 - P_B))dt.$$

We can write the filter for the averaged system as

$$dP_B = \bar{p}_0(1 - P_B)dt + [E_t I_B(x(t))\bar{g}(x(t))dt - E_t I_B(x(t))E_t\bar{g}(x(t))]dv$$

$$= \bar{p}_0(1 - P_B)dt + P_B(1 - P_B)(\bar{g}(B) - \bar{g}(G))dv. \qquad (6.1)$$

If $\overline{P}_B^\epsilon(\cdot)$ denotes the filter (6.1), but with the physical observations as the input, we then have the representation

$$d\overline{P}_B^\epsilon = \overline{p}_0(1 - \overline{P}_B^\epsilon)dt + \overline{P}_B^\epsilon(1 - \overline{P}_B^\epsilon)(\overline{g}(B) - \overline{g}(G)) \times$$
$$[dy^\epsilon - (\overline{g}(B)\overline{P}_B^\epsilon + \overline{g}(G)(1 - \overline{P}_B^\epsilon))dt]. \tag{6.2}$$

The system (6.2) is the averaged filter of Section 2. It is not the optimal filter for $x^\epsilon(\cdot)$.

The cost function. The following is a common type of cost structure for inspection/repair problems. If the system operates in state B, then we suppose that a cost is due to the poorer quality of the production, and this cost is to be proportional (c_0) to the number of items produced while the machine is in state B. We also have the option of stopping the production and inspecting and repairing (if necessary) the machine. If, on inspection, the system is found to be G, then a cost of $c_1 \geq 0$ is realized. If the system is found to be B on inspection, then a cost of $c_2 > c_1$ is realized. It is assumed that the inspection and repair take a negligable amount of time. If we suppose otherwise, then it is easy to incorporate that extra time into the formulation. After repair, the system starts again in state G, and with a new "work process" $z^\epsilon(\cdot)$ which is independent of the past. Our cost function is

$$\limsup_T \frac{1}{T} E\left[\int_0^T c_0 P_B^\epsilon(t)dt + c_1(\text{number of repairs by } T \text{ when in } G)\right.$$

$$\left. + c_2(\text{number of repairs by } T \text{ when in } B)\right]. \tag{6.3}$$

We can suppose that the same strategy for observing and stopping will be used on each of the intervals between inspections. Then the time intervals between inspections are mutually independent and identically distributed. This would not be the case if the $z^\epsilon(\cdot)$ process did not start "anew" after each inspection. But owing to the effects of the "fast time scale" of $z^\epsilon(\cdot)$, the "memory" in $z^\epsilon(\cdot)$ between the inspection intervals can be shown to have no effect in the limit as $\epsilon \to 0$ under our assumptions, even if we did not start the $z^\epsilon(\cdot)$ anew.

In the development below, we follow the approach used in [K22, Section 5] for a related inspection and repair problem, but where the "limits" were taken in a "heavy traffic" sense. The details of proof use the averaging methods of Section 3. They also use the techniques of [K22] to bound the moments of and characterize the limits of the stopping times. Let T_ϵ denote the duration of the first renewal interval. We can suppose that $ET_\epsilon < \infty$. Otherwise, the cost will be c_0, and it will be cheaper to operate a system in the state B than to repair it. We assume that the cost function is such that this will not be the best solution. Let $P_B^\epsilon(0) = 0$. Then by the renewal theorem [K26], we can write (6.3) in the form

$$V^\epsilon(T_\epsilon) \equiv \frac{E\int_0^{T_\epsilon} c_0 P_B^\epsilon(s)ds + c_1(1 - P_B^\epsilon(T_\epsilon)) + c_2 P_B^\epsilon(T_\epsilon)]}{ET_\epsilon}. \tag{6.4}$$

The cost functional for the averaged system is

$$V(T) \equiv \frac{E[\int_0^T c_0 P_B(s)ds + c_1(1 - P_B(T)) + c_2 P_B(T)]}{ET}, \tag{6.5}$$

where T is the duration of the first inspection interval and $P_B(0) = 0$. The following result is well known.

Theorem 6.1 [S6]. *Let $P_B(0) = 0$. There is a $\Delta_1 \in (0,1)$ such that the stopping time $\overline{T} = \min\{t : P_B(t) = 1 - \Delta_1\}$ minimizes $V(T)$, with respect to all stopping times. Also $E\overline{T} < \infty$.*

In order to simplify the development in the remaining theorems of this section, we assume that there is some $0 < \Delta_0 < \Delta_1$ such that we always stop and inspect no later than the first time that $P_B^\epsilon(t) \geq 1 - \Delta_0$; i.e.,

$$T_s \leq \min\{t : P_B^\epsilon(t) = 1 - \Delta_0\}. \tag{6.6}$$

This condition implies the following results.

Theorem 6.2. *Any sequence of $\{T_\epsilon\}$ satisfying (6.6) is uniformly integrable.*

Theorem 6.3. *Assume the conditions of Theorem 3.1a, and let $\{\hat{T}_\epsilon\}$ be the optimal stopping time for the cost (6.4) subject to the constraint (6.6). Then $\{P_B^\epsilon(\cdot), x^\epsilon(\cdot), y^\epsilon(\cdot), \hat{T}_\epsilon, \epsilon > 0\}$ is tight. The limit of any weakly convergent subsequence can be written as $(P_B(\cdot), x(\cdot), y(\cdot), \hat{T})$, where $x(\cdot)$ is the averaged jump process. There is a filtration $\mathcal{F}_t$ such that $P_B(\cdot)$ satisfies (6.1), where $dy = \overline{g}(x) + dw_0$, and $w_0(\cdot)$ is a Wiener process with respect to $\mathcal{F}_t$ and is independent of $x(\cdot)$. The limit $\hat{T}$ is an $\mathcal{F}_t$-stopping time and satisfies $E\hat{T} < \infty$. Also*

$$V^\epsilon(\hat{T}_\epsilon) \to V(\hat{T}) \geq V_0 \equiv \inf_T V(T). \tag{6.7}$$

where the inf *is over all stopping times.*

The next two theorems are the actual approximate optimality result. Theorem 6.5 states that the averaged filter yields a nearly optimal control for small ϵ.

Theorem 6.4. *Assume the conditions of Theorem 3.1a. Define $\overline{T}_\epsilon = \min\{t : P_B^\epsilon(t) = 1 - \Delta_1\}$. Then*

$$V^\epsilon(\overline{T}_\epsilon) \to V(\overline{T}) = V_0. \tag{6.8}$$

Also the $\geq$ in (6.7) is an equality.

Proof. (6.8) implies the second assertion of the theorem. Assume that the Skorohod representation (Theorem 1.3.1) is used so that weak convergences are also convergences w.p.1 in the appropriate topology. Since $\{\overline{T}_\epsilon\}$ is uniformly integrable by Theorem 6.2, to prove (6.8), we need only to prove that $\overline{T}_\epsilon \to \overline{T}$, the first time that the process $P_B(\cdot)$ hits $1 - \Delta_1$. Let $\phi(\cdot)$ denote the canonical function in $D[0, \infty)$, and define $\tau(\phi) = \min\{t : \phi(t) = 1 - \Delta_1\}$. Note that the limit $P_B(\cdot)$ has continuous paths w.p.1. The function $\tau(\cdot)$ is not continuous at every point $\phi(\cdot)$. But it is continuous at each $\phi(\cdot)$ which is continuous and which *crosses* the level $1 - \Delta_1$ at its first hitting time. The paths of $P_B(\cdot)$ do cross the level $1 - \Delta_1$ (w.p.1) at any hitting time. Thus, the weak limit of $\{\overline{T}_\epsilon\}$ must be (w.p.1) the first hitting time of $1 - \Delta_1$ by the limit process $P_B(\cdot)$. Q.E.D.

In general, the filter $P_B^\epsilon(\cdot)$ cannot be constructed, since its construction requires getting the conditional distribution of $z^\epsilon(\cdot)$. Here is where the averaged filter comes in. The next theorem says that if ϵ is small, then the use of the averaged filter (in lieu of $P_B^\epsilon(\cdot)$) gives us a cost which is close to minimal.

Theorem 6.5. *Assume the conditions of Theorem 3.1a, and let $\{T_\epsilon\}$ be any sequence of stopping times which satisfy (6.6). Then the conclusions of Theorem 6.3 hold with $\overline{P}_B^\epsilon(\cdot)$ replacing $P_B^\epsilon(\cdot)$. Define the stopping time $\tilde{T}_\epsilon = \min\{\overline{P}_B^\epsilon(t) = 1 - \Delta_1\}$. Then Theorem 6.4 holds for $\tilde{T}_\epsilon$ replacing $\overline{T}_\epsilon$.*

7. Robustness of the Averaged Filters

The outputs of the filters which are represented by (1.3) or (2.7) are obviously functions of the observations $y(\cdot)$ or $y^\epsilon(\cdot)$, resp. For practical purposes, it is useful to have some representation of the filter with which these outputs are continuous functions of the observations, and there is a considerable theory devoted to such continuity or robustness results [C1, D2]. The desire for this continuity property raises a number of interesting questions of both philosophical and practical importance. In particular, there is the implicit assumption that the sample paths of the observation process $y(\cdot)$ or $y^\epsilon(\cdot)$ might be corrupted by some type of noise other than white Gaussian noise, or that the signal model which is used is inexact, or that the filter processor would introduce additional errors. But then the very meaning of the filters (1.3) or (2.7) is called into question, since they are derived under certain explicit assumptions concerning the signal and the observation noise processes. They are not optimal under the variations just mentioned. Nevertheless, a filter would not be readily acceptable in applications if its output is not representable as a smooth function of the observational data. Actually, filters such as (1.3) and (2.7) can not usually be constructed by practical devices. Useful approximations are available, provided that the dimension of the signal is not too large. The "robustness" of the approximations are probably of greater interest than the analogous properties of (1.3) or (2.7), simply because it is the approximations which will be constructed. This "robustness" or "continuity" must be uniform in the approximation parameter, if it is to be useful. In this section, we will prove the robustness (uniformly in ϵ) of the averaged filter (2.7). In the next section we describe one widely used class of approximations (taken from [K13, K18]) and prove their robustness (continuity), uniformly in the approximation parameter. Reference [K21] contains a treatment of non-linear filtering problems for wide band noise driven systems, and where the observational noise is also wide band noise. Adaptations of the filters for the limit process (the weak limit of the physical processes, as bandwidth $\to \infty$) are used, and the "approximate optimality" properties are shown (with respect to an appropriate class of alternative estimators). Also, the robustness properties, uniformly in the data and in the bandwidth are shown. Since "typical" systems are driven by wide band noise and have wide band observation noise such robustness results as in [K21] are of particular importance. These cited results can be extended to the singular perturbation problem.

First, we prove the robustness of (2.7) via a classical argument [C1, D2]. By a stochastic "integration by parts", we can write $\overline{R}^\epsilon(\cdot)$ and define $\overline{R}_0(\cdot)$ as follows:

$$\overline{R}^\epsilon(t) = \exp\left[\overline{g}'(\hat{x}(t))y^\epsilon(t) - \int_0^t y^\epsilon(s)'d\overline{g}(\hat{x}(s)) - \frac{1}{2}\int_0^t |\overline{g}(\hat{x}(s))|^2 ds\right]$$

$$\equiv \overline{R}_0(t, y^\epsilon(\cdot)). \tag{7.1}$$

Note that $\overline{R}_0(\cdot)$ does not depend on ϵ. Let $t \leq T < \infty$ and let $\psi(\cdot)$ denote the canonical function in $C^{r''}[0,T]$, the path space for $y^\epsilon(\cdot)$ or $y(\cdot)$. The function with values $\overline{R}_0(t, \psi(\cdot))$ is well defined. Using the representation (7.1), we can write the function $\overline{E}_t^\epsilon \phi(x^\epsilon(t))$ defined in (2.7) as an explicit function of $y^\epsilon(\cdot)$ and define the function $\overline{F}_t(\cdot)$ as follows:

$$\overline{E}_t^\epsilon \phi(x^\epsilon(t)) = \frac{E_t^\epsilon \phi(\hat{x}(t))\overline{R}_0(t, y^\epsilon(\cdot))}{E_t^\epsilon \overline{R}_0(t, y^\epsilon(\cdot))} \equiv \overline{F}_t(y^\epsilon(\cdot)), \tag{7.2}$$

where $\overline{F}_t(\cdot)$ does not depend on ϵ.

Similarly, we define a function $R_0^\epsilon(\cdot)$ by another integration by parts:

$$R^\epsilon(t) \equiv R_0^\epsilon(t, y^\epsilon(\cdot)) = \exp[g'(\hat{x}^\epsilon(t), \hat{z}^\epsilon(t))y^\epsilon(t)$$

$$- \int_0^t y^\epsilon(s)'dg(\hat{x}^\epsilon(s), \hat{z}^\epsilon(s)) - \frac{1}{2}\int_0^t |g(\hat{x}^\epsilon(s), \hat{z}^\epsilon(s))|^2 ds]. \tag{7.3}$$

Now, we can write (2.5) as an explicit function of $y^\epsilon(\cdot)$ and define the function $F_t^\epsilon(\cdot)$ by:

$$E_t^\epsilon \phi(x^\epsilon(t)) = \frac{E_t^\epsilon \phi(\hat{x}^\epsilon(t))R_0^\epsilon(t, y^\epsilon(\cdot))}{E_t^\epsilon R_0^\epsilon(t, y^\epsilon(\cdot))} = F_t^\epsilon(y^\epsilon(\cdot)), \tag{7.4}$$

$F_t^\epsilon(\cdot)$ is well defined on $C^{r''}[0,T]$.

We now show that the function $\overline{F}_t(\cdot)$ is continuous in the canonical element of $C^{r''}[0,T]$. The form of the averaged filter defined by (7.2) is called the *robust form*. Robust filters were first investigated in [C1, D2], and various stochastic partial differential equations satisfied by the "robust" form of the conditional density were also obtained in these papers. The proof of the following theorem also yields a similar robustness result for (7.4) for each ϵ, but that robustness is unfortunately not uniform in ϵ.

Theorem 7.1. *Assume that (2.3) has a unique weak sense solution for each initial condition and has bounded and continuous coefficients. Let $\overline{g}(\cdot)$ and its partial derivatives up to order two be bounded and continuous. Then for each bounded set S in $C^{r''}[0,T]$, there is a $K(S) < \infty$ such that for all $\psi(\cdot)$ and $\tilde{\psi}(\cdot)$ in S and all $t \leq T$,*

$$|\overline{F}_t(\psi(\cdot)) - \overline{F}_t(\tilde{\psi}(\cdot))| \leq K(S)|\psi(\cdot) - \tilde{\psi}(\cdot)| = K(S)\sup_{s \leq t}|\psi(s) - \tilde{\psi}(s)|.$$

Proof. Let S be an arbitrary bounded set in $C^{r''}[0,T]$, with $\psi(\cdot)$ and $\tilde{\psi}(\cdot)$ being arbitrary elements of S. We need only show that there is a constant $K_1(S) < \infty$ such that

$$E|\overline{R}_0(t, \psi(\cdot)) - \overline{R}_0(t, \tilde{\psi}(\cdot))| \leq K_1(S)|\psi(\cdot) - \tilde{\psi}(\cdot)|, \text{ all } t \leq T, \qquad (7.5)$$

and that

$$\inf_{t \leq T} \inf_{\psi(\cdot) \in S} E|\overline{R}_0(t, \psi(\cdot))| > 0. \qquad (7.6)$$

We can (and will) drop the $\int_0^t |\overline{g}(\hat{x}(u))|^2 du/2$ term in the $\overline{R}_0(t, \psi(\cdot))$ without causing any problems, since $\overline{g}(\cdot)$ is bounded. By the inequality

$$|e^x - e^y| \leq |x - y|(e^x + e^y), \qquad (7.7)$$

we can bound the left side of (7.5) above by

$$E\left|(\psi(t) - \tilde{\psi}(t))'\overline{g}(\hat{x}(t)) - \int_0^t (\psi(s) - \tilde{\psi}(s))'d\overline{g}(\hat{x}(s))\right|$$

$$\cdot\left\{\exp\left[\psi'(t)\overline{g}(\hat{x}(t)) - \int_0^t \psi'(s)d\overline{g}(\hat{x}(s))\right]\right. \qquad (7.8)$$

$$\left. + \exp\left[\tilde{\psi}'(t)\overline{g}(\hat{x}(t)) - \int_0^t \tilde{\psi}'(s)d\overline{g}(\hat{x}(s))\right]\right\}.$$

By Itô's Formula

$$d\overline{g}(\hat{x}(t)) = \overline{A}\,\overline{g}(\hat{x}(t))dt + \overline{g}_x'(\hat{x}(t))\overline{\sigma}(\hat{x}(t))dw + [\overline{g}(\hat{x}(t)) - \overline{g}(\hat{x}(t^-))].$$

Let $v(\cdot)$ and $f(\cdot)$ be bounded and measurable functions. There are functions $K_i(\cdot)$ which are bounded on bounded sets and such that

$$E \exp \int_0^t v'(\hat{x}(s))dw(s) \leq \exp K_2(v(\cdot))t$$

$$E \exp \sum_{s \leq t} f(s)[\overline{g}(\hat{x}(s)) - \overline{g}(\hat{x}(s^-))] \leq \exp[\lambda t \exp K_3(f(\cdot))]. \qquad (7.9)$$

Using (7.9) and analogous estimates for

$$E \left| \int_0^t (\psi(s) - \tilde{\psi}(s))' d\overline{g}(\hat{x}(s)) \right|^2,$$

and Schwarz's inequality in (7.8) yields a $K_1(s)$ such that (7.5) holds. The inequality (7.6) follows by a similar calculation and the details are omitted. Q.E.D.

8. A Robust Computational Approximation to the Averaged Filter

In this section, we will describe an approximation to the non-linear filter which has been found to work well in practice. The approximation can be used for the filter (1.3) applied to the system (1.1), (1.2), or for the averaged filter (2.7) for the singular perturbation problem (2.1), (2.2). We will do the latter application only. For simplicity in the development, the jump term will be dropped. The approximate filter is of the dynamical systems or recursive form and was devised initially for the optimal filtering problem for a diffusion problem in [K13, K18]. The same idea was extended to various jump process and other models in [D3]. Owing to the problem of "dimensional explosion," the method seems to be practical at present only if the dimension r of the state x is no greater than three. The procedure exploits the facts that a filter can be built or well approximated for models where the signal process is a finite state Markov chain, and where the observation noise is "white Gaussian", and that a diffusion process can be well approximated by an appropriately chosen Markov chain. The time parameter for the approximating Markov chain can be either discrete or continuous. If the time parameter is discrete, then the approximate recursive filter just is a simple realization of Bayes' rule. If the time parameter of the approximating Markov chain is continuous, then the approximating filter is a simple finite set of stochastic differential equations which are obtainable from (1.4), and which will be written below.

The output of the approximate recursive filter can be thought of as a measure valued process which is actually an approximation to the conditional distribution of the signal given the observations, in that it converges weakly to that latter quantity as the approximation parameter goes to zero. The discrete parameter approximation was dealt with in [K13] and the continuous parameter form of the approximation in [K18], where the robustness property was also shown. The choice among the cases depends solely on computational convenience. We will adapt the line of development in [K18] to get an approximation to the averaged filter for the singularly perturbed systems (2.1), (2.2) (with the jump term deleted). The continuous parameter form of the approximating Markov chain will be used, but the entire development can be carried over to the "singular perturbation" analog of the discrete parameter form used in [K13]. The "approximation" filter depends on two parameters; h, a "computational" approximation, and ϵ, the singular perturbation parameter. It is important that the filter be robust, uniformly in both parameters.

The optimal filter for a continuous parameter Markov chain. For use below, we now write the expression for the filter when the signal is a continuous parameter Markov chain. Let $\xi(\cdot)$ be a finite state Markov chain with infinitesimal transition probabilities $p_{xy}dt$ that do not depend on time. Let $y(\cdot)$ denote the observation process, where $dy = k(\xi)dt + dw_0$, and where $w_0(\cdot)$ is a standard vector-valued Wiener process which is independent of $\xi(\cdot)$. Let $\hat{\xi}(\cdot)$ have the same probability law that $\xi(\cdot)$ has, but be independent of $(\xi(\cdot), w_0(\cdot))$. Let (as in Section 1) E_t denote the expectation conditioned on $\{y(u), u \leq t\}$. The process $\xi(\cdot)$ is just a special case of the $x(\cdot)$ process used in Section 1. We use the representations of the filter which are given in that section: In particular, we have

$$E_t\phi(\xi(t)) = \frac{E_t\phi(\hat{\xi}(t))R_\xi(t)}{E_tR_\xi(t)}, \qquad (8.1)$$

where

$$R_\xi(t) = \exp\left[\int_0^t k(\hat{\xi}(s))'dy(s) - \frac{1}{2}\int_0^t |k(\hat{\xi}(s))|^2 ds\right].$$

Define $P_x(\cdot)$ by $P_x(t) = P(\xi(t) = x\,|\,y(u), u \leq t)$, and define $\phi(\cdot)$ by $\phi(\xi) = I_{\{\xi=x\}}$. Then we can write (1.4) as

$$dP_x(t) = \sum_{y \neq x} p_{yx} P_y(t)dt - \sum_{y \neq x} p_{xy} P_x(t)dt$$

$$+ P_x(t)\left[k(x) - \sum_y P_y(t)k(y)\right]'\left[dy(t) - \sum_y P_y(t)k(y)dt\right]. \qquad (8.2)$$

Approximating $x(\cdot)$ by a Markov chain. We next define a class of continuous parameter Markov chain approximations $\xi^h(\cdot)$ to the averaged process $x(\cdot)$, which has been found to be very useful for computations in stochastic control [K13, K24, K25]. We work with the case where (2.3) and $x^\epsilon(\cdot)$ have no jump component, but the extension to the jump diffusion case is straightforward. Let $h > 0$ denote a scalar approximation parameter and suppose, for simplicity, that the $x(t)$ take their values in a compact set $\overline{S}$. Let R_h be a finite set of points in $\overline{S}$ such that the distance of any point x in $\overline{S}$ to R_h is $O(h)$. R_h will be the state space for the chain. Often R_h is just a "grid" in $\overline{S}$ with the grid lines spaced h units apart. The "approximating chain" $\xi^h(\cdot)$ will have the "local" properties of the averaged process $x(\cdot)$. The constructed process $\xi^h(\cdot)$ will be "close" to $x(\cdot)$ the sense that $\xi^h(\cdot) \Rightarrow x(\cdot)$, as $h \to 0$. To construct the actual approximate filter we first construct a filter for the signal process $\xi^h(\cdot)$ and observation process defined by $dy^h = \overline{g}(\xi^h(t))dt + dw_0$. Then we use the actual physical observations $y^\epsilon(\cdot)$ in place of $y^h(\cdot)$, analogously to what was done in the construction of the averaged filter in Section 2.

Let $\Delta t^h(x)$ be a function which goes to zero as $h \to 0$, uniformly in x. Let $p_{xy}^h \geq 0$ be defined for $x, y \in R_h$ and satisfy $\sum_y p_{xy}^h = 1$. The infinitesimal transition probabilities of the chain $\xi^h(\cdot)$ are defined to be $p_{xy}^h dt / \Delta t^h(x)$. Hence, the mean sojourn time of $\xi^h(\cdot)$ at state x is $\Delta t^h(x)$. Let $\{\tau_n^h, n < \infty\}$ denote the jump times of $\xi^h(\cdot)$. Then, we require the following properties, which guarantee that $\xi^h(\cdot) \Rightarrow x(\cdot)$, the averaged process:

$$|\xi^h(\tau_{n+1}^h) - \xi^h(\tau_n^h)| = O(h),$$

$$E[\xi^h(\tau_{n+1}^h) - x \,|\, \xi^h(\tau_n^h) = x] = \overline{G}(x)\Delta t^h(x) + o(\Delta t^h(x)), \qquad (8.3)$$

$$\text{covar}[\xi^h(\tau_{n+1}^h) - x \,|\, \xi^h(\tau_n^h) = x] = \overline{a}(x)\Delta t^h(x) + o(\Delta t^h(x)).$$

A digression on the construction of $\xi^h(\cdot)$. There are several convenient and easily programmable ways to construct the probabilities p_{xy}^h and mean sojourn times $\Delta t^h(x)$ which satisfy our needs. The reader is referred to the references for further details. Some useful methods are described in [K13, K24], for the discrete parameter analog of $\xi^h(\cdot)$. In the reference [K13] a discrete parameter process satisfying (8.3) is used to get the filter. In [K18], it is shown how to use the transition probabilities for such discrete parameter processes to get the transition probabilities for the continuous parameter process $\xi^h(\cdot)$ used here. The procedure is roughly as follows. Let $\{\xi_n^h\}$ be a discrete parameter process with transition probability $p^h(x, y)$, $x, y \in R_h$, and such that for some "interpolation time" $\Delta t^h(x)$

$$|\xi_{n+1}^h - \xi_n^h| = O(h),$$

$$E[\xi_{n+1}^h - x \,|\, \xi_n^h = x] = \overline{G}(x)\Delta t^h(x) + o(\Delta t^h(x)),$$

$$\text{covar}[\xi_{n+1}^h - x \,|\, \xi_n^h = x] = \overline{a}(x)\Delta t^h(x) + o(\Delta t^h(x)),$$

where $\Delta t^h(x) \to 0$ as $h \to 0$, uniformly in x. The process $\xi^h(\cdot)$ with infinitesimal transition probabilities (from x to y) $p_{xy}^h dt = p^h(x, y)dt / \Delta t^h(x)$

will satisfy (8.3). Given that $\xi^h(t) = x$, $\Delta t^h(x)$ is the mean time required for $\xi^h(\cdot)$ to leave x. It can be shown that $\xi^h(\cdot) \Rightarrow x(\cdot)$ [K13, K25].

The filter for the process $\xi^h(\cdot)$. Let the continuous parameter Markov chain $\xi^h(\cdot)$ satisfying (8.3) be given. Suppose that the noise corrupted observations $\{y^h(s), s \leq t\}$ of the $\xi^h(\cdot)$ process are available at time t, where $y^h(0) = 0$ and $dy^h = \overline{g}(\xi^h)dt + dw_0$, where $w_0(\cdot)$ is a standard vector-valued Wiener process independent of $\xi^h(\cdot)$. Let $\hat{\xi}^h(\cdot)$ have the same probability law as $\xi^h(\cdot)$, but be independent of $(\xi^h(\cdot), w_0(\cdot), y^\epsilon(\cdot))$. Define

$$R^h(t) = \exp\left[\int_0^t \overline{g}(\hat{\xi}^h(s))dy^h(s) - \frac{1}{2}\int_0^t |\overline{g}(\hat{\xi}^h(s))|^2 ds\right],$$

and let E_t^h denote the expectation conditioned on $\{y^h(s), s \leq t\}$. We continue to use E_t^ϵ to denote the expectation conditioned on $\{y^\epsilon(s), s \leq t\}$. Then, by (8.1) we have

$$E_t^h \phi(\xi^h(t)) = \frac{E_t^h \phi(\hat{\xi}^h(t)) R^h(t)}{E_t^h R^h(t)}. \tag{8.4}$$

The conditional probabilities $P_x^h(t) \equiv P(\xi^h(t) = x \,|\, y^h(s), s \leq t)$ satisfy (8.2) where $(k(\cdot), y(\cdot), \xi(\cdot), p_{xy})$ are replaced by $(\overline{g}(\cdot), y^h(\cdot), \xi^h(\cdot), p_{xy}^h)$.

The construction of the approximate averaged filter. We are now prepared to define an approximation to the averaged filter. It is defined by the right side of (2.7) with $\hat{\xi}^h(\cdot)$ replacing $\hat{x}(\cdot)$, and we will write out the evolution equations in detail. Define the function

$$\overline{P}_x^{\epsilon,h}(t) = \frac{E_t^\epsilon I_{\{\hat{\xi}^h(t)=x\}} \overline{R}^{\epsilon,h}(t)}{E_t^\epsilon \overline{R}^{\epsilon,h}(t)}, \tag{8.5}$$

where

$$\overline{R}^{\epsilon,h}(t) = \exp\left[\int_0^t \overline{g}(\hat{\xi}^h(s))' dy^\epsilon(s) - \frac{1}{2}\int_0^t |\overline{g}(\hat{\xi}^h(s))|^2 ds\right]. \tag{8.6}$$

Then, analogously to (8.2) we have

$$d\overline{P}_x^{\epsilon,h}(t) = \left[\sum_{y \neq x} p_{yx}^h \overline{P}_y^{\epsilon,h}(t) - \sum_{y \neq x} P_{xy}^h \overline{P}_x^{\epsilon,h}(t)\right] dt$$

$$+ \overline{P}_x^{\epsilon,h}(t) \left[\overline{g}(x) - \sum_y \overline{P}_y^{\epsilon,h}(t)\overline{g}(y)\right]' \left[dy^\epsilon(t) - \sum_y \overline{P}_y^{\epsilon,h}(t)\overline{g}(y)dt\right]. \tag{8.7}$$

The $\overline{P}_x^{\epsilon,h}(\cdot)$ are *not* actually conditional probabilities, despite their formal similarilty to the solution to (8.2), since the observation process is

$y^\epsilon(\cdot)$ and *not* $y^h(\cdot)$. From the $\overline{P}_x^{\epsilon,h}(t)$ we can define an "Ersatz conditional expectation" $\overline{E}_t^{\epsilon,h}$ by:

$$\overline{E}_t^{\epsilon,h}\phi(x^\epsilon(t)) = \sum_x \overline{P}_x^{\epsilon,h}(t)\phi(x). \qquad (8.8)$$

We use the overbar $-$ on $\overline{E}_t^{\epsilon,h}$ as a reminder that we are approximating the $\overline{E}_t^\epsilon$ of the averaged filter. The filter (8.8) is constructed by first solving (8.7), and then doing the calculation (8.8). It has the advantage that it retains the structure of the averaged filter (2.7), since $\hat{\xi}^h(\cdot)$ is an approximation to $\hat{x}(\cdot)$ and is constructed from an optimal filter for $\xi^h(\cdot)$. We can write (note the formal similarity to (2.7))

$$\overline{E}_t^{\epsilon,h}\phi(x^\epsilon(t)) = \frac{E_t^\epsilon\phi(\hat{\xi}^h(t))\overline{R}^{\epsilon,h}(t)}{E_t^\epsilon\overline{R}^{\epsilon,h}(t)}. \qquad (8.9)$$

Let $\overline{A}^h$ denote the differential operator of $\xi^h(\cdot)$. Then we can rewrite (8.9) in the 'moment equation' form

$$d\overline{E}_t^{\epsilon,h}\phi(x^\epsilon(t)) = \overline{A}^h\phi(x^\epsilon(t))dt + [\overline{E}_t^{\epsilon,h}\phi(x^\epsilon(t))\overline{g}(x^\epsilon(t))$$
$$- \overline{E}_t^{\epsilon,h}\phi(x^\epsilon(t))\overline{E}_t^{\epsilon,h}\overline{g}(x^\epsilon(t))][dy^\epsilon - \overline{E}_t^{\epsilon,h}\overline{g}(x^\epsilon(t))dt]. \qquad (8.10)$$

We can use the discrete parameter approximation $\{\xi_n^h\}$ cited above in lieu of $\xi^h(\cdot)$ to construct the filter also. The differential equations (8.7) would then be replaced by a difference equation [K13] which is usually easier to calculate with. The convergence and approximation results are the same as obtained here.

Let $\phi_j(\cdot), j \leq n$, be bounded and continuous real valued functions. Define the functions,

$$F_j(t) = E_t\phi_j(x(t)), \quad F_j^\epsilon(t) = E_t^\epsilon\phi_j(x^\epsilon(t)),$$

$$F_j^{\epsilon,h}(t) = \overline{E}_t^{\epsilon,h}\phi_j(x^\epsilon(t)).$$

The following convergence theorem can be proved by the weak convergence arguments which show $\xi^h(\cdot) \Rightarrow x(\cdot)$, together with the method of Theorem 3.1 for averaging out the $z^\epsilon(\cdot)$ in the filter equations.

Theorem 8.1. *Assume the conditions of Theorem 3.1a (except that we omit the "jump" components of $x(\cdot), x^\epsilon(\cdot)$), and let $\xi^h(\cdot)$ satisfy (8.3). Let $P(0)$ denote the distribution of $x^\epsilon(0)$ and let the initial condition $\{\overline{P}_x^{\epsilon,h}(0), x \in R_h\}$ converge weakly to $P(0)$ as $\epsilon \to 0$ and $h \to 0$. Then*

$$(\overline{F}_j^{\epsilon,h}(\cdot), j \leq n, x^\epsilon(\cdot), y^\epsilon(\cdot)) \Rightarrow (F_j(\cdot), j \leq n, x(\cdot), y(\cdot)), \qquad (8.11)$$

where $x(\cdot)$ and $y(\cdot)$ are defined by (2.3), (2.4). *Also*

$$(\overline{F}_j^{\epsilon,h}(\cdot), j \leq n) - (F_j^{\epsilon}(\cdot), j \leq n) \Rightarrow \text{zero process}, \qquad (8.12)$$

as $\epsilon \to 0$, $h \to 0$.

The robustness of the computational approximation (8.9). We now show that the filter defined by (8.9) is "robust" uniformly in ϵ and h. The uniform robustness is essential if the approximation is to be of practical use. The procedure used in Section 7 will be followed. By an integration by parts, we can rewrite $\overline{R}^{\epsilon,h}(t)$ as an explicit function of $y^{\epsilon}(\cdot)$, and define the function $\overline{R}_0^{\epsilon,h}(\cdot)$ by

$$\overline{R}_0^{\epsilon,h}(t) = \exp\left[\overline{g}'(\hat{\xi}^h(t))y^{\epsilon}(t) - \int_0^t y^{\epsilon}(s)'d\overline{g}(\hat{\xi}^h(s)) - \frac{1}{2}\int_0^t |\overline{g}(\hat{\xi}^h(s))|^2 ds\right]$$

$$\equiv \overline{R}_0^{\epsilon,h}(t, y^{\epsilon}(\cdot)).$$

Then, for each $\phi(\cdot) \in C_0(R^r)$, we can write $\overline{E}_t^{\epsilon,h}\phi(x^{\epsilon}(t))$ and define the function $\overline{F}_t^h(\cdot)$ by

$$\overline{E}_t^{\epsilon,h}\phi(x^{\epsilon}(t)) = \frac{E_t^{\epsilon}\phi(\hat{\xi}^h(t))\overline{R}_0^{\epsilon,h}(t, y^{\epsilon}(\cdot))}{E_t^{\epsilon}\overline{R}_0^{\epsilon,h}(t, y^{\epsilon}(\cdot))} = \overline{F}_t^h(y^{\epsilon}(\cdot)).$$

Analogously to Theorem 7.1, we have the following result.

Theorem 8.2. *Assume the conditions of Theorem 7.1 (for the case where $x(\cdot)$, $x^{\epsilon}(\cdot)$ are diffusions) and 8.1. Then for each bounded set S in $C^{r''}[0,T]$, there is a constant $K(S) < \infty$ such that*

$$|\overline{F}_t^h(\psi(\cdot)) - \overline{F}_t^h(\tilde{\psi}(\cdot))| \leq K(S)|\psi(\cdot) - \tilde{\psi}(\cdot)|, \qquad (8.13)$$

for all functions $\psi(\cdot)$ and $\tilde{\psi}(\cdot)$ in S and all small $h > 0$ and $t \leq T$.

Proof. The proof is an adaptation of that in [K18], where a specific Markov chain which satisfies our requirements is used. We need to prove the analog of (7.5) and (7.6). Following the method of proof of Theorem 7.1, in order to get the analog of (7.5) we need to show that for each bounded set S in $C^{r''}[0,T]$, there is a constant $K_1(S) < \infty$ such that (again dropping the $\exp\int_0^t |\overline{g}(\hat{\xi}^h(s))|^2 ds/2$ term, w.l.o.g.)

$$E\left|(\psi(t) - \tilde{\psi}(t))'\overline{g}(\hat{\xi}^h(t)) - \int_0^t (\psi(s) - \tilde{\psi}(s))'d\overline{g}(\hat{\xi}^h(s))\right| \times$$

$$\left\{\exp\left[\psi'(t)\overline{g}(\hat{\xi}^h(t)) - \int_0^t \psi'(s)d\overline{g}(\hat{\xi}^h(s))\right]\right\}$$

$$+ \exp\left[\tilde{\psi}'(t)\overline{g}(\hat{\xi}^h(t)) - \int_0^t \tilde{\psi}'(s)d\overline{g}(\hat{\xi}^h(s))\right]\right\} \leq K_1(S)\,|\,\psi(\cdot) - \tilde{\psi}(\cdot)|.$$

Equivalently, we only need to show that there is a $K_2 < \infty$ and a K_q which is bounded for $q(\cdot)$ in any bounded set in $C^{r''}[0,T]$ and such that

$$E\left|\int_0^t q'(s)d\overline{g}(\hat{\xi}^h(s))\right|^2 \leq K_2|q(\cdot)|^2 \qquad (8.14)$$

$$E\exp\int_0^t q'(s)d\overline{g}(\hat{\xi}^h(s)) \leq K_q. \qquad (8.15)$$

To get these estimates, we decompose the process $\overline{g}(\hat{\xi}^h(\cdot))$ into a sum of a martingale and a "compensator". The compensator can be defined roughly as the integral of the "conditional mean rate of change of $\overline{g}(\hat{\xi}^h(\cdot))$." Write

$$\overline{g}(\hat{\xi}^h(t)) = M^h(t) + \Gamma^h(t),$$

where $M^h(\cdot)$ is a martingale and $\Gamma^h(\cdot)$ is the "compensator". By a direct calculation we have

$$\Gamma^h(t) = \int_0^t \gamma^h(\hat{\xi}^h(s))ds,$$

where

$$\gamma^h(x) = \sum_{y \in R_h} [\overline{g}(y) - \overline{g}(x)]p_{xy}^h/\Delta t^h(x).$$

Note that $\gamma^h(x)$ is the mean rate of change of $\overline{g}(\xi^h(\cdot))$ at t, when $\xi^h(t) = x$. By a Taylor expansion and use of (8.3), it can be shown that

$$\gamma^h(x) = \overline{A}g(x) + o(\Delta t^h(x))/\Delta t^h(x).$$

The quadratic variation of $M^h(\cdot)$ is

$$\langle M^h \rangle(t) = \int_0^t v^h(\hat{\xi}^h(s))ds,$$

where

$$v^h(x) = \sum_y [\overline{g}(y) - \overline{g}(x)][\overline{g}(y) - \overline{g}(x)]'p_{xy}^h/\Delta t^h(x).$$

By a Taylor expansion and use of (8.3), we have

$$v^h(x) = \overline{g}_x'(x)\overline{a}(x)\overline{g}_x'(x) + o(\Delta t^h(x))/\Delta t^h(x).$$

Let $q(\cdot) \in C^{r''}[0,T]$. Then using the properties of $\Gamma^h(\cdot)$ and $\langle M^h \rangle(\cdot)$, we have

$$E \left| \int_0^t q'(s) d\overline{g}(\hat{\xi}^h(s)) \right|^2 = E \left| \int_0^t q'(s) d\Gamma^h(s) \right|^2$$

$$+ E \left| \int_0^t q'(s) dM^h(s) \right|^2 \leq \text{constant} \int_0^t |q(s)|^2 ds,$$

which yields (8.14).

To get (8.15), we use an approximation procedure. We first prove it assuming that $\hat{\xi}^h(\cdot)$ is stopped after the nth jump. The K_q will not depend on n, and then we let $n \to \infty$. Let $\hat{\xi}^{h,n}(\cdot)$ denote the $\hat{\xi}^h(\cdot)$ process stopped after the nth jump. Then, we can write

$$\overline{g}(\hat{\xi}^{h,n}(t)) = \Gamma^{h,n}(t) + M^{h,n}(t). \tag{8.16}$$

where $\Gamma^{h,n}(\cdot)$ and $M^{h,n}(\cdot)$ are just $\Gamma^h(\cdot)$ and $M^h(\cdot)$, resp., stopped at the time of the nth jump of $\hat{\xi}^h(\cdot)$. Let $E_t^{h,n}$ denote the expectation given $\{\hat{\xi}^{h,n}(s), s \leq t\}$, the path up to the nth jump n or to time t, whichever comes first. Note that $d\overline{g}(\hat{\xi}^{h,n}(s)) = 0$ for s larger than the nth jump time of $\hat{\xi}^h(\cdot)$. For notational simplicity, we henceforth consider the scalar case only. Then for small $\Delta > 0$, we can write

$$E \exp \int_0^t q(s) d\overline{g}(\hat{\xi}^{h,n}(s))$$

$$= E \exp \int_0^{t-\Delta} q(s) d\overline{g}(\hat{\xi}^{h,n}(s)) E_{t-\Delta}^{h,n} \exp \int_{t-\Delta}^t q(s) d\overline{g}(\hat{\xi}^{h,n}(s)). \tag{8.17}$$

Using the representation (8.16), we have

$$E_{t-\Delta}^{h,n} \exp \int_{t-\Delta}^t q(s) d\overline{g}(\hat{\xi}^{h,n}(s))$$

$$\leq \exp(K_1 \Delta \cdot |q(\cdot)|) \cdot E_{t-\Delta}^{h,n} \exp \int_{t-\Delta}^t q(s) dM^{h,n}(s), \tag{8.18}$$

where K_1 does not depend on n. By a truncated Taylor expansion there are C and C_n, which are bounded on each bounded $q(\cdot)$-set such that for $h \leq 1$,

$$E_{t-\Delta}^{h,n} \exp \int_{t-\Delta}^t q(s) dM^{h,n}(s) \leq 1 + E_{t-\Delta}^{h,n} \left\{ \int_{t-\Delta}^t q(s) dM^{h,n}(s) \right.$$

$$+ C \left| \int_{t-\Delta}^t q(s) dM^{h,n}(s) \right|^2 I_{\{\leq 1 \text{ jump on } (t-\Delta, t]\}} + C_n I_{\{>1 \text{ jump on } (t-\Delta, t]\}} \right\}$$

$$\leq 1 + C E_{t-\Delta}^{h,n} \int_{t-\Delta}^t q^2(s) d\langle M^{h,n} \rangle(s) + o_n(\Delta). \tag{8.19}$$

The $o_n(\Delta)$ might depend on $q(\cdot)$ also, but it is $o(\Delta)$ for each n and $q(\cdot)$. Then using an upper bound on the derivative of the quadratic variation, "iterating backwards" via (8.17), using the bounds (8.18) and (8.19) and then letting $\Delta \to 0$ yields

$$E \exp \int_0^t q(s) d\overline{g}(\hat{\xi}^{h,n}(s)) \leq E \exp(\hat{K}_q)t,$$

for some $\hat{K}_q$ which does not depend on n or on $h \leq 1$. The $\hat{K}_q$ might depend on $q(\cdot)$ but it is bounded on each bounded $q(\cdot)$-set. Thus (8.15) is proved. The analog of (7.6) follows by a similar calculation and we omit the details.   Q.E.D.

9. The Averaged Filter on the Infinite Time Interval

In this section, we treat the filtering problem when it is of interest on a time interval of arbitrary length, and we are interested in the average filtering errors per unit time. The results of Sections 3 and 5 will be extended and it will be shown that the "averaged" filter (2.7), (2.7′) is often a good approximation to the optimal filter, even when the filtering errors are of interest for a very long time. The probability measures $P(t)$, $P^\epsilon(t)$ and $\overline{P}^\epsilon(t)$ will all take values in $\mathcal{P}(R^r)$, where the Prohorov topology is used. In the ensuing development, it will be supposed that the outputs of the filters of interest do not "blow up" as time goes to infinity. If the $\{x^\epsilon(t), \epsilon > 0, t < \infty\}$ is tight, then so is $\{P^\epsilon(t), \epsilon > 0, t < \infty\}$. But it is not always true that the outputs of the averaged filter $\{\overline{P}^\epsilon(t), \epsilon > 0, t < \infty\}$ are tight, unless the state space is compact. We will simply assume that "tightness." One would not usually use an averaged filter on an arbitrary long time interval unless its outputs did not "blow up," but the point needs further work.

Notation and consistency of initial conditions. The measure valued processes $Z^\epsilon(\cdot)$ and $\overline{Z}^\epsilon(\cdot)$ of Section 3 cannot be used when working on the infinite time interval, since they are unnormalized and do not have limits in any useful sense as $t \to \infty$. We must work with the normalized measures. In preparation for the assumptions, we first make some remarks concerning the initial conditions of the process $P(\cdot)$.

In defining the $P(\cdot)$, $P^\epsilon(\cdot)$ and $\overline{P}^\epsilon(\cdot)$ of (2.6′), (2.5′) and (2.7′), resp., we assumed that the $\hat{x}(\cdot)$ and $\hat{x}^\epsilon(\cdot)$ processes had the same probability law as $x(\cdot)$ and $x^\epsilon(\cdot)$, resp. The $x(\cdot)$ and $\hat{x}(\cdot)$ always refer to the averaged system (2.3), and the $y(\cdot)$ process is always defined by (2.4). Henceforth, for simplicity, let $x(0) = x^\epsilon(0)$. Let the SDE's (1.1) and (2.1) have unique weak sense solutions for each initial condition. Suppose that $\hat{x}(\cdot)$ and $\hat{x}^\epsilon(\cdot)$ satisfy the SDE's of the types (2.3) and (2.1), resp., and are independent of the other processes, as required in Section 3, but that the distributions of $\hat{x}(0)$ and $x(0)$ are not the same. Then the $P(\cdot)$ and $P^\epsilon(\cdot)$ which are defined in Section 2 are no longer the asserted "conditional probability" processes, but they (and $\overline{P}^\epsilon(\cdot)$) are still well defined processes with paths in $D[\mathcal{P}(R^r); 0, \infty)$. The moment equations remain the same ((1.4) for $P(\cdot)$ and (2.8) for $\overline{P}^\epsilon(\cdot)$), but the bracketed term on the right of (1.4) is no longer an innovations process.

Let $P(0, B)$ denote the value of $P(0)$ on the set $B \in \mathcal{B}(R^r)$. Let the initial condition $P(0)$ be deterministic (i.e., it is a measure and not a measure-valued random variable) and let $P(0, B)$ equal the probability that $x(0)$ is in B for all $B \in \mathcal{B}(R^r)$. Then $P(0)$ and $x(0)$ are said to be *consistent*. This is the case if $x(0)$ and $\hat{x}(0)$ have the same distribution. The initial condition $P(0)$ might be a random variable. This would be the case if it were obtained from estimates made on "prior data." Suppose that $P(0)$ is random and let the probability that $x(0) \in B$ conditioned on $P(0)$ be $P(0, B)$ w.p.1. Then the initial conditions are also said to be *consistent*. When working on arbitrarily large time intervals, processes with inconsistent initial conditions arise since we need to take limits of the sequences $(x^\epsilon(t), P^\epsilon(t), \overline{P}^\epsilon(t))$, and whatever the properties of these sequences over a finite time interval, consistency can be lost in the limit. For example, let $\overline{P}^\epsilon(0) = P^\epsilon(0)$ and let $(x(0), P^\epsilon(0))$ be consistent. We know that, under appropriate conditions, $(x^\epsilon(\cdot), \overline{P}^\epsilon(\cdot)) \Rightarrow (x(\cdot), P(\cdot))$, where again $x(0)$ and $P(0)$ are consistent. Now, let $(x^\epsilon(t + \cdot), \overline{P}^\epsilon(t + \cdot)) \Rightarrow (x(\cdot), P(\cdot))$ as $\epsilon \to 0$ and $t \to \infty$. It is no longer guaranteed that $(x(0), P(0))$ are consistent. Hence, even though the moments of this limit measure valued process $P(\cdot)$ might obey (1.4), $P(\cdot)$ might not be a conditional probability for the limit process $x(\cdot)$. We will need some assumption which guarantees that this loss of consistency cannot occur. This will be an assumption on the uniqueness of the invariant measure of the pair of processes $(x(\cdot), P(\cdot))$. If we define $P(\cdot)$ via (1.4) but let $P(0)$ be arbitrary (i.e., not necessarily the distribution of $x(0)$) then the pair $(x(\cdot), P(\cdot))$ is a Markov–Feller process with continuous paths. The paths are elements of $D^r[0, \infty) \times D[\mathcal{P}(R^r); 0, \infty)$. Let $\Pi(x, v, t, \cdot)$ denote the transition function of the pair, where the variable v is the canonical variable of $\mathcal{P}(R^r)$.

The following notation will be used in Theorem 9.1. Assume the conditions of Theorem 3.1. If we allow arbitrary initial conditions, then the two sets of processes

$$x^\epsilon(\cdot), z^\epsilon(\cdot), \overline{P}^\epsilon(\cdot), P^\epsilon(\cdot)$$

and

$$x^\epsilon(\cdot), z^\epsilon(\cdot), \overline{P}^\epsilon(\cdot)$$

are also Markov–Feller processes with paths in $D^{r+r'}[0, \infty) \times D^2[\mathcal{P}(R^r); 0, \infty)$ and $D^{r+r'}[0, \infty) \times D[\mathcal{P}(R^r); 0, \infty)$, resp. Let $\hat{\Pi}^\epsilon(x, z, \overline{v}, v, t, \cdot)$ and $\overline{\overline{\Pi}}^\epsilon(x, z, \overline{v}, t, \cdot)$, resp., denote the transition functions for the two sets of processes.

We will need the following conditions.

A9.1. *There is a unique invariant measure* $\Pi_0(\cdot)$ *for the transition function* $\Pi(x, v, t, \cdot)$.

A9.2. *Let* $P(\cdot)$ *and* $P'(\cdot)$ *be processes which are obtained from* (2.6) *but with arbitrary initial conditions* $P(0)$, $P'(0)$ *(which need not be the same or consistent with* $x(0)$*). The triple* $(x(\cdot), P(\cdot)), P'(\cdot))$ *has a unique invariant measure* $\Pi_0'(\cdot)$ *which is concentrated on the set where* $P = P'$.

Assumption (A9.2) essentially implies that the initial condition of the filter is unimportant if the process is "run" for a long time. In a sense, the assumption allows us to make errors in the assignment of the initial distribution, and the effects of the errors decrease as $t \to \infty$. Since we usually do not know the initial distribution of the signal process, one usually makes the assumption implicitly. Unfortunately, there is little work on the issue. In [K7], the asymptotics of the optimal filter $P(\cdot)$ for $x(\cdot)$ is studied. Then $P(\cdot)$ is itself a Markov–Feller process, and the driving innovations process is "white noise." Under a "strictly non-deterministic process" condition on $x(\cdot)$, it is shown that this $P(\cdot)$ has a unique invariant measure. But the methods cannot be carried over to our case. A "wide band noise" version of [K7] is in [K21].

We will also need

A9.3. (a) $\sup_{t, \epsilon} \overline{P}^\epsilon(t, \{|x| \geq N\}) \to 0$ *as* $N \to \infty$.
(b) *The set* $\{x^\epsilon(t), \epsilon > 0, t < \infty\}$ *is tight.*

Theorem 9.1. *Assume* (A2.1), (A3.1), (A3.3) *(with* $T = \infty$*),* (A3.4), (A9.1) *and* (A9.3), *and let the fixed* x *process* $z_0(\cdot | x)$ *have a unique invariant measure for each* x. *Let* $\phi(\cdot)$ *be bounded, real valued and continuous. Then*

$$\lim_{T, \epsilon} \frac{1}{T} \int_0^T E[\phi(x^\epsilon(t)) - (\overline{P}^\epsilon(t), \phi)]^2 dt = \int [\phi(x) - (v, \phi)]^2 \Pi_0(dx dv). \quad (9.1)$$

If (A9.3) *is satisfied for the* $P(t)$ *and* $x(t)$ *replacing the* $\overline{P}^\epsilon(t)$ *and* $x^\epsilon(t)$, *then*

$$\lim_T \frac{1}{T} \int_0^T E[\phi(x(t)) - (P(t), \phi)]^2 dt = \int [\phi(x) - (v, \phi)]^2 \Pi_0(dx dv), \quad (9.2)$$

where in (9.2) $P(\cdot)$ *is the conditional probability process for the averaged process* $x(\cdot)$ *of* (2.3) *with observations* $y(\cdot)$ *of* (2.4). *If, in addition,* (A3.2) *and* (A9.2) *also hold, then*

$$\lim_{T, \epsilon} \frac{1}{T} \int_0^T E[(P^\epsilon(t), \phi) - (\overline{P}^\epsilon(t), \phi)]^2 dt = 0. \quad (9.3)$$

Remark. If for the stationary process $(x(\cdot), P(\cdot))$, $P(\cdot)$ is the conditional distribution process for $x(\cdot)$, then $\Pi_0(x \in B \mid v) = v(B)$ and (9.1) and

(9.2) can be simplified. Via the functional occupation measure method of Chapter 5, the limits in (9.1) and (9.2) (with the E deleted) can be shown to hold in probability.

Proof. By (A9.3) and (A3.3), (with $T = \infty$ there), in the argument below we can suppose that $\phi(\cdot) \in C_b(\overline{R}^r)$. Let $\overline{\Pi}^{\epsilon,t}(\cdot)$ and $\overline{\Pi}_m^{\epsilon,t}(\cdot)$ denote the distribution of $(x^\epsilon(t), z^\epsilon(t), \overline{P}^\epsilon(t))$ and of $(x^\epsilon(t), \overline{P}^\epsilon(t))$, resp. Define the measure

$$\overline{\Pi}_T^\epsilon(\cdot) = \frac{1}{T} \int_0^T \overline{\Pi}^{\epsilon,t}(\cdot)dt,$$

and define $\overline{\Pi}_{m,T}^\epsilon(\cdot)$ analogously.

By the definition of $\overline{\Pi}_{m,T}^\epsilon(\cdot)$ we can write

$$\frac{1}{T} \int_0^T E[\phi(x^\epsilon(t)) - (\overline{P}^\epsilon(t), \phi)]^2 dt = \int [\phi(x) - (v, \phi)]^2 \overline{\Pi}_{m,T}^\epsilon(dx\,dv).$$

By (A9.3) and (A3.3) (using $T = \infty$ there), $\{\overline{\Pi}_T^\epsilon(\cdot), \overline{\Pi}_{m,T}^\epsilon(\cdot), \epsilon > 0, T < \infty\}$ is tight in $\mathcal{P}^2(R^r)$. Thus, to prove (9.1), we need only characterize appropriately the limits of $\{\overline{\Pi}_{m,T}^\epsilon(\cdot), \epsilon > 0, T < \infty\}$.

Let $f(\cdot)$ be a real valued, bounded and continuous function on $R^r \times \mathcal{P}(R^r)$. Then the function $f_\delta^\epsilon(\cdot)$ defined by

$$\int \overline{\Pi}^\epsilon(x, z, v, \delta, dx'dz'dv')f(x', v') = f_\delta^\epsilon(x, z, v) \tag{9.4}$$

is continuous in (x, z, v). Let $\overline{P}^\epsilon(\cdot)$ be defined by (2.7') but with arbitrary (i.e., not necessarily consistent) initial condition $\overline{P}^\epsilon(0) = v$. I.e., v is the distribution of the $\hat{x}(0)$ which is used in the calculation of $\overline{P}^\epsilon(\cdot)$, and it need not be the distribution of $x(0)$. Then, by the arguments of Theorem 3.1, $(x^\epsilon(\cdot), \overline{P}^\epsilon(\cdot)) \Rightarrow (x(\cdot), P(\cdot))$, where $P(\cdot)$ satisfies (2.6') with initial condition $P(0) = v$. This weak convergence is uniform in any compact (x, z, v)-set, and the limit does not depend on z. Thus, as $\epsilon \to 0$, $f_\delta^\epsilon(\cdot)$ converges to the continuous function $\overline{f}_\delta(\cdot)$ which is defined by

$$\overline{f}_\delta(x, v) = \int \Pi(x, v, \delta, dx'dv')f(x', v'),$$

and the convergence is uniform in (x, z, v) in any compact set.

By this convergence result together with (A9.3) and (A3.3), (with $T = \infty$ there), we can write

$$\lim_{\epsilon,T} \int f(x,v) \overline{\Pi}^\epsilon_{m,T}(dxdv) =$$

$$\lim_{\epsilon,T} \frac{1}{T} \int_\delta^T dt \int \overline{\Pi}^{\epsilon,t-\delta}(dxdzdv)\overline{\Pi}^\epsilon(x,z,v,\delta,dx'dz'dv')f(x',v')$$

$$= \lim_{\epsilon,T} \int \overline{\Pi}^\epsilon_{m,T}(dxdv)\overline{f}_\delta(x,v). \tag{9.5}$$

Let $\tilde{\Pi}(\cdot)$ denote a weak limit of $\{\overline{\Pi}^\epsilon_{m,T}(\cdot), \epsilon > 0, T < \infty\}$, with $\epsilon \to 0$, $T \to \infty$. Then (9.5) implies that

$$\int f(x,v)\tilde{\Pi}(dxdv) = \int \tilde{\Pi}(dxdv)\overline{f}_\delta(x,v)$$

$$= \int \tilde{\Pi}(dxdv)\left[\int \Pi(x,v,\delta,dx'dv')f(x',v')\right], \tag{9.6}$$

which implies that $\tilde{\Pi}(\cdot)$ is an invariant measure for $(x(\cdot), P(\cdot))$. By the uniqueness of the invariant measure, $\Pi_0(\cdot) = \tilde{\Pi}(\cdot)$. Hence the chosen subsequence is irrelevant, and (9.1) follows. Eqn. (9.2) is proved in the same way. We just drop the ϵ in the above argument. The proof of the result (9.3) follows similar lines, but also uses the 'averaging' results of Theorem 3.1 concerning the $E_t^\epsilon \phi(x^\epsilon(t))$ as well as the fact that (A9.3b) implies that $\{P^\epsilon(t), \epsilon > 0, t < \infty\}$ satisfies (A9.3a). The details are omitted.   Q.E.D.

7

Weak Convergence: The Perturbed Test Function Method

0. Outline of the Chapter

The methods which were used in Chapters 3–6 to characterize the limits of $\{x^\epsilon(\cdot), m^\epsilon(\cdot)\}$ as solutions to the averaged system are very well suited to the types of "white noise driven" Itô equations which were used there. For the singular perturbation problem of Chapter 4, with model (4.1.1), (4.1.2) with control $m^\epsilon(\cdot)$, the martingale method which was used to characterize the limit process can be roughly described as follows (see Theorem 4.1.2): We apply the differential operator A^ϵ of the $(x^\epsilon(\cdot), z^\epsilon(\cdot))$ process to a nice test function $f(\cdot)$. Then for continuous and bounded functions $h(\cdot)$ and $\phi_j(\cdot)$, and for positive times $t_i \leq t \leq t + s$, Itô's formula was used to show that for each ϵ

$$Eh(x^\epsilon(t_i), (\phi_j, m^\epsilon)_{t_i}, i \leq q, j \leq p)[f(x^\epsilon(t+s)) - f(x^\epsilon(t))$$

$$- \int_t^{t+s} (A^\epsilon f)(x^\epsilon(u), z^\epsilon(u), u)du] = 0. \tag{0.1}$$

Tightness of $\{x^\epsilon(\cdot), \epsilon > 0\}$ was a simple consequence of the assumed growth conditions on the coefficients $G(\cdot)$ and $\sigma(\cdot)$ in (4.1.1). We used this tightness together with the assumed tightness of $\{z^\epsilon(t), t \leq T, \epsilon > 0\}$ and an averaging method to show that, as $\epsilon \to 0$ the $z^\epsilon(\cdot)$ could be averaged out in (0.1) and that the limits of $\int_t^{t+s} (A^\epsilon f)(x^\epsilon(u), z^\epsilon(u), u)du$ were also those of $\int_t^{t+s} \int_U \overline{A}^\alpha f(x^\epsilon(u))m_u^\epsilon(d\alpha)du$. Then a weakly convergent subsequence of $\{x^\epsilon(\cdot), m^\epsilon(\cdot)\}$ was extracted and limits taken in (0.1). Finally, it was shown that the limit pair $(x(\cdot), m(\cdot))$ solves the martingale problem for the averaged system (4.1.7).

The situation is not so simple if the "white noise" driving (4.1.1) or (4.1.2) is replaced by a "wide-band" noise $\xi(\cdot)$. In this book, the descriptor "wide-band" is used loosely to describe a "time scale" separation. It *does not imply* the existence of a bandwidth in the usual sense. E.g., for the singularly perturbed model of Chapter 4, $z^\epsilon(\cdot)$ can be said to be of wide-band with respect to $x^\epsilon(\cdot)$ even if $z^\epsilon(\cdot)$ is nonstationary. Singularly perturbed systems driven by wide-band noise will be discussed in the next chapter. In this chapter we will develop a powerful method for proving the tightness of a sequence of wide-band noise driven processes, and for characterizing the limits of the weakly convergent subsequences as particular diffusions or jump-diffusions. A *simple form* of the model that will be considered is the process $x^\gamma(\cdot)$ defined by

$$\dot{x}^\gamma = G(x^\gamma) + F(x^\gamma)\xi^\gamma/\gamma, \qquad (0.2)$$

where $\int_0^t \xi^\gamma(s)ds/\gamma$ converges weakly to a Wiener process as $\gamma \to 0$.

In general, in applications one does not have white noise driven systems. Nevertheless, the white noise driven systems are very useful and important models since so much can be done with them from both the analytical and computational point of view, and since they can be used to approximate an enormous variety of processes which occur in nature. But generally they must be seen as approximations only. We need to understand the sense in which they approximate the physical process of interest: E.g., in the example (0.2), which particular Itô process $x(\cdot)$ best represents the physical process $x^\gamma(\cdot)$ when the "bandwidth" of $\xi^\gamma(\cdot)$ is large, and what is the sense of the approximation (which functionals of $x^\gamma(\cdot)$ can be well approximated by the same functionals of $x(\cdot)$)? Such questions are particularly important when dealing with control problems, since there we are not dealing with just one process but rather with a family of processes indexed by the control, and it is useful to know that the approximating Itô process is a controlled process whose behavior is similar to that of the wide-band noise driven process under similar controls. I.e., the approximation must be uniformly good over a large class of controls. An analogous situation occured in the singular perturbation problem, where it was essential that the averaged system (4.1.7) was a good approximation to (4.1.1), (4.1.2), in the sense that the minimum value functions were close, and that controls which made sense (or were nearly optimal) for the "limit" also made sense (or were nearly optimal) for the physical process (4.1.1), (4.1.2). It was also important that the "rate of convergence" of any weakly convergent subsequence of $\{x^\epsilon(\cdot), m^\epsilon(\cdot)\}$ to the limit $(x(\cdot), m(\cdot))$ not depend on the chosen controls $\{m^\epsilon(\cdot)\}$. The "uniformity" of the rate was just a consequence of the tightness of $\{x^\epsilon(\cdot), \epsilon > 0,$ all controls of interest$\}$. We will be concerned with similar problems for the wide-band noise case.

In applications, the precise nature of the wide-band noise $\xi^\gamma(\cdot)$ (used, for example in (0.2)) is not usually known. Hence, it is important that the approximations of processes such as (0.2) (whether or not there is added a control term or singular perturbation) by an appropriate Itô process be robust, in the sense that the limit process depends only on a few general properties of the noise processes, and that the limit is attained uniformly over an appropriately large class of the noise processes.

In this chapter we will describe a powerful and readily usable method for proving the desired tightness and weak convergence, when the driving noise is wide-band. We are not concerned with the singular perturbation problem now, but with the general problem of approximation of "wideband" noise driven systems by Itô processes. In the next chapter, the method will be applied to the singularly perturbed system. The method to be introduced is an extension and refinement of the so-called "direct averaging-perturbed test function" method [K20, Chapter 5]. The method uses relatively simple test function perturbations together with laws of large numbers to accomplish the desired averaging. For the problems of interest to us in this book, as well as for typical problems in control and communication theory with either wide band noise driven systems or discrete systems driven by correlated noise, this method seems to be the method of choice now and the possibly most powerful of those currently available. In Chapter 8, the method will be used to extend the results of Chapter 4. Reference [K20] contains an extensive development of the ideas of the perturbed test function method as well as many examples of their application to problems of interest in control and communication theory. In Section 1, we give an example in order to motivate some of the later constructions. In Sectins 2 and 3 we will review the main ideas of the perturbed test function method. In Section 2, we state a general theorem for the perturbed test function method and exhibit the particular perturbation which is needed to get a convergence or approximation result for the example of Section 1. In Section 3, the general criterion is made more exploit and a useful criterion for tightness of a sequence of "wide band noise driven systems" is given. In Section 4, we develop the direct averaging-perturbed test function method, and show how to characterize the limit process.

In order not to overburden the text and to provide a reasonably explicit development, the processes of concern in Section 4 are restricted to a certain (broad) class of wide-band noise driven differential equations. The basic method is applicable to all the examples in [K20] as well, and to many other classes of processes as well.

1. An Example

Let $\xi(\cdot)$ be a scalar valued Markov jump process with values ± 1 and jump rate $\lambda > 0$ from either state. For small $\gamma > 0$, define the process $\xi^\gamma(\cdot)$ by $\xi^\gamma(t) = \xi(t/\gamma^2)$. Then, it can be shown that the sequence of processes defined by $\int_0^t \xi^\gamma(s)ds/\gamma$ converges weakly to a Wiener process with variance $2\int_0^\infty E\xi(s)\xi(0)ds = \int_{-\infty}^\infty E\xi(s)\xi(0)ds = 1/\lambda$, where the expectation is with respect to the distribution of the particular $\xi(\cdot)$ process where $\xi(0)$ has the stationary distribution. This result will be proved later in the Chapter. Note that to get $\xi^\gamma(\cdot)$ from $\xi(\cdot)$, time is compressed via the transformation $t \rightarrow t/\lambda^2$, and the amplitude is rescaled via the $1/\gamma$ coefficient. This rescaling is a widely used method of modelling wide-band noise. If $\xi(\cdot)$ has the spectral density $S(\omega)$, then $\xi(\cdot/\gamma^2)$ has a spectral density $S(\gamma^2\omega)\gamma^2$ and $\xi(\cdot/\gamma^2)/\gamma$

has a spectral density $S(\gamma^2\omega)$. If $S(\cdot)$ is continuous and nonzero at $\omega = 0$, then as $\gamma \to 0$ the spectral density of the normalized and rescaled process $\xi(\cdot/\gamma^2)/\gamma$ converges in an obvious sense to the constant spectral density with value $S(0)$, which is that of white noise with "intensity" $S(0)$.

Define the process $x^\gamma(\cdot)$ by $\dot{x}^\gamma(\cdot) = \sigma(x^\gamma)\xi^\gamma/\gamma$, where we suppose that $\sigma(\cdot) \in C_b^1(R)$. The pair $(x^\gamma(\cdot), \xi^\gamma(\cdot))$ is a Markov process with differential operator A^γ defined by

$$A^\gamma f(x, \xi^\gamma) = f_x(x, \xi^\gamma)\sigma(x)\xi^\gamma/\gamma + \lambda[f(x, -\xi^\gamma) - f(x, \xi^\gamma)]. \tag{1.1}$$

Due to the presence of the $1/\gamma$ term (hence the "asymptotic unboundedness" of the "driving noise"), the characterization of the weak limit of $\{x^\gamma(\cdot)\}$ (or even the proof of tightness) is more complicated than for the analogous problem for the singular perturbation case of Chapter 4.

Since we are seeking a limit Markov process $x(\cdot)$, we would like to use test functions $f(\cdot)$ which depend on x only. Then $A^\gamma f(x) = f_x'(x)\xi^\gamma/\gamma$, a possibly troublesome expression. Let $f(\cdot)$ be a function of x only. It turns out that there is a "perturbation" $f^\gamma(\cdot)$ of the test function $f(\cdot)$ such that (in a sense to be defined in Section 3)

$$
\begin{aligned}
f^\gamma(\cdot) - f(x^\gamma(\cdot)) &\to 0, \\
A^\gamma f^\gamma(\cdot) - Af(x^\gamma(\cdot)) &\to 0,
\end{aligned}
\tag{1.2}
$$

where the differential operator A is defined by

$$Af(x) = [f_x(x)\sigma(x)]_x \sigma(x)/2\lambda. \tag{1.3}$$

The operator A in (1.3) is the differential operator of the Itô process $x(\cdot)$ defined by

$$dx = \sigma(x)\sigma_x(x)dt/2\lambda + \sigma(x)dw/\sqrt{\lambda}. \tag{1.4}$$

In fact, as shown in Section 3, $x^\gamma(\cdot) \Rightarrow x(\cdot)$. See also [K20, Chapter 4]. The idea of the perturbed test function method is to find small perturbations $f^\gamma(\cdot)$ of the test function $f(\cdot)$ such that (1.2) holds. With such perturbed test functions $f^\gamma(\cdot)$ available, simple adaptations of the martingale method of Theorem 3.5.1 can be used. Since the driving noise process $\xi^\gamma(\cdot)$ will not be Markov in general, we need a suitable replacement for the differential generator A^γ. This will be an operator of the type of the $\hat{A}$ introduced in the next section. The idea of the perturbed test function seems to have originated in the works of Rishel [R1], Kurtz [K10], and Papanicolaou, Stroock and Varadhan [P1]. The method was developed extensively in [B7, K19, K20], in order to deal with weak convergence problems of the sort that arise in applications to control and communication systems. See [K20, Chapter 3] for more motivation and discussion. In the next section, we give some general theorems concerning the perturbed test function method, and exhibit the perturbation which is needed to deal with the above example.

2. The Perturbed Test Function Method: Introduction

The operator $\hat{A}$. Since the wide band noise driven systems of interest are not Markovian, we define an operator $\hat{A}$ which is a generalization of the differential operator of a Markov process, and which can be used for the necessary extensions of the martingale problem arguments we used in the previous chapter. Let $(\Omega, P, \mathcal{F})$ denote the probability space, let $\mathcal{F}_t$ be a filtration on the space , and let $f(\cdot)$ and $g(\cdot)$ denote measurable processes which are $\mathcal{F}_t$-adapted and progressively measurable and satisfy the following condition: For each $T_1 < \infty$

$$\sup_{t \leq T_1} E|f(t)| < \infty, \quad \sup_{t \leq T_1} E|g(t)| < \infty,$$

$$\sup_{0 < \delta, t \leq T_1} \frac{E|E[f(t + \delta)|\mathcal{F}_t] - f(t)|}{\delta} < \infty, \tag{2.1}$$

$$E\left|\frac{E[f(t + \delta)|\mathcal{F}_t] - f(t)}{\delta} - g(t)\right| \to 0, \text{ almost all } t.$$

Then we say that $f(\cdot) \in \mathcal{D}(\hat{A})$, the *domain of the operator* $\hat{A}$, and write $\hat{A}f = g$. If the filtration $\mathcal{F}_t$ (or the process which induces it) is indexed by a parameter (say, γ), then we write $\hat{A}^\gamma$ for the operator. Note that, if $\mathcal{F}_t$ is induced by a Markov process $x(\cdot)$ with paths in $D^k[0, \infty)$ for some k, then the operator $\hat{A}$ is just the weak infinitesimal operator of $x(\cdot)$, or the differential operator if $x(\cdot)$ is a jump diffusion and the operator acts on smooth functions. For our purposes, the most important property of $\hat{A}$ is given by Theorem 2.1.

Theorem 2.1. *Let $f(\cdot) \in \mathcal{D}(\hat{A})$. Then the process defined by*

$$M_f(t) = f(t) - \int_0^t \hat{A}f(s)ds$$

is an $\mathcal{F}_t$-martingale and, w.p.1,

$$E[f(t + s)|\mathcal{F}_t] = f(t) + \int_t^{t+s} E[\hat{A}f(u)|\mathcal{F}_t]du, \tag{2.2}$$

Equation (2.2) continues to hold if the t and $t + s$ are replaced by any bounded stopping times τ_1 and τ_2 with $\tau_1 \leq \tau_2$, and which take countably many values. If $f(\cdot)$ is right continuous, then the τ_i can be any bounded stopping times with $\tau_1 \leq \tau_2$.

Remarks. The proof is in [K20, Theorem 3.1], where the third line of (2.1) is assumed to hold for all $t \leq T_1$. But that proof carries over to our case. We say that the filtration $\mathcal{F}_t$ is *right continuous* if $\cap_{\delta > 0} \mathcal{F}_{t+\delta} \equiv \mathcal{F}_{t+} = \mathcal{F}_t$. If $\mathcal{F}_t$ is right continuous, then there is a right continuous version of any $\mathcal{F}_t$-martingale. Thus, if $\mathcal{F}_t$ is right continuous, then there is a version of the $M_f(\cdot)$ which is right continuous. This, in turn, implies that there is a version of $f(\cdot)$ which is right continuous. In the applications in the sequel, we can often use $\mathcal{F}_{t+}$ in lieu of $\mathcal{F}_t$ even if we do not know a priori that $\mathcal{F}_t$ is right continuous. Suppose that there are $\mathcal{F}_t$-adapted and progressively measurable functions $f(\cdot)$ and $g(\cdot)$ which satisfy: $\sup_{t \leq T_1} E|g(t)| < \infty$, $\sup_{t \leq T_1} E|f(t)| < \infty$ for each $T_1 < \infty$, and for each t the set

$$\left\{ \frac{1}{\delta} \int_t^{t+\delta} g(s)ds, \ \delta > 0 \right\}$$

is uniformly integrable, and where the process defined by $M(t) = f(t) - \int_0^t g(s)ds$ is an $\mathcal{F}_t$-martingale. Then $g = \hat{A}f$.

An example. Let us return to the example of Section 1. As pointed out in Section 1, the martingale method which was used to characterize the weak limits of $\{x^n(\cdot)\}$ in Theorem 3.5.1 or of the $\{x^\epsilon(\cdot)\}$ in Theorem 4.1.2 cannot be used here without some modification. We now use the example to illustrate how perturbed test functions provide an appropriate and easily usable modification.

Suppose that $\{x^\gamma(\cdot)\}$ is tight (we will show how to prove this later) so that our only problem is the characterization of the limits of weakly convergent subsequences. Define the filtration $\mathcal{F}_t^\gamma = \mathcal{B}(\xi^\gamma(s), x^\gamma(s), s \leq t)$, let E_t^γ denote the expectation conditioned on $\mathcal{F}_t^\gamma$, and let $\hat{A}^\gamma$ be defined as $\hat{A}$ was above but with $\mathcal{F}_t^\gamma$ replacing $\mathcal{F}_t$. Let E_t denote the expectation conditioned on $\{\xi(s), s \leq t\}$. Let $f(\cdot) \in C_0^2(R)$ be a given test function. Then $f(x^\gamma(\cdot)) \in \mathcal{D}(\hat{A}^\gamma)$ and

$$\hat{A}^\gamma f(x^\gamma(t)) = f_x(x^\gamma(t))\sigma(x^\gamma(t))\xi(t/\gamma^2)/\gamma. \tag{2.3}$$

I.e., acting on such $f(x^\gamma(\cdot))$, $\hat{A}^\gamma$ is just a "right" differentiation operator. The "right derivative" characterization follows from the definition of $\hat{A}^\gamma$ and the right continuity of $\xi(\cdot)$ and $x^\gamma(\cdot)$. For each large T (or for $T = \infty$, if the corresponding integral is well defined), define the function

$$f_1^\gamma(x,t) = \int_t^T f_x(x)\sigma(x)E_t^\gamma \xi(s/\gamma^2)ds/\gamma. \tag{2.4}$$

Note that the integrand in (2.4) is just a conditional expectation of (2.3) with x treated *as a parameter*. By a change of the time scale $s \rightarrow s\gamma^2$ (which will be used very often in the sequel), we "stretch out" the noise process $\xi^\gamma(\cdot)$ and rewrite (2.4) as

$$f_1^\gamma(x,t) = \gamma \int_{t/\gamma^2}^{T/\gamma^2} f_x(x)\sigma(x)E_t^\gamma\xi(s)ds. \qquad (2.5)$$

Note that $E_t^\gamma\xi(s) \to 0$ exponentially as $s - t \to \infty$. Thus the integral (2.5) is defined for $T \leq \infty$ and $f_1^\gamma(x,t) = O(\gamma)$. In fact, for $T = \infty$ (2.5) can be readily evaluated due to the simple Markovian structure of $\xi(\cdot)$ and $f_1^\gamma(x,t) = \gamma\xi^\gamma(t)f_x(x)\sigma(x)/2\lambda$. Next, define the function $f_1^\gamma(\cdot)$ by: $f_1^\gamma(t) = f_1^\gamma(x^\gamma(t),t)$. Define the *perturbed test function* $f^\gamma(\cdot)$ by $f^\gamma(t) = f(x^\gamma(t)) + f_1^\gamma(t)$. We have $f_1^\gamma(\cdot) \in \mathcal{D}(\hat{A}^\gamma)$ and (writing x^γ for $x^\gamma(t)$)

$$\hat{A}^\gamma f_1^\gamma(t) = -f_x(x^\gamma)\sigma(x^\gamma)\xi(t/\gamma^2)/\gamma$$

$$+ \frac{1}{\gamma}\int_t^T [f_x(x^\gamma)\sigma(x^\gamma)]_x E_t^\gamma\xi(s/\gamma^2)ds \cdot \dot{x}^\gamma(t). \qquad (2.6)$$

Calculations of the type needed to get (2.6) are discussed in detail in [K20, Chapters 3,4]. The action of $\hat{A}^\gamma$ on $f_1^\gamma(x^\gamma(t),t)$ is essentially a (right) differentiation operation. The variable t (with x^γ replaced by $x^\gamma(t)$) occurs in three places in (2.4): as a lower index in the integral, as an argument of $x^\gamma(t)$, and in E_t^γ. The two terms on the right of (2.6) are due to the first two cited occurances of t. The t in E_t^γ plays no role due to the fact that the calculations which are used to get $\hat{A}^\gamma f_1^\gamma$ are all conditioned on $\mathcal{F}_t^\gamma$. Note that the first term on the right of (2.6) is the *negative* of (2.3). Thus we can write (using the abbreviation $x^\gamma = x^\gamma(t)$)

$$\hat{A}^\gamma f^\gamma(t) = \hat{A}^\gamma f(x^\gamma(t)) + \hat{A}^\gamma f_1^\gamma(t)$$

$$= \frac{1}{\gamma^2}\int_t^T [f_x(x^\gamma)\sigma(x^\gamma)]_x E_t^\gamma\xi(s/\gamma^2)\sigma(x^\gamma)\xi(t/\gamma^2)ds. \qquad (2.7)$$

Important remark. The perturbation $f_1^\gamma(\cdot)$ was constructed as it was just so the first term on the right side of (2.6) (the derivative of (2.4) with respect to the lower limit of integration) would be the negative of the "$1/\gamma$-term" in (2.3). The perturbation $f_1^\gamma(\cdot)$ is always constructed with this 'cancellation' in mind.

With a change of scale, $s \to s\gamma^2$, (2.7) can be rewritten and $Q(\cdot)$ defined as

$$\int_{t/\gamma^2}^{T/\gamma^2} [f_x(x^\gamma)\sigma(x^\gamma)]_x \sigma(x^\gamma)E_t^\gamma\xi(s)\xi(t/\gamma^2)ds$$

$$\equiv Q(x^\gamma, \xi(t/\gamma^2)) = O(1). \qquad (2.8)$$

For notational simplicity, set $T = \infty$. Then $Q(\cdot)$ doesn't depend on T. Write

$$\int_0^t \hat{A}^\gamma f^\gamma(x^\gamma(s))ds = \int_0^t Q(x^\gamma(s), \xi(s/\gamma^2))ds$$

$$= \gamma^2 \int_0^{t/\gamma^2} Q(x^\gamma(\gamma^2 s), \xi(s))ds. \tag{2.9}$$

By the assumed tightness of $\{x^\gamma(\cdot)\}$, for small γ the process $x^\gamma(\gamma^2 \cdot)$ varies "slowly" compared with $\xi(\cdot)$. By using this fact, together with an ergodic theorem for the $\xi(\cdot)$ process, we can average out the $\xi(\cdot)$ in (2.9) and get that

$$\int_0^t Q(x^\gamma(s), \xi(s/\gamma^2))ds - \int_0^t Af(x^\gamma(s))ds \Rightarrow 0, \tag{2.10}$$

where the operator A is defined by (1.3). The asymptotic equivalence between $\int_0^t \hat{A}^\gamma f^\gamma(x^\gamma(s))ds$ and $\int Af(x^\gamma(s))ds$ and the fact that the perturbation $f_1^\gamma(\cdot)$ is $O(\gamma)$ is essentially what is meant by (1.2). The precise sense in which (1.2) holds is given in the next section. Using the property cited in the last sentence and a "martingale problem type" argument yields that the weak limits of $\{x^\gamma(\cdot)\}$ solve the martingale problem for operator A.

The full details are in Sections 3 and 4. The fact that (2.10) holds can be seen most easily (and formally) by fixing $x^\gamma(s) \equiv x$. In the rest of this section, we give a formal argument for (2.10). Due to the separation of the time scales of the processes $x^\gamma(\gamma^2 \cdot)$ and $\xi(\cdot)$, our formal argument can be justified by an approximation argument over small time intervals, and will be developed in Section 4. Fix $x^\gamma(s) \equiv x$ and define $[f_x(x)\sigma(x)]_x\sigma(x) = b(x)$. Then we can write the left-hand term of (2.10) (with $T = \infty$) as

$$b(x) \int_0^t ds \int_{s/\gamma^2}^\infty E_{s/\gamma^2}\xi(u)\xi(s/\gamma^2)du$$

$$= b(x)\gamma^2 \int_0^{t/\gamma^2} ds \int_s^\infty du E_s\xi(u)\xi(s). \tag{2.11}$$

As $\gamma \to 0$, the right side of (2.11) converges in probability to

$$b(x)t \int_0^\infty E\xi(u)\xi(0)du = b(x)t/2\lambda = tAf(x), \tag{2.12}$$

where $\xi(\cdot)$ is the *stationary process*.

3. The Perturbed Test Function Method: Tightness and Weak Convergence

In this section, the convergence and approximation theorems are stated in a general setting. Specific and verifiable criteria are given in Section 4. For $\gamma > 0$, let $x^\gamma(\cdot)$ be a process with paths in $D^k[0, \infty)$. Let $\mathcal{F}_t^\gamma$ be a filtration such that $\mathcal{F}_t^\gamma \supset \mathcal{B}(x^\gamma(s), s \leq t)$ and define the operator $\hat{A}^\gamma$ by (2.1) where the filtration $\mathcal{F}_t^\gamma$ is used. Let E_t^γ denote the expectation conditioned on $\mathcal{F}_t^\gamma$. The next theorem gives a basic method for characterizing the limit of a weakly convergent subsequence of $\{x^\gamma(\cdot)\}$. The method defined by the theorem will be elaborated upon in the next section, where its convenience of use and power will be seen. The theorem is essentially an adaptation of the martingale problem method to non-Markov processes.

Theorem 3.1. *Let A denote the differential operator of a diffusion (2.4.1) or a jump-diffusion process (2.7.7) (with u deleted) with coefficients that are continuous functions of x, and let the sequence $\{x^\gamma(\cdot)\}$ with paths in $D^r[0,\infty)$ be weakly convergent to a process $x(\cdot)$. Let C_1 be a dense set in $C_0(R^r)$ such that for each $T_1 < \infty$ and $f(\cdot) \in C_1$, there are $f^\gamma(\cdot) \in \mathcal{D}(\hat{A}^\gamma)$ which satisfy (3.1) and either (3.2a) or (3.2b) for $t \leq T_1$ (typically $C_1 = C_0^2(R^r)$):*

$$E|f^\gamma(t) - f(x^\gamma(t))| \xrightarrow[\gamma]{} 0, \tag{3.1}$$

$$E\left| \int_0^t [\hat{A}^\gamma f^\gamma(s) - Af(x^\gamma(s))ds \right| \xrightarrow[\gamma]{} 0, \tag{3.2a}$$

$$E\left| \int_0^t \left[\frac{(E_s^\gamma f^\gamma(s+\delta) - f^\gamma(s))}{\delta} - Af(x^\gamma(s)) \right] ds \right| \xrightarrow[\gamma,\delta]{} 0. \tag{3.2b}$$

Then $x(\cdot)$ solves the martingale problem for operator A and the initial condition $\lim_\gamma x^\gamma(0) = x(0)$.

Proof. We do the proof under (3.2a) only and adapt the martingale problem method of Theorem 3.5.1. Let $h(\cdot)$ be a bounded and continuous real valued function, q an integer, $t_i \leq t \leq t+s$, $i \leq q$, and let $f(\cdot) \in C_1$, with $f^\gamma(\cdot)$ being the sequence associated with $f(\cdot)$ in the theorem statement. It is assumed without loss of generality that the probability is zero that $x(\cdot)$ has a discontinuity at the t_i or t or $t+s$. (There are at most a countable set of points that this criteria will exclude.) Since $f^\gamma(\cdot) \in \mathcal{D}(\hat{A}^\gamma)$, the martingale property in Theorem 2.1 implies that

$$Eh(x^\gamma(t_i), i \leq q) \left[f^\gamma(t+s) - f^\gamma(t) - \int_t^{t+s} \hat{A}^\gamma f^\gamma(u)du \right] = 0.$$

By (3.1) and (3.2a),

$$\lim_\gamma Eh(x^\gamma(t_i), i \leq q) \left[f(x^\gamma(t+s)) - f(x^\gamma(t)) - \int_t^{t+s} Af(x^\gamma(u))du \right] = 0.$$

Thus, upon taking the weak limit,

$$Eh(x(t_i), i \leq q) \left[f(x(t+s)) - f(x(t)) - \int_t^{t+s} Af(x(u))du \right] = 0. \tag{3.3}$$

Equation (3.3), the arbitrariness of $h(\cdot)$, q, t and $t + s$ implies that the bracketed term in (3.3) is a martingale with respect to the filtration $B(x(s)$, $s \leq t)$ for all $f(\cdot) \in C_1$, hence for all $f(\cdot) \in C_0^2(R^r)$. Now, Theorem 2.8.1 imply that $x(\cdot)$ solves the martingale problem for the operator A and the cited initial condition.   Q.E.D.

A slight modification of Theorem 3.1 is needed for the control problem. In Theorem 3.2, $m^\gamma(\cdot)$ is a relaxed control for the process $x^\gamma(\cdot)$ and the filtration $\mathcal{F}_t^\gamma$ includes $B(m^\gamma(s)$, $x^\gamma(s)$, $s \leq t)$. The operator $\hat{A}^\gamma$ is defined with respect to $\mathcal{F}_t^\gamma$. We do not need to specify the form of the process $x^\gamma(\cdot)$. The proof of Theorem 3.2 is almost identical to that of Theorem 3.1 and is omitted.

Theorem 3.2. *Let A^α denote the differential operator of a controlled diffusion or jump-diffusion process (3.1.1) with control parameter α and coefficients that are continuous functions of (x, α). Let $\{x^\gamma(t), m^\gamma(\cdot)\}$ denote a weakly convergent sequence with limit $(x(\cdot), m(\cdot))$. Let T_1, C_1 and $f^\gamma(\cdot)$ be defined as in Theorem 3.1, and assume that (3.1) and either (3.4a) or (3.4b) are satisfied:*

$$E \left| \int_0^t \left[\hat{A}^\gamma f^\gamma(s) - \int_{\mathcal{U}} A^\alpha f(x^\gamma(s)) m_s^\gamma(d\alpha) \right] ds \right| \xrightarrow[\gamma]{} 0, \quad t \leq T, \quad (3.4a)$$

$$E \left| \int_0^t \left[\frac{(E_s^\gamma f^\gamma(t + s) - f^\gamma(s))}{\delta} \right. \right.$$

$$\left. \left. - \int_{\mathcal{U}} A^\alpha f(x^\gamma(s)) m_s^\gamma(d\alpha) \right] ds \right| \xrightarrow[\gamma, \delta]{} 0, \quad t \leq T_1. \quad (3.4b)$$

Then $(x(\cdot), m(\cdot))$ solves the martingale problem associated with the operator A^α, and initial condition $\lim_\gamma x^\gamma(0)$.

Tightness. Truncated Processes. Let $\{x^\gamma(\cdot)\}$ be a sequence of processes with paths in $D^r[0, \infty)$. It is easier to prove tightness if the $x^\gamma(\cdot)$ are bounded uniformly in γ. In order to take advantage of this, one often works with *truncated processes*. For $N > 0$, a process $y^N(\cdot)$ is said to be an N-*truncation* of a process $y(\cdot)$ (both with paths in $D^r[0, \infty)$), if $y^N(\cdot) = y(\cdot)$ until at least the first time that $|y(\cdot)|$ reaches or exceeds N, but $|y^N(t)| \leq N +$ constant for all t. Suppose that for each N we can prove tightness of the sequence of N-truncations $\{x^{\gamma,N}(\cdot), \gamma > 0\}$. To simplify notation, let γ index a sequence such that for each N, $\{x^{\gamma,N}(\cdot), \gamma > 0\}$ converges weakly to a process denoted by $x_N(\cdot)$. Suppose that for each $T_1 < \infty$,

$$\sup_{M \geq N} P\big(\sup_{t \leq T_1} |x_M(t)| \geq N \big) \to 0$$

as $N \to \infty$. Then for any $\delta > 0$,

$$\lim_{M,N} P\left(\sup_{t \le T_1} |x^{\gamma,N}(t) - x^{\gamma,M}(t)| \ge \delta\right) = 0.$$

In this case, there is a process $x(\cdot)$ such that the $x_N(\cdot)$ are N-truncations of $x(\cdot)$ and $x^\gamma(\cdot) \Rightarrow x(\cdot)$. This is, in fact, the typical case in applications, unless the limit process has a finite explosion time.

In order to minimize detail when proving tightness, we will always *implicitly* assume that the processes are bounded or truncated, and that the limits (as the truncation parameter N goes to ∞) are taken, as appropriate, after the weak convergence of each N-truncation $\{x^{\gamma,N}(\cdot), \gamma > 0\}$ is shown. We will generally *omit* the truncation argument, since it can *always* be added under our conditions without the need for any additional conditions. Simply for completeness in this discussion, the truncation method is illustrated in the example after the theorem.

The next theorem gives a criterion for tightness that is widely used and often easily verifiable.

Theorem 3.3. *Let $\{x^\gamma(\cdot)\}$ be sequence of processes with paths in $D^r[0,\infty)$. For each $\epsilon > 0$ and $T_1 < \infty$, let there be a compact set $K_\epsilon(T_1)$ such that*

$$\sup_\gamma \sup_{t \le T_1} P(x^\gamma(t) \notin K_\epsilon(T_1)) \le \epsilon. \tag{3.5}$$

Let C_2 be a dense set in $C_0(R^r)$ which contains the square of each function in it. Typically, $C_2 = C_0^2(R^r)$.) For each $T_1 < \infty$ and $f(\cdot) \in C_2$, let there be a sequence $f^\gamma(\cdot) \in \mathcal{D}(\hat{A}^\gamma)$ such that the set

$$\{\hat{A}^\gamma f^\gamma(t), t \le T_1, \gamma > 0\} \tag{3.6}$$

is uniformly integrable, and that for each $\delta > 0$ we have

$$\lim_\gamma P\left(\sup_{t \le T_1} |f^\gamma(t) - f(x^\gamma(t))| \ge \delta\right) = 0. \tag{3.7}$$

Then $\{x^\gamma(\cdot)\}$ is tight in $D^r[0,\infty)$.

The proof is given in [K19] and in [K20, Theorem 3.4], and it shows that the conditions of the theorem imply those of Theorem 1.5.1.

Return to the example. We now illustrate Theorem 3.3 and the truncation argument for the example of Section 1. Unless (3.5) is known to hold, we need to truncate and work with the truncation $x^{\gamma,N}(\cdot)$. We will see how that truncation goes for this example, but will generally omit such "truncation" considerations in the sequel, since they can always be carried out and are only of technical importance. We can define a convenient truncation as follows. Let $q_N(\cdot)$ be a real-valued and continuously differentiable function which takes values in the interval $[0,1]$ and satisfies $q_N(x) = 0$ for $|x| \geq N + 1$, and $q_N(x) = 1$ for $|x| \leq N$. Define the truncation $x^{\gamma,N}(\cdot)$ by

$$\dot{x}^{\gamma,N} = \sigma(x^{\gamma,N})q_N(x^{\gamma,N})\xi^{\gamma}/\gamma. \tag{3.8}$$

Suppose that $\sup_{\gamma} |x^{\gamma}(0)| \leq N$ for large enough N. Now apply Theorem 3.3 to $\{x^{\gamma,N}(\cdot), \gamma > 0\}$ for each N. Suppose that we prove tightness via the theorem for each N. Thus for each N, there is a weakly convergent subsequence. We can suppose (via a use of the diagonal method and an abuse of the terminology concerning the index) that γ indexes a sequence $\{x^{\gamma,N}(\cdot), \gamma > 0\}$, which converges weakly to a process $x_N(\cdot)$ for each N. If this limit can be shown to be a truncation of a process $x(\cdot)$ with paths in $D^r[0,\infty)$, then $x^{\gamma}(\cdot) \Rightarrow x(\cdot)$.

To apply Theorem 3.3 to the example, let $x^{\gamma}(0) \Rightarrow x(0)$, let $f(\cdot) \in C_0^2(R) = C_2$, and define $f_1^{\gamma}(\cdot)$ and $f^{\gamma}(\cdot)$ as above (2.6). Since $f_1^{\gamma}(t) = O(\gamma)$ and $\hat{A}^{\gamma} f^{\gamma}(t) = O(1)$, the conditions of Theorem 3.3 are satisfied for each N-truncation. Thus, $\{x^{\gamma,N}(\cdot), \gamma > 0\}$ is tight for each N. With the truncation defined by (3.8), the limit of any weakly convergent subsequence $x_N(\cdot)$ has the differential operator A of (1.3), but with $\sigma(x)$ replaced by $\sigma(x)q_N(x)$. Since $\sigma(x)q_N(x)$ and $[\sigma(x)q_N(x)]_x$ are bounded, for any $T_1 < \infty$ we have

$$\sup_{M \geq N} P\big(\sup_{t \leq T_1} |x_M(t)| \geq N\big) \xrightarrow{N} 0.$$

Thus, as discussed above, there is a process $x(\cdot)$ such that $x_N(\cdot)$ is an N-truncation of $x(\cdot)$. The process $x(\cdot)$ solves the martingale problem for the differential operator A.

Now, let us return to the *original* sequence $\{x^{\gamma}(\cdot)\}$. Let $x^{\epsilon}(0) \Rightarrow x(0)$. Suppose that there is a differential operator A with continuous coefficients such that the limit of any weakly convergent subsequence solves the martingale problem for operator A. If the solution to that martingale problem is unique, then the original sequence converges weakly.

4. Characterization of the Limits

After stating the general Theorem 4.1, we will work with the processes $x^{\gamma}(\cdot)$ defined by the differential equation

$$\dot{x}^{\gamma} = \overline{G}(x^{\gamma}) + \tilde{G}(x^{\gamma}, \xi^{\gamma}) + F(x^{\gamma}, \xi^{\gamma})/\gamma, \quad x \in R^k, \tag{4.1}$$

where $\xi^{\gamma}(\cdot)$ is a "noise" process. Loosely speaking, we require that the "time scale" of the noise process $\xi^{\gamma}(\cdot)$ is "faster" than that of $x^{\gamma}(\cdot)$. Typically, $\xi^{\gamma}(\gamma^2 t)$ and $x^{\gamma}(t)$ are on the same "scale". For one example, let $\xi(\cdot)$ be a right continuous process and define $\xi^{\gamma}(t) = \xi(t/\gamma^2)$. The form (4.1) is chosen for convenience only, in order to simplify the presentation. In the next chapter, we restrict our attention to singularly perturbed forms of

(4.1). The techniques of the weak convergence proofs here are applicable much more generally.

Theorem 4.1 gives a general prototype result, and is a minor modification of [K20, Theorem 5.8]. That reference also treats the discrete parameter case. Theorem 4.1 and Theorem 3.1 are similar, but Theorem 4.1 is stated in a way which is easier to use with the ergodic assumptions, which will be used below. The $x^\gamma(\cdot)$, $\hat{A}^\gamma$ and $\mathcal{F}_t^\gamma$ are defined as in Section 2. If $\xi^\gamma(t) = \xi(t/\gamma^2)$ for some process $\xi(\cdot)$, then we use E_t for the expectation conditioned on $\{\xi(s),\ s \leq t\}$. Theorems 4.2 to 4.4 are applications of Theorem 4.1 to more specific cases.

Theorem 4.1. *Assume that $\{x^\gamma(t)\}$ is tight in $D^r[0,\infty)$. For each $T < \infty$ and $f(\cdot) \in C_0^2(R^r)$, let there be $f^\gamma(\cdot) \in \mathcal{D}(\hat{A}^\gamma)$ such that*

$$E|f^\gamma(t) - f(x^\gamma(t))| \to 0 \qquad (4.2)$$

for each $t \leq T$. Let A denote the differential operator of a jump-diffusion process with continuous coefficients. For each $T < \infty$, suppose that there are $\delta_\gamma \to 0$ such that

$$\sup_{t \leq T} E \frac{1}{\delta_\gamma} \left| \int_t^{t+\delta_\gamma} E_t^\gamma [\hat{A}^\gamma f^\gamma(u) - Af(x^\gamma(u))] du \right| \to 0 \qquad (4.3)$$

as $\gamma \to 0$. Then the limit of any weakly convergent subsequence of $\{x^\gamma(t)\}$ solves the martingale problem associated with the operator A.

Proof. We follow the lines of the usual martingale method (say, of Theorem 3.5.1) for proving weak convergence and identifying the limit process. For $t_i \leq t \leq t+s$, bounded continuous $h(\cdot)$, and $f(\cdot)$ in $C_0^2(R^r)$, the martingale property in Theorem 2.1 implies that

$$Eh(x^\gamma(t_i), i \leq q) \left[f^\gamma(t+s) - f^\gamma(t) - \int_t^{t+s} \hat{A}^\gamma f^\gamma(u) du \right] = 0. \qquad (4.4)$$

Let s be an integral multiple of δ_γ. Then (4.4) implies that

$$Eh(x^\gamma(t_i), i \leq q) \left[f(x^\gamma(t+s)) - f(x^\gamma(t)) - \int_t^{t+s} Af(x^\gamma(u)) du \right]$$

$$= O(1)[E|f^\gamma(t+s) - f(x^\gamma(t+s))| + E|f^\gamma(t) - f(x^\gamma(t))|]$$

$$+ Eh(x^\gamma(t_i), i \leq q) \left[\delta_\gamma \sum_{i=0}^{s/\delta_\gamma - 1} \frac{1}{\delta_\gamma} \int_{t+i\delta_\gamma}^{t+i\delta_\gamma+\delta_\gamma} E_{t+i\delta_\gamma}^\gamma [\hat{A}^\gamma f^\gamma(u) \right.$$

$$\left. - Af(x^\gamma(u))] du \right].$$

By (4.2)–(4.4), the right-hand side goes to zero as $\gamma \to 0$. The proof then follows by the argument of Theorem 3.5.1.   Q.E.D.

The perturbed test functions $f^\gamma(\cdot)$ which are required in Theorem 4.1 are generally the same as those used earlier in this chapter, say in Theorem 3.3. The perturbed test functions which will be used for the tightness proofs will usually be the same as those used for this purpose in [K20], and will generally be what are referred to in [K20] as first order perturbed test functions, since the perturbation is just a simple integral. An example is the perturbation (2.4) used in our illustrative example. In some cases, we will have to use the "second order" perturbed test functions. That name arises from the fact that the "second order" perturbation is the sum of a single and a double integral. We will need such perturbations when dealing with the wide bandwidth noise problem when the noise and the "fast" variable are "coupled." It will also be the case when dealing with the so-called parametric singularities in Chapter 10. The first order perturbed test functions will be used for the applications of Theorem 4.1 for all cases except those just cited. In [K20], much of the theoretical effort was devoted to the more complicated second order perturbed test function case, although the use of the first order case was introduced and made much use of in the examples and in many of the applications in Chapters 8–10 there. In the treatment of the example, we saw (formally) how the first order perturbed test function $f^\gamma(\cdot)$ could be used to average out the noise, when used together with an ergodic theorem for $\xi(\cdot)$. In some cases, the operator $\hat{A}^\gamma$, when applied to a first order perturbed test function still has terms of the order of $1/\gamma$. Such terms cannot be averaged easily by use of ergodic theorems, and we then introduce the second order perturbation to help average them.

We now apply the tightness Theorem 3.3 and the weak convergence Theorem 4.1 to the model (4.1). We will use the following conditions.

A4.1. $\xi^\gamma(\cdot)$ is right continuous and bounded uniformly in $\gamma > 0$.

A4.2. $\overline{G}(\cdot)$, $\tilde{G}(\cdot)$, $F(\cdot)$ and $F_x(\cdot)$ are continuous.

The boundedness assumption in (A4.1) is made simply to avoid difficulties in the calculation of the $\hat{A}^\gamma$, and to be able to express the basic ideas in a relatively simple way. There are numerous other possibilities: approximations to random impulses, etc. For large T, define $f_1^\gamma(\cdot)$ by $f_1^\gamma(t) = f_1^\gamma(x^\gamma(t), t)$ and

$$f_1^\gamma(x, t) = \frac{1}{\gamma} \int_t^T f_x(x)' E_t^\gamma F(x, \xi^\gamma(s)) ds.$$

We can now state the general Theorem 4.2. The theorem might seem a little hard to use at first sight. But, as illustrated by the commonly occuring cases covered by Theorems 4.3 and 4.4 and by the development in the next chapter, it is often very easy to use. The methods given by the theorems in this section represent very natural and possibly the easiest to use methods for getting weak convergence results for wide band noise driven systems. When we write for t in a compact set, we always mean a compact set in $[0, T)$, but the value of T is arbitrarily large, and can equal infinity if the $f_1^\gamma(x, t)$ is still well defined.

Theorem 4.2. *Let $\{x^\epsilon(0)\}$ be tight. Assume (A4.1), (A4.2), that T_1 is arbitrary (with $T_1 < T$ arbitrary) and that*

$$\lim_\gamma E |f_1^\gamma(x, t)| = 0, \quad t \leq T_1, \tag{4.5a}$$

uniformly in x in any compact set. Let

$$\lim_\gamma P \left(\sup_{t \leq T_1} |f_1^\gamma(x^\gamma(t), t)| \geq \delta \right) = 0 \tag{4.5b}$$

for any $\delta > 0$. Suppose that the set (recall that the support of $f(\cdot)$ is compact)

$$\left\{ \sup_x \frac{1}{\gamma^2} \left| \int_t^T E_t^\gamma [f_x(x)' F(x, \xi^\gamma(s))]_x ds \right|, t \leq T, \gamma > 0 \right\} \tag{4.6}$$

is uniformly integrable and that the term

$$D^\gamma(x, t) = \frac{1}{\gamma^2} \int_t^T E_t^\gamma [f_x(x)' F(x, \xi^\gamma(s))]_x ds \tag{4.7}$$

is continuous in x in the sense that if $\{x^\gamma, y^\gamma\}$ are bounded $\mathcal{F}_t$-measurable sequences such that $|x^\gamma - y^\gamma| \to 0$ w.p.1, then

$$|D^\gamma(x^\gamma, t) - D^\gamma(y^\gamma, t)| \xrightarrow{P} 0$$

as $\gamma \to 0$, uniformly in t in any compact set. Let B_0 be a second order operator with continuous coefficients which are bounded by $K(1 + |x|)$ for some $K < \infty$. Suppose that there are $\delta_\gamma \to 0$ such that for each x

$$\frac{1}{\delta_\gamma} \int_t^{t+\delta_\gamma} ds E_t^\gamma \int_s^T E_s^\gamma [f_x(x) F(x, \xi^\gamma(u))/\gamma]_x' F(x, \xi^\gamma(s)) du/\gamma \to B_0 f(x), \tag{4.8}$$

$$\frac{1}{\delta_\gamma} \int_t^{t+\delta_\gamma} E_t^\gamma \tilde{G}(x, \xi^\gamma(s)) ds \to 0, \tag{4.9}$$

as $\gamma \to 0$, where the convergence is in the mean and is uniform in t in any compact set. Then $\{x^\gamma(\cdot)\}$ is tight and the limit of any weakly convergent subsequence solves the martingale problem associated with the operator A defined by

$$Af(x) = f_x(x)' \overline{G}(x) + B_0 f(x). \tag{4.10}$$

Remark on the conditions. Conditions (4.8) and (4.9) are essentially ergodic assumptions. To see this, let $T = \infty$ and rescale ("stretch out") time by letting $s \to \gamma^2 s$ and $u \to \gamma^2 u$. Let $\gamma^2/\delta_\gamma \to 0$. Write the left side of (4.8) as

$$\frac{\gamma^2}{\delta_\gamma} \int_{t/\gamma^2}^{(t+\delta_\gamma)/\gamma^2} ds E_t^\gamma \left\{ \int_s^\infty du E_{s\gamma^2}^\gamma [f_x'(x)F(x,\xi^\gamma(\gamma^2 u))]_x' \cdot F(x,\xi^\gamma(\gamma^2 s)) \right\}$$

$$\equiv \frac{\gamma^2}{\delta_\gamma} \int_{t/\gamma^2}^{(t+\delta_\gamma)/\gamma^2} E_t^\gamma H^\gamma(s,x)ds \qquad (4.11)$$

where $H^\gamma(s,x)$ is the bracketed term on the left side of (4.11) and it is a functional of $\{\xi^\gamma(u), u \in (-\infty, \gamma^2 s]$. Let $\xi^\gamma(\cdot)$ take the frequently used form $\xi^\gamma(t) = \xi(t/\gamma^2)$ where $\xi(\cdot)$ is a stationary process. Then $H^\gamma(x,s)$ does not depend on γ, and we write it as $H(x,s)$. The "ergodic" nature of the condition (4.8) is then apparant. In this case, the right side of (4.11) can be written as

$$\frac{1}{T_\gamma} \int_{t_\gamma}^{T_\gamma+t_\gamma} E_{t_\gamma} H(x,s)ds$$

where T_γ and t_γ are sequences which go to ∞ as $\gamma \to 0$. It is often the case that $E_t H(x,s) \to EH(x,0)$ in the mean as $s - t \to \infty$. The conditions (4.8) and (4.9) can be quite useful and verifiable even in the absence of stationarity and for processes which are not necessarily defined by the scaling $\xi^\gamma(t) = \xi(t/\gamma^2)$. Some examples are in [K20]. See also the additional remarks below Theorems 4.3 and 4.4.

Continue to suppose that $\xi^\gamma(s) = \xi(s/\gamma^2)$ **for some process** $\xi(\cdot)$. A scale change $s \to \gamma^2 s$ shows that the term in (4.6) equals (with $T = \infty$)

$$\int_{t/\gamma^2}^\infty E_t^\gamma [f_x'(x)F(x,\xi(s))]_x ds. \qquad (4.6')$$

Thus the uniform integrability required in (4.6), holds if $E_t F(x,\xi(s)) \to 0$ fast enough as $s - t \to \infty$. Usually, the upper limit T is unimportant and can be replaced by the value $T = \infty$. Sometimes we use the finite value T since then the integrals are defined over a bounded time interval, and there is less detail to check when working with the perturbations to the test functions. If $E| \int_t^\infty E_s f_x'(x)F(x,\xi(u))du| \to 0$ uniformly in x as $t - s \to \infty$, then the continuity of $f(\cdot)$ and $F(\cdot)$ imply the continuity of (4.7). The conditions of the theorem can be readily verified in many applications: ϕ-mixing processes, stable Gauss-Markov processes (see below or [K20, Chapter 4]; periodic or almost periodic functions replacing the processes, scaled "physical" impulses [K20, Chapter 4.2], etc.

Proof. We have $f^\gamma(\cdot) \in \mathcal{D}(\hat{A}^\gamma)$ and (write $x^\gamma = x^\gamma(t)$ for simplicity)

$$\hat{A}^\gamma f_1^\gamma(t) = -f_x'(x^\gamma)F(x^\gamma,\xi^\gamma(t))/\gamma$$

$$+ \frac{1}{\gamma}\int_t^T ds[E_t^\gamma f_x'(x^\gamma)F(x^\gamma,\xi^\gamma(s))]_x' \dot{x}^\gamma(t). \tag{4.12}$$

Define the perturbed test function $f^\gamma(\cdot) = f(x^\gamma(\cdot)) + f_1^\gamma(\cdot)$. Then, using $\hat{A}^\gamma f(x^\gamma) = f_x'(x^\gamma)\dot{x}^\gamma$ and (4.12), we have

$$\hat{A}^\gamma f^\gamma(t) = \frac{1}{\gamma}\int_t^T ds[E_t^\gamma f_x'(x^\gamma)F(x^\gamma,\xi^\gamma(s))]_x' \left[\frac{F(x^\gamma,\xi^\gamma(t))}{\gamma} + \overline{G}(x^\gamma)\right.$$

$$\left. + \tilde{G}(x^\gamma,\xi^\gamma(t))\right] + f_x'(x^\gamma)[\overline{G}(x^\gamma) + \tilde{G}(x^\gamma,\xi^\gamma(t))]. \tag{4.13}$$

Note that in the calculation of $\hat{A}^\gamma f^\gamma(t)$, the first term on the right side of (4.12) cancels the (negative of the) same term appearing in $f_x'(x^\gamma)\dot{x}^\gamma$. Tightness of $\{x^\gamma(\cdot)\}$ then follows from Theorem 3.3 and (A4.2), (4.5b), (4.6), (and the implied truncation argument if $x^\gamma(t)$ is not known to be uniformly bounded a priori). Note that the limit processes are continuous w.p.1, since the $x^\gamma(\cdot)$ are.

For the characterization of the limits we use Theorem 4.1. With an abuse of notation, let γ index a subsequence which converges weakly. By (4.6) and the scale change $s \to \gamma^2 s$, we see that the term in (4.13) which is the product of the integral and $(\overline{G}+\tilde{G})$ is asymptotically negligible, being $O(\gamma)$ times a uniformly (in γ) integrable sequence. Given $\epsilon > 0$, let $d_n^\epsilon(\cdot)$, $n = 1, 2, \ldots$, be a finite collection of continuous non-negative real valued functions with support denoted by S_n^ϵ, whose union contains the support of $f(\cdot)$. We suppose that the diameters of the sets S_n^ϵ are no larger than ϵ and $\sum_n d_n^\epsilon(x) = 1$ for x in the support of $f(\cdot)$, and $\sum_n d_n^\epsilon(x) \leq 1$ otherwise. Choose any sequence $\{t_\gamma\}$ such that $t_\gamma \to t$ and let x_n^ϵ be any point in S_n^ϵ. By the continuity condition (A4.2), the continuity of (4.7), and the tightness of $\{x^\gamma(\cdot)\}$, the limits of (4.14) (as $\gamma \to 0$) and (4.15) (as $\gamma \to 0$, and then $\epsilon \to 0$) are the same:

$$\frac{1}{\delta_\gamma}\int_{t_\gamma}^{t_\gamma+\delta_\gamma} ds E_{t_\gamma}^\gamma \int_s^T du E_s^\gamma [f_x'(x^\gamma(s))F(x^\gamma(s),\xi^\gamma(u))]_x' \frac{F(x^\gamma(s),\xi^\gamma(s))}{\gamma^2}$$

$$\tag{4.14}$$

$$\sum_n d_n^\epsilon(x^\gamma(t)) \left\{ \frac{1}{\delta_\gamma}\int_{t_\gamma}^{t_\gamma+\delta_\gamma} ds E_{t_\gamma}^\gamma \int_s^T du E_s^\gamma [f_x'(x_n^\epsilon)F(x_n^\epsilon,\xi^\gamma(u))]_x' \cdot \right.$$

$$\left. \cdot \frac{F(x_n^\epsilon,\xi^\gamma(s))}{\gamma^2} \right\}. \tag{4.15}$$

Thus it is sufficient to take limits of (4.14) for fixed x replacing $x^\gamma(\cdot)$, and then to replace x in the limit by the limit of $x^\gamma(t)$. Now (4.8) implies that the difference of the limits of the bracketed term in (4.15) and $B_0 f(x_n^\epsilon)$ goes to zero in the mean, uniformly in n and ϵ, as $\gamma \to 0$. Thus, the difference between (4.14) and $B_0 f(x^\gamma(t))$ goes to zero in the mean (uniformly in t on each compact set) as $\gamma \to 0$. A similar argument and (4.9) can be used to eliminate the $f'_x \tilde{G}$ term in (4.13). Thus (4.3) holds. Since (4.2) also holds, the theorem now follows from Theorem 4.1.   Q.E.D.

An important special case. For processes $\xi^\gamma(\cdot)$ of a more specific form, Theorem 4.2 can usually be rewritten so that it looks considerably simpler. In the next two theorems, we will use the special form $\xi^\gamma(t) = \xi(t/\gamma^2)$ for some given process $\xi(\cdot)$ and the following conditions.

A4.3. $\xi^\gamma(t) = \xi(t/\gamma^2)$. *The process $\xi(\cdot)$ is bounded and right continuous.*

A4.4. *For each $f(\cdot) \in C_0^2(R^k)$, either (a): the processes defined by*

$$\int_t^\infty E_t F(x, \xi(s))ds,$$

$$\int_t^\infty E_t[f'_x(x)F(x, \xi(s))]'_x F(x, \xi(t))ds \tag{4.16}$$

are bounded and x-continuous, uniformly in each compact x-set and in (t, ω); or (b):

$$E \left| \int_t^\infty E_s F(x, \xi(u))du \right| \to 0$$

$$E \left| \int_t^\infty E_s F_x(x, \xi(u))du \right| \to 0 \tag{4.17}$$

uniformly in x in each compact set as $t - s \to \infty$.

A4.5.

$$\frac{1}{\tau} \int_t^{t+\tau} E_t \tilde{G}(x, \xi(s))ds \to 0$$

in the mean for each x as t and τ go to infinity.

A4.6. *There are continuous $\overline{F}(\cdot)$ and $a(\cdot) = \{a_{ij}(\cdot)\}$ (a symmetric matrix) such that with B_0 defined by*

$$B_0 f(x) = f'_x(x)\overline{F}(x) + \frac{1}{2} \sum_{i,j} a_{ij}(x) f_{x_i x_j}(x),$$

we have

$$\frac{1}{\tau} \int_t^{t+\tau} ds \int_s^\infty du E_t[f'_x(x)F(x, \xi(u))]'_x F(x, \xi(s)) \to B_0 f(x)$$

in the mean for each x as t and τ go to infinity. Also the coefficients of B_0 are bounded by $K(1 + |x|)$ for some $K < \infty$.

Theorem 4.3. *Assume* (A4.2) *to* (A4.6) *and let* $\{x^\gamma(0)\}$ *be tight. Then* $\{x^\gamma(\cdot)\}$ *is tight and the limit of any weakly convergent subsequence solves the martingale problem associated with the operator A defined by*

$$Af(x) = f_x'(x)\overline{G}(x) + B_0 f(x).$$

Example. Before giving the proof, let us return to the example of Section 1. Write $b(x) = [f_x(x)\sigma(x)]_x\sigma(x)$. For $u \geq t$, we have $E_s\xi(u) = [\exp -2\lambda(u - t)]\xi(t)$. Thus for $s \geq t$,

$$\int_s^\infty b(x)E_t\xi(u)\xi(s)du = b(x)E_t\xi^2(s)/2\lambda = b(x)/2\lambda,$$

which is also the value of the right-hand term in the second expression of (A4.6). This example illustrates the advantages of having the E_t in (A4.6).

Proof. Define

$$f_1^\gamma(x,t) = \frac{1}{\gamma}\int_t^\infty E_t^\gamma f_x'(x)F(x,\xi^\gamma(s))ds = \gamma \int_{t/\gamma^2}^\infty E_t^\gamma f_x'(x)F(x,\xi(s))ds.$$

Recall that $E_t^\gamma = E_{t/\gamma^2}$ here, by our notation. Use the perturbed test function $f^\gamma(t) = f(x^\gamma(t)) + f_1^\gamma(x^\gamma(t),t)$ of the last theorem. Then the proof follows from Theorem 4.2 when $T = \infty$ is used in the definition of the perturbation to the test function.   Q.E.D.

Unbounded noise. If the driving noise is unbounded, one must specify the functional form of $\tilde{G}(x,\xi)$ and $F(x,\xi)$ more carefully. We will do one important case and assume the following.

A4.7. $\xi(\cdot)$ *is a stable Gauss-Markov process with a stationary transition function,* $\xi^\gamma(t) = \xi(t/\gamma^2)$, *and* $F(x,\xi) = F(x)\xi$, $\tilde{G}(x,\xi) = \tilde{G}(x)\xi$, *where* $F(\cdot)$, $\overline{G}(\cdot)$, *and* $\tilde{G}(\cdot)$ *have the x-smoothness required in* (A4.2).

By a stable Gauss-Markov process with a stationary transition function, we mean the solution to a stable linear stochastic differential equation with constant coefficients. Hence the correlation matrix $E\xi(t)\xi(0)'$ goes to zero exponentially as $t \to \infty$.

Theorem 4.4. *Under* (A4.7) *and the tightness of the set of initial conditions,* $\{x^\gamma(\cdot)\}$ *is tight and the limit of any weakly convergent subsequence solves the martingale problem associated with the operator A defined in Theorem 4.3, and B_0 is defined by*

$$B_0 f(x) = \int_0^\infty duE[f_x'(x)F(x,\xi(u))]_x' F(x)\xi(0), \qquad (4.18)$$

where the stationary $\xi(\cdot)$ process is used in (4.18).

Proof. The perturbed test function $f^\gamma(\cdot)$ defined in Theorem 4.3 will be used. We have

$$\frac{1}{\gamma} \int_t^\infty E_t^\gamma \xi^\gamma(s) ds = \gamma \int_{t/\gamma^2}^\infty E_t^\gamma \xi(s) ds = O(\gamma)|\xi(t/\gamma^2)|.$$

Thus (4.5a) holds. Since, for each $T_1 < \infty$,

$$\gamma \sup_{t \le T_1/\gamma^2} |\xi(t)| \to 0$$

in probability as $\gamma \to 0$, (4.5b) holds. In Theorem 4.2, we supposed that the $\xi^\gamma(\cdot)$ was bounded. But, with only minor modifications, the same proof carries over to the present case. Since

$$\frac{1}{\gamma^2} \int_t^\infty ds E_t^\gamma [f_x(x)'F(x,\xi^\gamma(s))]_x' F(x,\xi^\gamma(t)) = O(|\xi^\gamma(t)|^2),$$

the set in (4.6) as well as $\{|\tilde{G}(x,\xi^\gamma(t))|, \gamma > 0, t \le T_1\}$ are both uniformly integrable for each $T_1 < \infty$. Using this and the perturbed test functions $f^\gamma(\cdot)$ yields the tightness.

The proof is completed by noting that the following limits hold in the mean as t and τ go to infinity:

$$\frac{1}{\tau} \int_t^{t+\tau} E_t \xi(s) ds \to 0$$

$$\frac{1}{\tau} \int_t^{t+\tau} E_t \xi(s) ds \int_s^\infty E_s \xi(u)' du \to \int_0^\infty E\xi(0)\xi(u)' du, \qquad (4.19)$$

where the expression on the right of (4.19) is for the stationary $\xi(\cdot)$ process. Q.E.D.

8

Singularly Perturbed Wide-Band Noise Driven Systems

0. Outline of the Chapter

In this chapter, the ideas of Chapter 7 will be used to extend the results in Chapters 4 and 5 to the singular perturbation problem for wide band noise driven systems. The actual physical models for many practical systems are driven by wide band noise and it is important to know how to approximate these models and the associated control problems by simple diffusion models and control problems.

The basic problem is defined in Section 1. We choose one particular form for the model, due to lack of space and to avoid overburdening the text with detail. But the methods used are of general interest—for many types of approximation problems, where the system is driven by non-white noise.

There are also extensions to the discrete parameter case. In Section 2, the weak convergence of the "stretched out" fast system is proved. The method uses the perturbed test function technique introduced in the last chapter. The convergence of the wide band noise driven controlled "slow" system to a controlled "averaged" system is dealt with in Section 3. The situation here is more complex than that in Chapter 4, due to the presence of the wide band noise. But, the tightness and averaging methods of Chapter 7 can be combined with the basic method of averaging out the $z^\epsilon(\cdot)$ process in Theorem 4.1.2 to yield the correct result. The limit averaged system is not quite so obvious as that in Chapter 4, due to the occurrance of a "correction term."

The "almost" optimality theorems are given in Section 4, and in Section 5, we do the average cost per unit time problem. Here, there are three parameters to be dealt with; the time interval and the "fast scale" also occured in Chapters 4 and 5. The "band width" parameter is new here. It is shown that the correct limits hold no matter how these parameters go to their limits.

We do not treat the non-linear filtering problem for the wide band noise case. Interesting relevant results are in [K21]. That reference concerns filtering problems where both the system and the observations are "driven" by wide band noise. It is shown that an optimal filter for a "limit" system, but with the physical observations used, is nearly optimal with respect to a large and natural class of alternative filters. The results can be extended to the singularly perturbed model.

The results of Section 5 can also be extended to the case where the cost is of the discounted type and the discount factor goes to zero, analogously to the situation in Chapter 5.7. See also [K23] for some results on pathwise convergence of ergodic and discounted cost functions for (non-singularly perturbed) wide band noise driven systems.

1. The System and Noise Model

For simplicity of development, we will work with a particular structure for the system and noise. However, the structure is "canonical" in the sense that the techniques which are to be used for the averaging and weak convergence arguments can be used for a wide variety of cases, including discrete parameter analogs. Let $\psi(\cdot)$ and $\xi(\cdot)$ be processes with paths in $D^k[0,\infty)$ for some k. For small $\rho > 0$ and small $\beta > 0$, define the rescaled processes $\psi^\rho(t) = \psi(t/\rho^2)$ and $\xi^\beta(t) = \xi(t/\beta^2)$. These will be the wide-band "driving" noise processes.

The "fast" scale is parametrized by $\epsilon > 0$, as in (4.1.2). Define $\gamma = (\epsilon, \rho, \beta)$. The statement $\gamma \to 0$ implies that all components of γ go to zero. The control system model which we will use is defined by (1.1) and (1.2):

$$\dot{x}^\gamma = \int_{\mathcal{U}} G(x^\gamma, z^\gamma, \alpha) m_t^\gamma(d\alpha) + \tilde{G}(x^\gamma, z^\gamma, \psi^\rho)$$

$$+ F(x^\gamma, z^\gamma, \psi^\rho)/\rho, \quad x \in R^r, \tag{1.1}$$

$$\epsilon \dot{z}^\gamma = H(x^\gamma, z^\gamma) + \tilde{H}(x^\gamma, z^\gamma, \xi^\beta) + \sqrt{\epsilon}\, J(x^\gamma, z^\gamma, \xi^\beta)/\beta, \quad z \in R^{r'}. \tag{1.2}$$

It will be shown that if $\beta^2/\epsilon \to 0$ and $\rho^2/\epsilon \to 0$ as $\gamma \to 0$, then the system (1.1) and (1.2) is the natural "wide-band" noise analog of (4.1.1), (4.1.2). To get a preliminary idea of why this is so, see equation (2.2) in the next section. This is the equation for the "stretched out" fixed-x wide band noise driven fast process $z_0^\gamma(t \mid x)$. Now, recall the discussion in Chapter 7.4 where, under appropriate conditions, a process represented by an equation such as (2.2) converged to a diffusion process as the scale parameter β^2/ϵ (called γ^2 there) tended to zero.

Analogously to (4.1.3), we use the cost functional

$$V_{T_0}^\gamma(x, z, m^\gamma) = E_{x,z}^{m^\gamma} \int_0^{T_0} \int_{\mathcal{U}} k(x^\gamma(s), z^\gamma(s), \alpha) m_s^\gamma(d\alpha) ds$$

$$+ E_{x,z}^{m^\gamma} g(x^\gamma(T_0)). \tag{1.3}$$

The cost functional and the expectation depend on the joint distribution of $(m^\gamma, \psi^\rho, \xi^\beta)$, although we omit ψ^ρ and ξ^β from the notation for simplicity.

Notation and Admissible controls. Define $\mathcal{F}_t^\gamma = B(\psi^\rho(s), \xi^\beta(s), x^\gamma(s),$ $z^\gamma(s), s \leq t)$. Let E_t^γ denote expectation conditioned on $\mathcal{F}_t^\gamma$. Recall the definition of the space $\mathcal{R}(\mathcal{U} \times [0, \infty))$ of relaxed control values given in Chapter 3.2, where $\mathcal{U}$ is compact. The definition of admissible controls will be the obvious analog of that used in Chapter 3, in that the values at each time can depend only on the "past data," and that certain measurability properties are required. Let $m^\gamma(\cdot)$ be an $\mathcal{R}(\mathcal{U} \times [0, \infty))$-valued random variable. For $B \in \mathcal{B}(\mathcal{U})$, write $m^\gamma(B \times [0, t]) = m^\gamma(B, t)$, for simplicity. Suppose that $m^\gamma(B, \cdot)$ is a measurable process and is $\mathcal{F}_t^\gamma$-adapted for each $B \in \mathcal{B}(\mathcal{U})$. Then $m^\gamma(\cdot)$ is said to be an *admissible relaxed control* for (1.1), (1.2).

2. Weak Convergence of the Fast System

In this section and in the next, we show that the system (1.1), (1.2) can be well approximated (for small γ) by an appropriate "averaged" system, analogous to what was done in Theorem 4.1.2. But, due to the presence of "non-white" noise terms, the proof requires additional steps in order to accomplish the averaging. We will need conditions which guarantee that the process defined by the "stretched out" or "rescaled" fast process (2.1) converges to an appropriate diffusion process as $\gamma \to 0$. This diffusion will be (2.4), analogous to the fixed-x system (4.1.6), with $x = x(t)$, where $x(t) = \lim_\gamma x^\gamma(t)$. This part will be covered in this section. We will also need conditions which guarantee that $x^\gamma(\cdot)$ converges weakly to a controlled "averaged" diffusion. The appropriately averaged limit process (the weak limit of $\{x^\gamma(\cdot)\}$) will be obtained by a procedure which can be described simply by three steps which exploit the time scale differences between the three processes $x^\gamma(\cdot), z^\gamma(\cdot)$ and $\psi^\rho(\cdot)$. First the differential operator of $(x^\gamma(\cdot), z^\gamma(\cdot))$ acting on fuctions of $x^\gamma(\cdot)$ is "averaged" over the driving noise $\psi^\rho(\cdot)$ by a "perturbed test function method." Then the $z^\gamma(\cdot)$ is averaged out by integrating with respect to the invariant measure of (2.4). Then a final weak convergence argument identifies the limit $x(\cdot)$ as a certain controlled diffusion process. Each step requires its own conditions. This is done in the next section, where the limits of $\{x^\gamma(\cdot)\}$ will be dealt with. We will use various subsets of the following conditions (A2.1)–(A2.10). Theorem 2.1 is the analog of Theorem 4.1.2. In Theorem 4.1.2, tightness of $\{x^\epsilon(\cdot)$ followed directly from the growth properties of $b(\cdot)$ and $\sigma(\cdot)$. In Theorem 2.1, we assume this tightness. It is proved in Theorem 3.1 and 3.2, under additional conditions.

Assumptions.

A2.1.

$$\beta^2/\epsilon \to 0, \rho^2/\epsilon \to 0 \text{ as } \gamma \to 0,$$

$$0 < \liminf_{\gamma \to 0} \beta/\rho \le \limsup_{\gamma \to 0} \beta/\rho < \infty$$

(A2.1) implies the following formal time scale separation: $x^\gamma(\cdot)$ is "slower" than $z^\gamma(\cdot)$ which is "slower" than $\psi^\rho(\cdot)$ and $\xi^\beta(\cdot)$. Without (A2.1), the development is much harder, and there might not be a limit *diffusion* process which approximates the $x^\gamma(\cdot)$.

A2.2. *The $\psi(\cdot)$ and $\xi(\cdot)$ are bounded processes with paths in some $D^k[0,\infty)$, and $\xi^\beta(t) = \xi(t/\beta^2)$, $\psi^\rho(t) = \psi(t/\rho^2)$.*

A2.3. *The solution to (1.1), (1.2) exists and is unique on $[0,\infty)$ for each admissible relaxed control.*

Later, we will also use Gaussian noise processes.

The "stretched out" process. Define the processes $z_0^\gamma(t) = z^\gamma(\epsilon t)$ and $x_0^\gamma(t) = x^\gamma(\epsilon t)$. Then

$$\dot{z}_0^\gamma = H(x_0^\gamma, z_0^\gamma) + \tilde{H}(x_0^\gamma, z_0^\gamma, \xi(\epsilon t/\beta^2)) + J(x_0^\gamma, z_0^\gamma, \xi(\epsilon t/\beta^2)) / (\beta/\sqrt{\epsilon}), \quad (2.1)$$

$$\dot{x}_0^\gamma = \epsilon \int_{\mathcal{U}} G(x_0^\gamma, z_0^\gamma, \alpha) m_t^\gamma(d\alpha) + \epsilon \tilde{G}(x_0^\gamma, z_0^\gamma, \psi(\epsilon t/\rho^2))$$

$$+ \epsilon F(x_0^\gamma, z_0^\gamma, \psi(\epsilon t/\rho^2))/\rho$$

Define the fixed-x process $z_0^\gamma(\cdot|x)$ by

$$\dot{z}_0^\gamma(t \mid x) = H(x, z_0^\gamma(t \mid x)) + \tilde{H}(x, z_0^\gamma(t \mid x), \xi(\epsilon t/\beta^2))$$

$$+ J(x, z_0^\gamma(t \mid x), \xi(\epsilon t/\beta^2)) / (\beta/\sqrt{\epsilon}) \quad (2.2)$$

where x is treated as a parameter. Define $\lambda = \beta/\sqrt{\epsilon}$. If the value of x is obvious then we write $z_0^\gamma(\cdot \mid x)$ simply as $z_0^\gamma(\cdot)$; there should be minimal confusion with the process defined by (2.1).

Remark on weak convergence of (2.1), (2.2). If $\{x^\gamma(\cdot)\}$ can be proved to be tight, then inspecting (2.1) and recalling the discussion of the example in Chapter 7 suggests that the $x_0^\gamma(\cdot)$ varies "much more slowly" than either $z_0^\gamma(\cdot)$ or $\xi(\epsilon \cdot /\beta^2)$ does (for small $\beta^2/\epsilon = \lambda^2$), and that the fixed-x process $z_0^\gamma(\cdot|x)$ might be a good approximation to the solution to (2.1) over long time intervals, for appropriate values of x.

We now introduce conditions which guarantee that $z_0^\gamma(\cdot|x)$ converges weakly to a "fixed-x" diffusion process $z_0(\cdot|x)$ defined by (2.4). Define E_t^ξ and E_t^ψ to be the expectations conditioned on $\{\xi(s), s \le t\}$ and $\{\psi(s), s \le t\}$, resp.

Assumptions and the fixed-x limit process.

A2.4. $H(\cdot), \tilde{H}(\cdot), J(\cdot)$ *and* $J_z(\cdot)$ *are continuous.* $J_x(\cdot)$ *is bounded and continuous.*

A2.5. *The random functions* $\int_\tau^\infty E_t^\xi J(x, z, \xi(s))ds$, $\int_\tau^\infty E_t^\xi J_x(x, z, \xi(s))ds$ *and* $\int_\tau^\infty E_t^\xi J_z(x, z, \xi(s))ds$ *go to zero in the mean, uniformly on each compact* (x, z)-*set as* $\tau - t \rightarrow \infty$.

A2.6.

$$\frac{1}{\tau} \int_t^{t+\tau} E_t^\xi \tilde{H}(x, z, \xi(s))ds \xrightarrow{P} 0$$

for each x *and* z, *as* t, τ *go to infinity.*

A2.7. *Let* $f(\cdot) \in C_0^2(R^{r'})$. *There are continuous functions* $\bar{H}(\cdot)$ *and* $\theta(\cdot)$ *($\theta(\cdot)$ being a symmetric matrix) such that*

$$\frac{1}{\tau} \int_t^{t+\tau} ds \int_s^\infty du\, E_t^\xi [f_z'(z)J(x, z, \xi(u))]_z' J(x, z, \xi(s)) \xrightarrow{P} B_x^0 f(z)$$

for each x *and* z *as* t *and* τ *go to infinity, where we define*

$$B_x^0 f(z) = f_z'(z)\bar{H}(x, z) + \frac{1}{2}\sum_{i,j} \theta_{ij}(x, z)f_{z_i z_j}(z).$$

Define the operator A_x^0 *on* $C^2(R^{r'})$ *by*

$$A_x^0 f(z) = f_z'(z)H(x, z) + B_x^0 f(z). \tag{2.3}$$

The A_x^0 is the differential operator of a fixed-x process $z_0(\cdot|x)$ analogous to (4.1.6). The $z_0(\cdot|x)$ will turn out to be the weak limit of $\{z_0^\gamma(\cdot|x)\}$. We will assume, purely for notational convenience, that $\theta(\cdot)$ can be written as $\theta(\cdot) = v(\cdot)v'(\cdot)$, for some continuous matrix valued function $v(\cdot)$. Then, for given x, there is a standard vector-valued Wiener process $\tilde{w}_2(\cdot)$ such that $z_0(\cdot)$ satisfies the SDE

$$dz_0 = H(x, z_0)dt + \bar{H}(x, z_0)dt + v(x, z_0)d\tilde{w}_2, \tag{2.4}$$

where x is a parameter. The function $\bar{H}(\cdot)$ is known as the "correction" term [K20] and is a consequence of the correlation between $z_0^\gamma(t)$ and $\xi(\epsilon t/\beta^2)$.

A2.8. *For each* x *and initial condition* $z_0(0)$, *there is a unique solution on* $[0,\infty)$ *to the martingale problem associated with* A_x^0, *and that process has a unique invariant measure* $\mu_x(\cdot)$.

A2.9. *For each* $T_1 < \infty$, $\{z^\gamma(t), \gamma > 0, t \leq T_1\}$ *is tight.*

Criteria for the stability assumed in (A2.9) are in Chapter 9. Using the general ideas of the perturbed Liapunov function methods of Chapter 9, stability for a broad class of processes can be proved. Basically, the methods exploit the stability of the limit systems. We will also use the following alternative to (A2.2), (A2.4)–(A2.7).

A2.10. *Let* $\tilde{H}(x, z, \xi) = \tilde{H}_0(x, z)\xi$, $J(x, z, \xi) = J_0(x, z)\xi$, *where* $\tilde{H}_0(\cdot)$
and $J_0(\cdot)$ *satisfy the conditions in* (A2.4) *on* $\tilde{H}(\cdot)$ *and* $J(\cdot)$. *The process*
$\xi(\cdot)$ *is a stable Gauss-Markov process with a stationary transition function.*
$H(\cdot)$ *satisfies the conditions in* (A2.4).

When (A2.10) is used the operator B_x^0 can be also defined by: $B_x^0 f(z) = \int_0^\infty E[f_z'(z)J(x, z, \xi(u))]_z' J(x, z, \xi(0))du$, where $\xi(\cdot)$ is the stationary process. Note that we do not require the processes $\xi(\cdot)$ or $\psi(\cdot)$ to be stationary. We can get the same result if we use the Gaussian assumption (A3.7) in lieu of our assumptions on G, $\tilde{G}$, $F(\cdot)$ and $\psi(\cdot)$.

Theorem 2.1. *Assume* (A2.1)–(A2.9) *or* (A2.1), (A2.3), (A2.8)–(A2.10).
Let $\{x^\gamma(\cdot)\}$ *be tight. Then for any* $T_1 < \infty$, *the set of processes* $\{z_0^\gamma(t_0 + \cdot)$,
$t_0 \leq T_1/\epsilon$, $\gamma > 0\}$ *is tight. Let* $\{t_\gamma\}$ *be a sequence of real numbers such*
that $(z_0^\gamma(t_\gamma), x_0^\gamma(t_\gamma))$ *converges weakly to, say,* (z_0, x_0). *Then* $z_0^\gamma(t_\gamma + \cdot) \Rightarrow$
$z_0(\cdot|x_0)$, *where* $z_0(0|x_0) = z_0$.

Remark. Note that the set of processes in the theorem has the two parameters γ and t_0. In order to average the $z^\gamma(\cdot)$ out of (1.1) at say, real time t, we need to use the sequence $z_0^\gamma(t_\gamma + \cdot)$, where $\epsilon t_\gamma \to t$.

Proof. The proof will be done under (A2.1)–(A2.9), with only an occasional comment concerning (A2.10).

1. Tightness. We suppose, as usual, that the $z_0^\gamma(\cdot)$ are "truncated," as discussed in Chapter 7.3, so that we can suppose that they are bounded. If all the details of the truncation procedure were put in, we would actually prove that any sequence of N-truncations converges to an N-truncation of (2.4), with $x = x(t)$. By (A2.8) and the discussion in Chapter 7.3, this would yield the desired tightness and convergence. To prove the tightness, we use Theorem 7.3.3, but where the parameter is now (γ, t_0). Condition (7.3.5) holds by (A2.9). By the assumed tightness of $\{x^\gamma(\cdot)\}$, we can suppose (w.l.o.g.) that the $\{x^\gamma(\cdot)\}$ is uniformly bounded on each interval $[0, T_1]$. Let $f(\cdot) \in C_0^2(R^{r'})$, and let $\hat{A}_0^\gamma$ denote the $\hat{A}$ operator defined in Chapter 7.2, but with respect to the filtration $\mathcal{F}_{0,t}^\gamma = \mathcal{B}(x_0^\gamma(s), z_0^\gamma(s), \xi(\epsilon s/\beta^2), \psi(\epsilon s/\rho^2)$, $m_{\epsilon s}^\gamma(\cdot)$, $s \leq t)$. Note that $\mathcal{F}_{0,t}^\gamma = \mathcal{F}_{\epsilon t}^\gamma$, and it measures all the controls and noise which are used to drive $x_0^\gamma(\cdot)$ and $z_0^\gamma(\cdot)$ up to time t on their time scale. Let $E_{0,t}^\gamma$ denote the expectation conditioned on $\mathcal{F}_{0,t}^\gamma$ and define $\lambda = \beta/\sqrt{\epsilon}$. Then for each t_0, $f(z_0^\gamma(t_0 + \cdot)) \in \mathcal{D}(\hat{A}_0^\gamma)$ and (writing x_0^γ and z_0^γ for $x_0^\gamma(t_0 + t)$ and $z_0^\gamma(t_0 + t)$, resp., for simplicity) we have

$$\hat{A}_0^\gamma f(z_0^\gamma) = f_z'(z_0^\gamma)\dot{z}_0^\gamma = f_z'(z_0^\gamma)[H(x_0^\gamma, z_0^\gamma)$$

$$+ \tilde{H}(x_0^\gamma, z_0^\gamma, \xi((t_0 + t)/\lambda^2)) + J(x_0^\gamma, z_0^\gamma, \xi((t_0 + t)/\lambda^2))/\lambda]. \quad (2.5)$$

Following the idea behind the construction of the perturbation used in the example of Chapter 7 and in the proofs of Chapter 7.4, for each γ and

t_0 define the test function perturbation $f_1^\gamma(t_0 + t) = f_1^\gamma(t_0 + t, x_0^\gamma(t_0 + t), z_0^\gamma(t_0 + t))$ where

$$f_1^\gamma(v, x, z) = \int_v^\infty E_{0,v}^\gamma f_z'(z) J(x, z, \xi(s/\lambda^2))/\lambda ds. \qquad (2.6)$$

By a change of time scale $s \to \lambda^2 s$, we can write

$$f_1^\gamma(t_0 + t, x_0^\gamma, z_0^\gamma) = \lambda \int_{(t_0+t)/\lambda^2}^\infty E_{0,(t_0+t)}^\gamma f_z'(z_0^\gamma) J(x_0^\gamma, z_0^\gamma, \xi(s)) ds.$$

By (A2.4) and (A2.5), this expression is $O(\lambda)$ uniformly in t_0. Under the Gaussian assumption (A2.10), we have $f_1^\gamma(t) = O(\lambda)|\xi(t/\lambda^2)|$ and for any $\delta_0 > 0$, $T_0 < \infty$ and $T_1 < \infty$

$$\lim_\gamma \sup_{t_0 \leq T_1/\epsilon} P(\sup_{s \leq T_0} O(\lambda)|\xi((t_0 + s)/\lambda^2)| \geq \delta_0) = 0.$$

Analogous to the procedure in Theorem 7.4.3, define the perturbed test function $f^\gamma(t_0 + \cdot)$, parametrized by γ and t_0, by $f^\gamma(t_0 + t) = f(z_0^\gamma(t_0 + t)) + f_1^\gamma(t_0 + t)$. Then (7.3.7) holds, where (γ, t_0) replaces γ.

We have $f_1^\gamma(t_0 + \cdot) \in \mathcal{D}(\hat{A}_0^\gamma)$ and (again using the abbreviations $x_0^\gamma(t_0 + t) = x_0^\gamma, z_0^\gamma(t_0 + t) = z_0^\gamma$) we can write

$$\hat{A}_0^\gamma f_1^\gamma(t_0 + t) = -\frac{1}{\lambda} f_z'(z_0^\gamma) J(x_0^\gamma, z_0^\gamma, \xi((t_0 + t)/\lambda^2))$$

$$+ \frac{1}{\lambda^2} \int_{t_0+t}^\infty ds [E_{0,(t_0+t)}^\gamma f_z'(z_0^\gamma) J(x_0^\gamma, z_0^\gamma, \xi(s/\lambda^2))]_z' [J(x_0^\gamma, z_0^\gamma, \xi((t_0 + t)/\lambda^2)]$$

$$+ \frac{\epsilon}{\lambda \rho} \int_{t_0+t}^\infty ds [E_{0,(t_0+t)}^\gamma f_z'(z_0^\gamma) J(x_0^\gamma, z_0^\gamma, \xi(s/\lambda^2))]_x' F(x_0^\gamma, z_0^\gamma, \psi(\epsilon(t_0 + t)/\rho^2))$$

$$+ \text{ terms which are } O(\lambda). \qquad (2.7)$$

Via the scale change $s \to \lambda^2 s$, we see that the first integral on the right side of (2.7) is uniformly bounded, and the second is $O(\epsilon\lambda/\rho) = O(\sqrt{\epsilon}\beta/\rho) = O(\sqrt{\epsilon})$. Under the Gaussian assumption (A2.10), the two integrals on the right side of (2.7) are of the orders of $|\xi((t_0 + t)/\lambda^2)|^2$ and $|\xi((t_0 + t)/\lambda^2)| \cdot (\sqrt{\epsilon}\beta/\rho)$, resp., and the neglected terms are bounded by these factors times a function of γ which goes to zero as $\gamma \to 0$. Owing to the fact that the dominant "$1/\lambda$ term" in (2.5) is the negative of the first term of the right of (2.7) and to the bounds given earlier in this paragraph, for any $T_1 < \infty$ and $T_0 < \infty$ the set $\{\hat{A}_0^\gamma f^\gamma(t_0 + t), t_0 \leq T_1/\epsilon, t \leq T_0, \gamma \geq 0\}$ is uniformly integrable. Thus all the conditions of Theorem 7.3.3. are satisfied and the asserted tightness follows.

2. Weak convergence. We follow the method of Theorem 7.4.1. For bounded continuous $h(\cdot), t \geq 0, s \geq 0$ and $t_i \leq t \leq t + s$, and any sequence

$\{t_\gamma\}$ of positive real numbers with $\{\epsilon t_\gamma\}$ bounded, the martingale property (7.2.2) yields

$$Eh(z_0^\gamma(t_\gamma + t_i), x_0^\gamma(t_\gamma + t_i), i \leq q)\left[f^\gamma(t + t_\gamma + s) - f^\gamma(t + t_\gamma)\right.$$

$$\left. - \int_{t_\gamma + t}^{t_\gamma + t + s} \hat{A}_0^\gamma f^\gamma(u)du\right] = 0, \qquad (2.8)$$

and

$$Eh(z_0^\gamma(t_\gamma + t_i), x_0^\gamma(t_\gamma + t_i), i \leq q)\left[f^\gamma(t + t_\gamma + s) - f^\gamma(t + t_\gamma)\right.$$

$$\left. - E_{0,t_\gamma + t}^\gamma \int_{t_\gamma + t}^{t_\gamma + t + s} \hat{A}_0^\gamma f^\gamma(u)du\right] = 0. \qquad (2.9)$$

Abusing notation, let γ index a sequence such that $\{x_0^\gamma(t_\gamma + \cdot), z_0^\gamma(t_\gamma + \cdot)\}$ converges weakly, with the limit denoted by $(x_0(\cdot), z_0(\cdot))$. By the tightness of the original sequence $\{x^\gamma(\cdot)\}$, and the continuity of the limit process, it is not hard to show that $x_0(t) = x_0(0) = x_0$, all t, and we omit the details. Let us use the Skorohod representation so that we can suppose that $x_0^\gamma(t_\gamma + \cdot)$ and $z_0^\gamma(t_\gamma + \cdot)$ converge to their limits uniformly on each bounded interval, w.p.1. Now, as $\gamma \to 0$ we can replace the $(z_0^\gamma(t_\gamma + t_i), x_0^\gamma(t_\gamma + t_i))$ in (2.9) by the limit $(z_0(t_i), x_0(0))$. Also, we can replace $f^\gamma(t_\gamma + t + s)$ and $f^\gamma(t_\gamma + t)$ by their limits $f(z_0(t+s))$ and $f(z_0(t))$, resp. To complete the proof via the martingale method, we need to obtain the limits of the integral of $\hat{A}_0^\gamma f^\gamma(u)$ in (2.9). By the calculations in Part 1 of the proof, we have

$$\hat{A}_0^\gamma f^\gamma(u) = f_z'(z_0^\gamma(u))[H(x_0^\gamma(u), z_0^\gamma(u)) + \tilde{H}(x_0^\gamma(u), z_0^\gamma(u), \xi(u/\lambda^2))]$$

$$+ \frac{1}{\lambda^2}\int_u^\infty d\tau[E_{0,u}^\gamma f_z'(z_0^\gamma(u))J(x_0^\gamma(u), z_0^\gamma(u), \xi(\tau/\lambda^2))]_z'. \qquad (2.10)$$

$$\cdot J(x_0^\gamma(u), z_0^\gamma(u), \xi(u/\lambda^2)) + O(\sqrt{\epsilon}).$$

We will next "average out" the integral term in (2.9). Let $K^\gamma(u)$ denote the integral term on the right side of (2.10). For notational simplicity, let $t_\gamma = 0$. Let $\Delta \to 0$ as $\gamma \to 0$ such that $\lambda^2/\Delta \to 0$ and s/Δ is an integer. We can then write

$$E_{0,t}^\gamma\int_t^{t+s} K^\gamma(u)du = E_{0,t}^\gamma\sum_{i=0}^{s/\Delta - 1}\int_{t+i\Delta}^{t+i\Delta+\Delta} du\left\{\int_u^\infty d\tau[E_{0,u}^\gamma f_z'(z_0^\gamma(u))\cdot\right.$$

$$\left. \cdot J(x_0^\gamma(u), z_0^\gamma(u), \xi(\tau/\lambda^2))]_z' \cdot J(x_0^\gamma(u), z_0^\gamma(u), \xi(u/\lambda^2))/\lambda^2\right\} =$$

$$= \sum_{i=0}^{s/\Delta-1} \Delta \cdot \frac{\lambda^2}{\Delta} E_{0,t}^\gamma \int_{(t+i\Delta)/\lambda^2}^{(t+i\Delta+\Delta)/\lambda^2} du \left\{ \int_u^\infty d\tau [E_{0,\lambda^2 u}^\gamma f_z'(z_0^\gamma(u\lambda^2)) \cdot \right.$$

$$\left. \cdot J(x_0^\gamma(u\lambda^2), z_0^\gamma(u\lambda^2), \xi(\tau))]_z' \cdot J(x_0^\gamma(u\lambda^2), z_0^\gamma(u\lambda^2), \xi(u)) \right\}. \qquad (2.11)$$

Owing to the weak convergence of $z_0^\gamma(\cdot)$ and $x_0^\gamma(\cdot)$ and the continuity conditions (A2.4), (A2.5) and (A2.7), we can follow the procedure in Theorem 7.4.2 on each summand in (2.11) to get that the limit of (2.11) and that of

$$E_{0,t}^\gamma \sum_{i=0}^{s/\Delta-1} \Delta \cdot B_{x_0^\gamma(t+i\Delta)}^0 f(z_0^\gamma(t+i\Delta))$$

are the same.

By doing a similar averaging procedure on the integral of the term $f_z'(z_0^\gamma)\tilde{H}(x_0^\gamma, z_0^\gamma, \xi(u/\lambda^2))$ in (2.10), and using (A2.6) we can show that it disappears in the limit. Then taking limits in (2.9) as $\gamma \to 0$ yields

$$Eh(z_0(t_i), i \le q, x_0)[f(z_0(t+s)) - f(z_0(t)) - \int_t^{t+s} A_{x_0}^0 f(z_0(u))du] = 0.$$

Thus $z_0(\cdot)$ solves the martingale problem associated with A_x^0 with parameter $x = x_0$.

The calculations under the Gaussian assumption (A2.10) are similar and the details are omitted.   Q.E.D.

3. Convergence to the Averaged System

In this section, we show that the sequence $\{x^\gamma(\cdot), m^\gamma(\cdot)\}$ in (1.1) is tight and that the weak limits satisfy a type of "averaged" Itô equation. Theorem 3.1 is the wide band noise analog of Theorem 4.1.2.

Assumptions. We will use the following conditions on (1.1).

A3.1. $G(x, z, \alpha) = G_0(x, z) + G_1(x, \alpha)$. The $G_i(\cdot)$, $\tilde{G}(\cdot)$, $F(\cdot)$ and $F_x(\cdot)$ are continuous and are uniformly bounded on each (x, α)-set.

A3.2. The random functions

$$\int_\tau^\infty E_t^\psi F(x, z, \psi(s))ds, \int_\tau^\infty E_t^\psi F_x(x, z, \psi(s))ds$$

go to zero in the mean as $\tau - t \to \infty$, uniformly on each bounded (x, z)-set.

A3.3. For each x and z,

$$\frac{1}{\tau} \int_t^{t+\tau} E_t^\psi \tilde{G}(x, z, \psi(s))ds \xrightarrow{P} 0$$

as t and τ go to infinity.

A3.4. *For $f(\cdot) \in C_0^2(R^r)$, as t and τ go to infinity the random functions*

$$\frac{1}{\tau} \int_t^{t+\tau} ds \int_s^\infty du E_t^\psi [f_x'(x) F(x, z, \psi(u))]_x' F(x, z, \psi(s))$$

converge in the mean for each x, z.

Definitions. We denote the limit in (A3.4) by $(A^0 f)(x, z)$, and define $G_2(\cdot)$ and the symmetric matrix $a(\cdot) = \{a_{ij}(\cdot)\}$ by

$$(A^0 f)(x, z) = f_x'(x) G_2(x, z) + \frac{1}{2} \sum_{i,j} a_{ij}(x, z) f_{x_i x_j}(x).$$

By (A3.1), (A3.2) and the boundedness of $\psi(\cdot)$ in (A2.2), both $G_2(\cdot)$ and $a(\cdot)$ are continuous. Define the averaged differential operator $\bar{A}^0$ and the function $\bar{G}_2(\cdot)$ and the symmetric matrix $\bar{a}(\cdot)$ by

$$\bar{A}^0 f(x) = \int_{\mathcal{U}} (A^0 f)(x, z) \mu_x(dz) = f_x'(x) \bar{G}_2(x) + \frac{1}{2} \sum_{i,j} \bar{a}_{ij}(x) f_{x_i x_j}(x).$$

(3.1)

For notational convenience, we suppose that $\bar{a}(\cdot)$ has a factorization $\bar{a}(\cdot) = \bar{\sigma}(\cdot)\bar{\sigma}'(\cdot)$, where $\bar{\sigma}(\cdot)$ is continuous. Now, define the function $\overline{G}_0(x) = \int_{\mathcal{U}} G_0(x, z) \mu_x(dz)$ and $\bar{G}(x, \alpha) = \bar{G}_0(x) + G_1(x, \alpha) + \overline{G}_2(x)$. Define the operators $\bar{A}^\alpha$ and $\bar{A}^m$ acting on $C^2(R^r)$ by:

$$\bar{A}^\alpha f(x) = \bar{A}^0 f(x) + f_x'(x) G_1(x, \alpha) + f_x'(x) \bar{G}_0(x),$$

$$(\bar{A}^m f)(x, t) = \int_{\mathcal{U}} \overline{A}^\alpha f(x) m_t(d\alpha) = f_x'(x) \int_{\mathcal{U}} \bar{G}(x, \alpha) m_t(d\alpha)$$

$$+ \frac{1}{2} \sum_{i,j} \bar{a}_{ij}(x) f_{x_i x_j}(x).$$

(3.2)

The process $x(\cdot)$ associated with $\bar{A}^\alpha$ can be written as

$$dx = [\int_{\mathcal{U}} G_1(x, \alpha) m_t(d\alpha) + \bar{G}_0(x) + \overline{G}_2(x)] dt + \bar{\sigma}(x) dw, \qquad (3.3)$$

where $(m(\cdot), w(\cdot))$ is an admissible pair in the sense of Chapter 3. It will turn out that (3.3) represents the proper averaged system for (1.1) and (1.2).

Assumptions. We require a condition on the averaged system (3.3).

A3.5. *For each admissible pair $(m(\cdot), w(\cdot))$ and initial condition $x(0)$, there is a unique weak sense solution to (3.3).*

A3.6. *Let $k(\cdot)$ and $g(\cdot)$ be bounded and continuous real valued functions and let $k(\cdot)$ take the form $k(x, z, \alpha) = k_0(x, z) + k_1(x, \alpha)$.*

We also use the following "Gaussian" alternative to (A3.2)–(A3.4):

A3.7. $\psi(\cdot)$ *is a stable Gauss-Markov process with a stationary transition function and*

$$\tilde{G}(x, z, \psi) = \tilde{G}_0(x, z)\psi, \ F(x, z)\psi = F_0(x, z)\psi,$$

where $\tilde{G}_0(\cdot)$ and $F_0(\cdot)$ satisfy the smoothness requirements on $\tilde{G}(\cdot)$ and $F(\cdot)$ in (A3.1).

The cost function for the averaged system (3.3) is (4.1.8): i.e.,

$$V_{T_0}(x, m) = E_x^m \int_0^{T_0} \int_{\mathcal{U}} \bar{k}(x(t), \alpha)m_t(d\alpha)dt + E_x^m g(x(T_0)). \tag{3.4}$$

where $\bar{k}(x, \alpha) = \int_{\mathcal{U}} k(x, z, \alpha)\mu_x(d\alpha)$. The next theorem is the "wide-band" noise analog of Theorem 4.1.2. We prove it first when $F(\cdot)$ does not depend on z, although the z argument is retained in $F(\cdot)$ for reference in Theorem 3.2.

Theorem 3.1. *Assume either $\{(A2.1)–(A2.9), (A3.1)–(A3.6)\}$ or $\{(A2.1), (A2.3), (A2.8)–(A2.10), (A3.1), (A3.5)-(A3.7)\}$, and let $\{x^\gamma(0)\}$ be tight. Let $F(\cdot)$ not depend on z. Let $m^\gamma(\cdot)$ be an admissible relaxed control for (1.1), (1.2). Then $\{x^\gamma(\cdot), m^\gamma(\cdot)\}$ is tight. Let $(x(\cdot), m(\cdot))$ denote the limit of a weakly convergent subsequence (indexed by γ_n). Then there is a standard Wiener process $w(\cdot)$ such that $(m(\cdot), w(\cdot))$ is admissible, $(x(\cdot), m(\cdot), w(\cdot))$ satisfies (3.3) and*

$$V_{T_0}^{\gamma_n}(x, z, m^{\gamma_n}) \to V_{T_0}(x, m). \tag{3.5}$$

Remark on the Proof. We divide the proof into several sections in order to emphasize the several levels of averaging. The proof will be done under the first set of conditions, with only occasional remarks concerning the Gaussian condition.

Proof. Part 1. *Tightness of $\{x^\gamma(\cdot)\}$.* As usual in the proofs of tightness, we assume that the processes are truncated, without introducing the truncation notation. Let $f(\cdot) \in C_0^2(R^r)$. Then $f(x^\gamma(\cdot)) \in \mathcal{D}(\hat{A}^\gamma)$ and (we often write x^γ and z^γ for $x^\gamma(t)$ and $z^\gamma(t)$, resp., for simplicity)

$$\hat{A}^\gamma f(x^\gamma) = \dot{f}(x^\gamma) = f_x'(x^\gamma)[\int_{\mathcal{U}} G(x^\gamma, z^\gamma, \alpha)m_t^\gamma(d\alpha)$$

$$+ \tilde{G}(x^\gamma, z^\gamma, \psi^\rho) + F(x^\gamma, z^\gamma, \psi^\rho)/\rho]. \tag{3.6}$$

Define the test function perturbation $f_1^\gamma(t) = f_1^\gamma(x^\gamma, z^\gamma, t)$, where

$$f_1^\gamma(x^\gamma, z^\gamma, t) = \int_t^\infty f_x'(x^\gamma) E_t^\gamma F(x^\gamma, z^\gamma, \psi^\rho(s)) ds/\rho$$

$$= \rho \int_{t/\rho^2}^\infty f_x'(x^\gamma) E_t^\gamma F(x^\gamma, z^\gamma, \psi(s)) ds = O(\rho). \qquad (3.7)$$

The $O(\rho)$ value follows from (A3.2). (Under the Gaussian assumption, (3.7) equals $O(\rho)|\psi(t/\rho^2)|$). We have $f_1^\gamma(\cdot) \in \mathcal{D}(\hat{A}^\gamma)$ and

$$\hat{A}^\gamma f_1^\gamma(t) = -f_x'(x^\gamma) F(x^\gamma, z^\gamma, \psi^\rho(t))/\rho$$

$$+ \frac{1}{\rho^2} \int_t^\infty ds [E_t^\gamma f_x'(x^\gamma) F(x^\gamma, z^\gamma, \psi^\rho(s))]_x' \cdot F(x^\gamma, z^\gamma, \psi^\rho(t)) \qquad (3.8)$$

$$+ \text{ error terms.}$$

The error terms are

$$\frac{1}{\rho} \int_t^\infty ds [E_t^\gamma f_x'(x^\gamma) F(x^\gamma, z^\gamma, \psi^\rho(s))]_x' \cdot \left[\int_U G(x^\gamma, z^\gamma, \alpha) m_t^\gamma(d\alpha) \right.$$

$$\left. + \tilde{G}(x^\gamma, z^\gamma, \psi^\rho(t)) \right] = O(\rho). \qquad (3.9)$$

(In the "Gaussian" case, (3.9) equals $O(\rho)[1 + |\psi^\rho(t)|^2]$.) By the use of the scale change $s \to \rho^2 s$ and (A3.1) and (A3.2), we see that the main term on the right hand side term in (3.8) is bounded uniformly in γ, t. (It is $O(|\psi^\rho(t)|^2)$ in the "Gaussian" case.) Define the perturbed test function $f^\gamma(\cdot) = f(x^\gamma(\cdot)) + f_1^\gamma(\cdot)$. Then $f^\gamma(\cdot) \in \mathcal{D}(\hat{A}^\gamma)$ and $\hat{A}^\gamma f^\gamma(t)$ is bounded uniformly in γ, t. (It is $O(|\psi^\rho(t)|^2 + |\psi^\rho(t)| |\xi^\beta(t)| + 1)$ in the "Gaussian" case.) Thus, analogously to the case in Theorem 2.1, tightness of (the truncated) $\{x^\gamma(\cdot)\}$ follows from Theorem 7.3.3, since for small γ and each $T_0 < \infty$, $f_1^\gamma(\cdot)$ is "uniformly small" on $[0, T_0]$ and $\{\hat{A}^\gamma f^\gamma(t), t \leq T_0, \gamma > 0\}$ is uniformly integrable. The sequence $\{m^\gamma(\cdot)\}$ is always tight.

Part 2. Weak Convergence. Now, let us abuse notation and use γ to index a weakly convergent subsequence with limit denoted by $(x(\cdot), m(\cdot))$. Let $t \geq 0$, $\tau > 0$, $f(\cdot) \in C_0^2(R^r)$, and define $f_1^\gamma(\cdot)$ as in Part 1. In this part, we show that

$$\int_t^{t+\tau} E_t^\gamma \hat{A}^\gamma f^\gamma(s) ds - \int_t^{t+\tau} E_t^\gamma \int_U \overline{A}^\alpha f(x^\gamma(s)) m_s^\gamma(d\alpha) ds \xrightarrow{P} 0. \qquad (3.10)$$

as $\gamma \to 0$. Since the term is bounded, this is equivalent to convergence in the mean. For notational simplicity, we work only with the scalar case and the "second order" component of $\int_t^{t+\tau} E_t^\gamma \hat{A}^\gamma f^\gamma(s)ds$; namely, with the term

$$\frac{1}{\rho^2} \int_t^{t+\tau} E_t^\gamma f_{xx}(x^\gamma(s))ds \times$$

$$\int_s^\infty E_s^\gamma F(x^\gamma(s), z^\gamma(s)), \psi^\rho(u))F(x^\gamma(s), z^\gamma(s), \psi^\rho(s))du. \tag{3.11}$$

By the continuity condition in (A3.2) and the tightness of $\{x^\gamma(\cdot)\}$, the limits in (3.11) do not "change much" if we approximate $x^\gamma(\cdot)$ by piecewise constant functions if the "intervals" of constancy are small. More precisely, let $\Delta > 0$ be such that τ/Δ is an integer. Then the limits (in the mean) of (3.11) (as $\gamma \to 0$) and those of (3.12) (as $\gamma \to 0$, and then $\Delta \to 0$) are the same:

$$\sum_{i=0}^{\tau/\Delta-1} \frac{1}{\rho^2} \int_{(t+i\Delta)}^{(t+i\Delta+\Delta)} ds E_t^\gamma f_{xx}(x^\gamma(t+i\Delta)) \times$$

$$\int_s^\infty du E_s^\gamma F(x^\gamma(t+i\Delta), z^\gamma(s), \psi^\rho(u))F(x^\gamma(t+i\Delta), z^\gamma(s), \psi^\rho(s)). \tag{3.12}$$

Now, change scale $s \to \rho^2 s$, $u \to \rho^2 u$ in (3.12), and rewrite it as

$$\sum_{i=0}^{\tau/\Delta-1} \rho^2 \int_{(t+i\Delta)/\rho^2}^{(t+i\Delta+\Delta)/\rho^2} ds E_t^\gamma f_{xx}(x^\gamma(t+i\Delta)) \times \tag{3.13}$$

$$\int_s^\infty du E_{\rho^2 s}^\gamma F(x^\gamma(t+i\Delta), z^\gamma(\rho^2 s), \psi(u))F(x^\gamma(t+i\Delta), z^\gamma(\rho^2 s), \psi(s)).$$

Using (A2.1), there are $\delta > 0$ (which depend on γ) such that $\delta/\rho^2 \to \infty$ and $\delta = o(\epsilon)$ as $\gamma \to 0$. Since $\delta = o(\epsilon)$, it follows from Theorem 2.1 that

$$\sup_{s \le \delta} |z^\gamma(k\delta + s) - z^\gamma(k\delta)| \xrightarrow{P} 0 \tag{3.14}$$

as $\delta \to 0$ uniformly in k such that $k\delta \le (t + \tau)$. Because of (3.14) and the continuity conditions (A3.1) and (A3.2), for small Δ, δ and ρ, (3.13) is arbitrarily closely approximated by (3.15):

$$\sum_{i=0}^{\tau/\Delta-1} \sum_{k=(t+i\Delta)/\delta}^{(t+i\Delta+\Delta)/\delta-1} \delta\frac{\rho^2}{\delta} \int_{k\delta/\rho^2}^{(k\delta+\delta)/\rho^2} ds\Big\{ E_t^\gamma f_{xx}(x^\gamma(t+i\Delta)) \times$$

$$\int_s^\infty du E_{\rho^2 s}^\gamma F(x^\gamma(t+i\Delta), z^\gamma(k\delta), \psi(u))F(x^\gamma(t+i\Delta), z^\gamma(k\delta), \psi(s)) \Big\}.$$

$$\tag{3.15}$$

The approximation is in the sense that the limits in the mean of (3.13) as $\gamma \to 0$ and then $\Delta \to 0$ are the same as those of (3.15) as $\gamma \to 0$, then $\delta \to 0$ and then $\Delta \to 0$.

Given $\eta > 0$, there is $N_\eta < \infty$ such that $P(|z^\gamma(s)| \geq N_\eta) \leq \eta$, all $s \leq t + \tau$. Let $d_n^\eta(\cdot)$, $n = 1, 2, \ldots$, be a finite set of non-negative continuous functions with support having diameter $\leq \eta$ and such that $\sum_n d_n^\eta(x, z) = 1$ for $|z| \leq N_\eta$, and x in the support of $f(\cdot)$, with the sum being no larger than unity otherwise. Let (x_n^η, z_n^η) be in the support of $d_n^\eta(\cdot)$. Now, using the technique of the proof of Theorem 7.4.2 and the continuity conditions (A3.1) and (A3.2), for small enough η we can approximate (3.15) arbitrarily closely (in the mean) by using the sum

$$\sum_n d_n^\eta(x^\gamma(i\Delta + \Delta), z^\gamma(k\delta)) \int_{k\delta/\rho^2}^{(k\delta+\delta)/\rho^2} ds E_t^\gamma f_{xx}(x_n^\eta) \times$$

$$\int_s^\infty du E_s^\psi F(x_n^\eta, z_n^\eta, \psi(u)) F(x_n^\eta, z_n^\eta, \psi(s)) \qquad (3.16)$$

to replace the integral in (3.15) for each k and i.

By the approximation (3.16) together with (A3.4) and the definition of $a(\cdot)$ below (A3.4), the limits in (3.17) (as $\gamma \to 0$, then $\delta \to 0$, and then $\Delta \to 0$) are the same as those of (3.15) as $\gamma \to 0$, then $\delta \to 0$, and then $\Delta \to 0$:

$$\sum_{i=0}^{\tau/\Delta-1} \sum_{k=(t+i\Delta)/\delta}^{(t+i\Delta+\Delta)/\delta-1} \delta \frac{1}{2} E_t^\gamma f_{xx}(x^\gamma(t+i\Delta)) a(x^\gamma(t+i\Delta), z^\gamma(k\delta)). \qquad (3.17)$$

By (3.14), the limits (in the mean) of (3.17) and (3.18) (as $\gamma \to 0$, then $\Delta \to 0$) are the same:

$$\sum_{i=0}^{\tau/\Delta-1} \int_{t+i\Delta}^{t+i\Delta+\Delta} \frac{1}{2} E_t^\gamma f_{xx}(x^\gamma(s)) a(x^\gamma(s), z^\gamma(s)) ds$$

$$= \frac{1}{2} \int_t^{t+\tau} E_t^\gamma f_{xx}(x^\gamma(s)) a(x^\gamma(s), z^\gamma(s)) ds. \qquad (3.18)$$

We next show that the $z^\gamma(\cdot)$ process in (3.18) can be averaged out in essentially the same way that $z^\epsilon(\cdot)$ was in Theorem 4.1.2. Define the occupation measures $P_s^{\gamma, \Delta}(\cdot)$ by

$$P_s^{\gamma, \Delta}(B \times C) = \frac{1}{\Delta} \int_s^{s+\Delta} E_s^\gamma I_B(x^\gamma(u)) I_C(z^\gamma(u)) du$$

$$= \frac{\epsilon}{\Delta} \int_{s/\epsilon}^{(s+\Delta)/\epsilon} E_s^\gamma I_B(x_0^\gamma(u)) I_C(z_0^\gamma(u)) du$$

$$= \frac{\epsilon}{\Delta} \int_{s/\epsilon}^{(s+\Delta)/\epsilon} E_{0,s/\epsilon}^\gamma I_B(x_0^\gamma(u)) I_C(z_0^\gamma(u)) du,$$

where $E_{0,s}^\gamma = E_{\epsilon s}^\gamma$ is defined in Theorem 2.1, and $B \in \mathcal{B}(R^r)$ and $C \in \mathcal{B}(R^{r'})$. Then (3.18) can be written as

$$\frac{1}{2} \sum_{i=0}^{\tau/\Delta-1} \Delta E_t^\gamma \int f_{xx}(x)a(x,z)P_{t+i\Delta}^{\gamma,\Delta}(dxdz). \tag{3.19}$$

It will be shown in Part 3 below that if the $P_{(t+i\Delta)}^{\gamma,\Delta}(dxdz)$ in (3.19) is replaced by $\mu_{x^\gamma(t+i\Delta)}(dz)\delta_{x^\gamma(t+i\Delta)}(x)dx$, then the limits in the mean of (3.19) (as $\gamma \to 0$, then $\Delta \to 0$) do not change. This, together with an analogous averaging argument for the terms other than (3.11) which appear in the left side of (3.10) yields (3.10).

For the remainder of this section, suppose that (3.10) holds and let $h(\cdot), p, q, \varphi_j(\cdot)$ and t_i be as in Theorem 4.1.2. Then

$$Eh(x^\gamma(t_i), (m^\gamma, \varphi_j)_{t_i}, i \le q, j \le p)[f^\gamma(t+\tau) - f^\gamma(t) - \int_t^{t+\tau} \hat{A}^\gamma f^\gamma(s)ds] = 0.$$

Thus, by (3.10) and the weak convergence, the limit $(x(\cdot), m(\cdot))$ must satisfy

$$Eh(x(t_i), (m, \varphi_j)_{t_i}, i \le q, j \le p)\left[f(x(t+\tau)) - f(x(t))\right.$$

$$\left. - \int_t^{t+\tau} \int_U \bar{A}^\alpha f(x(s))m_s(d\alpha)ds\right] = 0. \tag{3.20}$$

Thus, $(x(\cdot), m(\cdot))$ solves the martingale problem for operator $\bar{A}^\alpha$. The rest of the proof follows the lines of Theorem 4.1.2. We need only prove that the occupation measure can be (asymptotically) replaced by the invariant measures.

Part 3. Convergence of the occupation measures. The proof parallels that of Part 4 of the proof of Theorem 4.1.2. Let $\{s_\gamma\}$ be such that $\epsilon s_\gamma \to s \le t + \tau$ as $\gamma \to 0$. For $f(\cdot) \in C_0^2(R^{r'})$, let $f^\gamma(\cdot)$ be defined as in Theorem 2.1 (not as previously defined in this proof); namely, we use $f^\gamma(t) = f(z_0^\gamma(t)) + f_1^\gamma(t, x_0^\gamma(t), z_0^\gamma(t))$ where $f_1^\gamma(\cdot)$ is defined by (2.6). Define the martingale $M_f^\gamma(\cdot)$ by

$$M_f^\gamma(v) = f^\gamma(s_\gamma + v) - f^\gamma(s_\gamma) - \int_{s_\gamma}^{s_\gamma+v} \hat{A}_0^\gamma f^\gamma(u)du,$$

where $\hat{A}_0^\gamma$ is defined in Theorem 2.1. By the tightness of $\{x^\gamma(\cdot)\}$, we can suppose w.l.o.g. that the $x^\gamma(\cdot)$ are uniformly bounded or truncated. Below, we will set $v = \Delta/\epsilon$. By (2.10) and the fact that $M_f^\gamma(\cdot)$ is a martingale, we can write

$$0 = E_{0,s_\gamma}^\gamma \frac{\epsilon}{\Delta} M_f^\gamma(\Delta/\epsilon) = \lim \frac{\epsilon}{\Delta} \int_{s_\gamma}^{s_\gamma + \Delta/\epsilon} E_{0,s_\gamma}^\gamma \hat{A}_0^\gamma f^\gamma(u))du$$

$$= \lim \frac{\epsilon}{\Delta} \int_{s_\gamma}^{s_\gamma + \Delta/\epsilon} E_{0,s_\gamma}^\gamma f_z'(z_0^\gamma(u))[H(x_0^\gamma(u), z_0^\gamma(u))$$

$$+ \tilde{H}(x_0^\gamma(u), z_0^\gamma(u), \xi(\epsilon u/\beta^2))]du +$$

$$+ \lim \frac{\epsilon}{\Delta} \int_{s_\gamma}^{s_\gamma + \Delta/\epsilon} du E_{0,s_\gamma}^\gamma \int_u^\infty dv [E_{0,u}^\gamma f_z'(z_0^\gamma(u)) J(x_0^\gamma(u), z_0^\gamma(u), \xi(\epsilon v/\beta^2))]_z' \times$$

$$[J(x_0^\gamma(u), z_0^\gamma(u), \xi(\epsilon u/\beta^2))/(\beta^2/\epsilon)]. \tag{3.21}$$

The limit in (3.21) is for $\gamma \to 0$ and then $\Delta \to 0$. The limit is uniform in the sequence $\{s_\gamma\}$ and in s in any compact set. By (3.21) and the last part of the proof of Theorem 2.1, we have

$$0 = \lim \frac{\epsilon}{\Delta} \int_{s_\gamma}^{s_\gamma + \Delta/\epsilon} E_{0,s_\gamma}^\gamma A_{x_0^\gamma(u)}^0 f(z_0^\gamma(u))du. \tag{3.22}$$

The limit in (3.22) is taken as $\gamma \to 0$ and then $\Delta \to 0$. It is also uniform in $\{s_\gamma\}$ and in s in any compact set. Equivalently to (3.22), we have

$$0 = \lim \int A_x^0 f(z) P_{\epsilon s_\gamma}^{\gamma, \Delta}(dxdz).$$

Now, let $\epsilon s_\gamma \to s$ and proceed as in Theorem 4.1.2 to show that all weak limits of $\{P_{\epsilon s_\gamma}^{\gamma, \Delta}(\cdot)\}$ are representable as $\mu_{x(s)}(dz)\delta_{x(s)}(x)dx$. Q.E.D.

$F(\cdot)$ **depending on** z. If $F(\cdot)$ depends on z, then we need to restrict the conditions somewhat. We will use:

A3.8. *Let* $H, \tilde{H}, J$, *and the first partial* x *and* z *derivatives of* F, F_z *and* J *be bounded and continuous on each bounded* x-*set. Let* $\rho/\epsilon \to 0$ *as* $\gamma \to 0$, *and suppose that* $\psi(\cdot)$ *and* $\xi(\cdot)$ *are mutually independent. For the Gaussian case, assume the analogous conditions, except let* $\beta/\sqrt{\epsilon} = O(\beta^\alpha)$ *for some* $\alpha > 0$.

Theorem 3.2. *Assume* (A3.8) *and the conditions of Theorem 3.1, except let* $F(\cdot)$ *depend on* z. *Then the conclusions of Theorem 3.1 hold.*

Proof. Only the "tightness" part of the proof of Theorem 3.1 will be modified here. The weak convergence argument uses analogous modifications. If $F(\cdot)$ depends on z, then we need to add the following term to (3.8):

$$\frac{1}{\rho} \int_t^\infty ds [E_t^\gamma f_x(x^\gamma) F(x^\gamma, z^\gamma, \psi^\rho(s))]_z' \times$$

$$\left[\frac{H(x^\gamma, z^\gamma)}{\epsilon} + \frac{\tilde{H}(x^\gamma, z^\gamma, \xi^\beta(t))}{\epsilon} + \frac{J(x^\gamma, z^\gamma, \xi^\beta(t))}{\sqrt{\epsilon}\beta} \right]. \qquad (3.23)$$

We do the non-Gaussian case only. The first two terms in (3.23) (due to H and $\tilde{H}$) are $O(\rho/\epsilon)$, as can be seen by use of the usual scale change $s \to \rho^2 s$. Thus, they are asymptotically negligable. To deal with the term containing the $J(\cdot)$, we need to use a "second order" test function perturbation. Define $f_2^\gamma(t) = f_2^\gamma(x^\gamma(t), z^\gamma(t), t)$, where

$$f_2^\gamma(x^\gamma, z^\gamma, t) = \frac{1}{\rho\beta\sqrt{\epsilon}} \int_t^\infty ds E_t^\gamma \int_s^\infty du [E_s^\gamma f_x'(x^\gamma) F(x^\gamma, z^\gamma, \psi^\rho(u))]_z' \times$$

$$J(x^\gamma, z^\gamma, \xi^\beta(s)).$$

By a change of variable $s \to \beta^2 s$, $u \to \rho^2 u$, we see that

$$f_2^\gamma(x^\gamma, z^\gamma, t) = O(\rho\beta/\sqrt{\epsilon}).$$

Also, $f_2^\gamma(\cdot) \in \mathcal{D}(\hat{A}^\gamma)$ and $\hat{A}^\gamma f_2^\gamma(t)$ equals the negative of the "bad" term in (3.23) (the one with the $J(\cdot)$) plus terms of order $O(\frac{\rho\beta}{\sqrt{\epsilon}}(\frac{1}{\epsilon} + \frac{1}{\beta\sqrt{\epsilon}} + \frac{1}{\rho}))$.

Defining $f^\gamma(\cdot) = f(x^\gamma(\cdot)) + f_1^\gamma(\cdot) + f_2^\gamma(\cdot)$, where $f_1^\gamma(\cdot)$ was defined in Part 1 of the proof of Theorem 3.1, and using Theorem 7.3.3 yields the tightness, since $\hat{A}^\gamma f^\gamma(t)$ is bounded and $f_1^\gamma(t) = O(\rho)$ and $f_2^\gamma(t) = O(\rho\beta/\sqrt{\epsilon})$. **End of comments on the proof.**

4. The Optimality Theorem

Theorem 4.1. *Assume the conditions of either Theorem 3.1 or Theorem 3.2 and either* (a) *or* (b) *below.*

(a) *For each $\delta > 0$, the averaged system (3.3) and cost (3.4) has a continuous δ-optimal feedback control $u^\delta(x, t)$ under which the solution to (3.3) is unique in the weak sense.*

(b) *$F(x, z, \psi) = F(x, z)\psi$ and, for each $\delta > 0$, (3.3), (3.4) has a δ-optimal relaxed control $m^\delta(\cdot)$ under which the solution to (3.3) is unique in the weak sense.*

Then

$$V_{T_0}^\gamma(x, z) = \inf_{m^\gamma adm.} V_{T_0}^\gamma(x, z, m^\gamma) \to V_{T_0}(x) = \inf_{m\, adm.} V_{T_0}(x, m).$$

Proof. Under (a), the proof follows from the asserted δ-optimality and uniqueness and the weak convergence argument of Theorems 3.1 or 3.2. The $u^\delta(\cdot)$ is used as the comparison control in lieu of the comparison controls of the type constructed in Theorem 4.2.1.

Under (b), the proof is similar to that of Theorem 4.2.1, except that the weak convergence arguments are those of Theorems 3.1 or 3.2. Define $W^\rho(t) = \int_0^t \psi^\rho(s)ds/\rho$. By appropriate normalizations, we can suppose w.l.o.g., that $W^\rho(\cdot) \Rightarrow w(\cdot)$, a standard vector-valued Wiener process. In fact it can be shown that $\{x^\gamma(\cdot), m^\gamma(\cdot), W^\rho(\cdot)\}$ is tight and any weak limit $(x(\cdot), m(\cdot), w(\cdot))$ satisfies (3.3). Now use the construction of Theorem 4.2.1, with $W^\rho(\cdot)$ replacing the $w^\epsilon(\cdot)$ there.   Q.E.D.

5. The Average Cost Per Unit Time Problem

In this section, we do the "wide band noise" analog of the results of Chapters 4.4 and 5.4, via the functional occupation measure approach. In analogy to (4.4.1) and (4.4.2), define

$$\Lambda_T^\gamma(x, z, m^\gamma) = \frac{1}{T} \int_0^T \int_\mathcal{U} k(x^\gamma(s), z^\gamma(s), \alpha)m_s^\gamma(d\alpha)ds,$$

$$\Lambda^\gamma(x, z, m^\gamma) = \overline{lim}_T E_{x,z}^{m^\gamma} \Lambda_T^\gamma(x, z, m^\gamma),$$

$$\Lambda(x, m) = \overline{lim}_T E_{x,z}^m \frac{1}{T} \int_0^T \int_\mathcal{U} \overline{k}(x(s), \alpha)m_s(d\alpha)ds,$$

where the averaged system $x(\cdot)$ is defined by (3.3). Define the functional occupation measures $\tilde{P}^{\gamma,t}(\cdot)$ and $\tilde{P}_T^\gamma(\cdot)$ as $\tilde{P}^{\epsilon,t}(\cdot)$ and $\tilde{P}_T^\epsilon(\cdot)$ were defined above Theorem 5.4.1, but with γ relacing ϵ. Recall the definition of a δ-optimal feedback control given above Theorem 5.4.2.

Theorem 5.1. *Assume the conditions of Theorems 3.1 or 3.2. Suppose that for each $\delta > 0$, there is an admissible δ-optimal control $m^\gamma(\cdot)$ for the cost functional $\Lambda^\gamma(x, z, m)$ such that the corresponding sequence $\{x^\gamma(t), z^\gamma(t), t < \infty, (small) \gamma > 0\}$ is tight. Then the conclusions of Theorem 5.4.1 hold (with γ replacing ϵ), where the limit averaged system is (3.3).*

Theorem 5.2. *Assume the conditions of Theorems 3.1 or 3.2. For each $\delta > 0$ suppose that there is a δ-optimal continuous feedback (a function of x only) control $u^\delta(\cdot)$ for (3.3) with cost $\Lambda(x, u^\delta) = \Lambda(u^\delta)$, under which (3.3) has a unique weak sense solution for each initial condition and a unique invariant measure, and when that control is applied to (1.1), (1.2), the set $\{x^\gamma(t), z^\gamma(t), (small) \gamma > 0, t < \infty\}$ is tight. Then the conclusions of Theorem 5.4.2 hold with γ replacing ϵ.*

Proof of Theorem 5.1. The details are essentially combinations of those in Theorems 2.1, 3.1, 3.2, and those in Chapter 5.4, and only a few details will be given. We basically follow the proof and terminology in Theorem 5.4.1, but with an "extra level" of averaging required due to the wide band noise. For simplicity, let $F(\cdot)$ not depend on z. Let $f(\cdot) \in C_0^2(R^r)$ and define

$f_1^\gamma(x, z, t)$ by (3.7) and set $f^\gamma(t) = f(x^\gamma(t)) + f_1^\gamma(x^\gamma(t), z^\gamma(t), t)$. Define the function $F_1(\cdot)$ as in Theorem 5.3.1 but where the $\overline{A}^\alpha$ operator is that for the process (3.3).

Define the martingale

$$M_f^\gamma(t) = f^\gamma(t) - f^\gamma(0) - \int_0^t \hat{A}^\gamma f^\gamma(s)ds.$$

Recall the definition of the shifted process $x_t^\gamma(\cdot) = x^\gamma(t + \cdot)$. Then we can write

$$\frac{1}{T}\int_0^T dt\left\{ h(x_t^\gamma(t_i), (\Delta_t m^\gamma, \psi_j)_{t_i}, i \leq q, j \leq p) \cdot \left[f^\gamma(t + \tau + s) \right.\right.$$

$$\left.\left. - f^\gamma(t + \tau) - \int_{t+\tau}^{t+\tau+s} du\, \hat{A}^\gamma f^\gamma(u) \right]\right\} \tag{5.1}$$

$$= \frac{1}{T}\int_0^T dt\{h(x_t^\gamma(t_i), (\Delta_t m^\gamma, \psi_j)_{t_i}, i \leq q, j \leq p) \cdot [M_f^\gamma(t+\tau+s) - M_f^\gamma(t+\tau)]\}.$$

For each $T_1 < \infty$, we have

$$\sup_{\gamma, t} \sup_{s \leq T_1} E[M_f^\gamma(t + s) - M_f^\gamma(t)]^2 < \infty. \tag{5.2}$$

By (5.2) and the martingale property of $M_f^\gamma(\cdot)$, the mean square value of the right side of (5.1) goes to zero as $T \to \infty$ and $\gamma \to 0$. Using this and the fact that $f_1^\gamma(\cdot) \to 0$ yields

$$\lim_T \frac{1}{T}\int_0^T dt\left\{ h(x_t^\gamma(t_i), (\Delta_t m^\gamma, \psi_j)_{t_i}, i \leq q, j \leq p) \times \right.$$

$$\left. \left[f(x_t^\gamma(\tau + s)) - f(x_t^\gamma(\tau)) - \int_{t+\tau}^{t+\tau+s} \hat{A}^\gamma f^\gamma(u)du \right] \right\} = 0, \tag{5.3}$$

where the limit is in the mean.

Now, rewrite the left side of (5.3) as

$$\int F_1(\phi, v)\tilde{P}_T^\gamma(d\phi dv) + \frac{1}{T}\int_0^T dt\left\{ h(x_t^\gamma(t_i), (\Delta_t m^\gamma, \psi_j)_{t_i}, i \leq q, j \leq p) \times \right.$$

$$\left. \int_{t+\tau}^{t+\tau+s} du\left[\int_{\mathcal{U}} \bar{A}^\alpha f(x^\gamma(u))m_u^\gamma(d\alpha) - \int_{\mathcal{U}} E_{t+\tau}^\gamma \bar{A}^\alpha f(x^\gamma(u))m_u^\gamma(d\alpha) \right] \right\}$$

$$- \frac{1}{T}\int_0^T dt\left\{ h(x_t^\gamma(t_i), (\Delta_t m^\gamma, \psi_j)_{t_i}, i \leq q, j \leq p) \times \right.$$

$$\left. E_{t+\tau}^\gamma \int_{t+\tau}^{t+\tau+s} du\left[\hat{A}^\gamma f^\gamma(u) - \int_{\mathcal{U}} \overline{A}^\alpha f(x^\gamma(u))m_u^\gamma(d\alpha) \right] \right\} -$$

$$- \frac{1}{T} \int_0^T dt \Big\{ h(x_i^\gamma(t_i), (\Delta_t m^\gamma, \psi_j)_{t_i}, i \leq q, j \leq p) \times$$

$$\int_{t+\tau}^{t+\tau+s} du [\hat{A}^\gamma f^\gamma(u) - E_{t+\tau}^\gamma \hat{A}^\gamma f^\gamma(u)] \Big\}. \qquad (5.4)$$

The second and the right hand term in (5.4) both go to zero in mean square as $T \to \infty$, because of the centering by the conditional expectation. We can also show that the third term goes to zero in the mean as $T \to \infty$ and $\gamma \to 0$ by using methods similar to those used in Theorems 2.1 and 3.1 to approximate the conditional expectation (given the past) of the integral of $\hat{A}^\gamma f^\gamma(u)$ by that of $\int_{\mathcal{U}} \overline{A}^\alpha f(x^\gamma(u)) m_u^\gamma(d\alpha)$ to average out the noise $\psi^\rho(\cdot)$ and the $z^\gamma(\cdot)$. The details are then completed as in Theorem 4.5.1.

We omit the details of the proof of Theorem 5.2.

9

Stability Theory

0. Outline of the Chapter

The tightness assumptions (A4.1.6) or (A4.4.2) on $\{z^\epsilon(t), t < \infty, \epsilon > 0\}$ or of (A4.4.1) on $\{x^\epsilon(t), t < \infty, \epsilon > 0\}$ are essentially questions of stochastic stability. Of course, if the state spaces are bounded, then the cited assumptions are automatically satisfied. The deterministic specialization of the above cited tightness requirements is that the trajectories of interest (those of $x^\epsilon(\cdot)$ and/or $z^\epsilon(\cdot)$) be bounded on the time interval of interest. To prove that boundedness in particular cases for the deterministic problem, some form of Liapunov function method is usually used. Stochastic "Liapunov function methods" are also very useful (if not indispensible at this time) to prove the required tightness for the stochastic problems, and we will discuss several approaches in this chapter.

The root of all the methods is the basic stochastic stability theory for diffusion or jump diffusions given in [K2, K15, K16], and in Section 1 we will state the few results which we will need from these references. Many of the basic methods have been extended to be of use for the wide band noise driven systems or for discrete parameter systems with correlated inputs. Via Liapunov function-type methods, properties such as recurrence and tightness can be obtained, as well as bounds on the probabilities of path excursions or on moments of recurrence times.

Next, let $x^\epsilon(\cdot)$ be a wide band noise driven system, where ϵ is the scale parameter, and let $x^\epsilon(\cdot) \Rightarrow x(\cdot)$, a diffusion or jump diffusion process, as $\epsilon \to 0$. I.e., for small ϵ, $x^\epsilon(\cdot)$ is "well approximated" by $x(\cdot)$ in the sense of weak convergence. Note that this "approximation" does not automatically imply that the "large time" or stability properties are similar. The desire is to prove stability or recurrence properties of the $x^\epsilon(\cdot)$ for small ϵ. This can often be done by exploiting the stability properties of $x(\cdot)$, and the "local similarity" of $x^\epsilon(\cdot)$ and $x(\cdot)$ for small ϵ. In [K20, Chapter 6], there is a development of the basic ideas—via a so-called *perturbed Liapunov function method*. The method is very useful for processes $x^\epsilon(\cdot)$ which are close to diffusions $x(\cdot)$ in the above sense. This method is briefly discussed in Section 5. The perturbed Liapunov function which is needed is constructed in two steps. First, one finds a stochastic Liapunov function $V(\cdot)$ for the "limit system" $x(\cdot)$. Then, the operator $\hat{A}^\epsilon$ (see Chapter 7) is applied to

$V(\cdot)$. As in Chapter 7 (e.g., see (7.2.3)) there is usually a "bad" term. A perturbation $V_1^\epsilon(\cdot)$ (analogous to the test function perturbation $f_1^\gamma(\cdot)$ of Chapter 7) is then added to $V(\cdot)$. The perturbed Liapunov function $V^\epsilon(\cdot)$ defined by $V^\epsilon(x^\epsilon(t)) = V(x^\epsilon(t)) + V_1^\epsilon(t)$ is then used to get recurrence and other properties for $x^\epsilon(\cdot)$ for small ϵ in much the same way that $V(\cdot)$ is used for that purpose for the process $x(\cdot)$. The reader is referred to [K20, Chapter 6] for full details and examples and to Section 5 below for a brief introduction.

The singularly perturbed system defined by the Itô equations (4.1.1), (4.1.2) requires somewhat a different technique. The two processes $x^\epsilon(\cdot)$ and $z^\epsilon(\cdot)$ must be treated together. One cannot usually treat the stability of each of the components separately, usless they are essentially uncoupled. Loosely speaking, suppose that the "fast" process $z_0^\epsilon(\cdot)$ is such that for each x and $t < \infty$, the "stretched-out" process $z_0^\epsilon(t/\epsilon + \cdot)$ converges weakly to the fixed-x process $z_0(\cdot|x)$ on the set where $x^\epsilon(t) \to x(t) = x$. Recall that $z_0^\epsilon(t/\epsilon + s) = z^\epsilon(t + \epsilon s)$. This does not necessarily imply that that the set

$$\{z^\epsilon(t), t \leq T, \epsilon > 0\} \tag{0.1}$$

is tight (i.e., that (A4.1.6) holds), since it is still possible that the slow variations of $x_0^\epsilon(\cdot)$ might "add enough energy" to the fast system to destabilize it. Thus, we need to work with the stability problems for both components simultaneously. To do this, we adapt the method which is used for the deterministic problem in [S1, S2]. These references (and the further references cited there) are generally concerned with the question of asymptotic stability, but boundedness in probability is enough for our purposes. The idea for the deterministic problem is developed in Section 2. The Liapunov function which is used for the pair $(x^\epsilon(\cdot), z^\epsilon(\cdot))$ is a suitable sum of a Liapunov function which can be used for the $x^\epsilon(\cdot)$ above with z fixed at a "stable point", and one which can be used for the "fixed-x" system $z^\epsilon(\cdot|x)$ alone.

In Section 3, the technique of Section 2 is altered appropriately so that it can be applied to the stochastic model (4.1.1), (4.1.2). In Section 4, we treat a linear case directly. Section 6 concerns a combination of the ideas of Sections 3 and 5, where one has a wide bandwidth driven singularly perturbed system of the form (8.1.1), (8.1.2). Here, there are two types of perturbations: (a) the type used for the singularly perturbed system in Section 3; (b) the "wideband noise type" of perturbation of Section 5. Only a brief outline is to be given, but the possiblities should be clear. Considerable further development is possible in many directions, as is usually the case with stability analyses of new classes of problems. An important class of examples appears in Chapter 10.

1. Stability Theory for Jump-Diffusion Processes of Itô Type

In this section, we discuss some classical stochastic stability ideas which will be useful from Section 3 on. For $p = 1, 2, \ldots$, let $x^p(\cdot)$ be processes which satisfy (1.1):

$$dx^p = b_p(x^p, t)dt + \sigma_p(x^p, t)dw + \int_\Gamma q_p(x^p, \gamma, t)N(dtd\gamma), x \in R^r. \quad (1.1)$$

Let $A^p(t)$ denote the differential operator of (1.1) at t. If $f(\cdot)$ is a smooth function of x, we use $A^p(t)f(x)$ to denote the value of $A^p(t)$ acting on $f(\cdot)$ at x, and not $(A^p f)(x, t)$ as used in previous chapters.

We will need the following conditions.

A1.1. $N(\cdot)$ *is a Poisson measure which is independent of the standard vector-valued Wiener process* $w(\cdot)$, *and has jump rate* $\lambda < \infty$ *and jump distribution* $\Pi(\cdot)$, *where* $\Pi(\cdot)$ *has compact support* Γ. *The* $q_p(\cdot)$ *are measurable and uniformly bounded and* $b_p(\cdot)$ *and* $\sigma_p(\cdot)$ *are measurable and satisfy* $|b_p(x, t)| + |\sigma_p(x, t)| \leq K(1 + |x|)$ *for some* $K < \infty$.

The $b_p(\cdot), \sigma_p(\cdot)$ and $q_p(\cdot)$ could be random functions, provided that they are non-anticipative with respect to $w(\cdot)$ and $N(\cdot)$.

A1.2. $V(\cdot) \in C^2(R^r)$ *and satisfies* $V(x) \geq 0$, *and* $V(x) \to \infty$ *as* $|x| \to \infty$.

A1.3. *There is* $\lambda_0 > 0$ *such that, for some* $c_0 > 0$ *and all* p, $A^p(t)V(x) \leq -c_0$ *for* $x \notin Q(\lambda_0) = \{x: V(x) \leq \lambda_0\}$ *and all* t.

Theorem 1.1. *Assume* (A1.1)–(A1.3). *Let* τ^p *be a (finite w.p.1) stopping time for* $x^p(\cdot)$, *and let* $x^p(\tau^p) \notin Q(\lambda_0)$. *Define the stopping time* $\tau_0^p = \inf\{t \geq \tau^p : x^p(t) \in Q(\lambda_0)\}$. *Then, for* $N > 0$ *we have*

$$P_{x^p(\tau^p)}\left(\sup_{\tau^p \leq s \leq \tau_0^p} V(x^p(s)) \geq N \right) \leq V(x^p(\tau^p))/N. \quad (1.2)$$

Also, with $x^p(0) = x$,

$$E_{x^p(\tau^p)}[\tau_0^p - \tau^p] \leq V(x^p(\tau^p))/c_0, \quad \text{w.p.1,}$$

$$E_x[\tau_0^p - \tau^p] \leq E_x V(x^p(\tau^p))/c_0. \quad (1.3)$$

Remark. By the theorem, the process $x^p(\cdot)$ will always return to the "recurrence set" $Q(\lambda_0)$, with the mean recurrence times and the upper bounds on the probability of large excursions (between recurrence times) being uniform in p. Such estimates are useful to get tightness. The proof of Theorem 1.1 is in [K2, K15, K16].

Theorem 1.2. *Assume* (A1.1). *Let* $\{\tau_k^p\}$ *be an arbitrary sequence of (finite w.p.1) stopping times such that all the* $x^p(\tau_k^p)$ *lie in some compact set. Suppose   that* $\{x^p(\tau_k^p + \cdot): p < \infty, \ k < \infty\}$ *is tight. Then, for each* $T < \infty$

$$P_k^p \left(\sup_{\tau_k^p \leq s \leq \tau_k^p + T} |x^p(s)| \geq N \right) \to 0, \quad \text{w.p.1} \tag{1.4}$$

as $N \to \infty$, *uniformly in* k, p, *where* P_k^p *denotes the probability conditioned on the data up to time* τ_k^p. *Assume* (A1.1)–(A1.3). *Then, for any compact set* S, *the set*

$$\{x^p(t), t \geq 0 : x(0) \in S, p < \infty\}$$

is tight.

Proof. The first tightness assumption and (A1.1) implies (1.4). For simplicity in the proof, we drop the jump term and let $\{|x^p(0)|\}$ be bounded. Let $\lambda_1 > \lambda_0$. Define the stopping times: $\sigma_1^p = min\{t : x^\epsilon(t) \in Q(\lambda_0)\}$, $\tau_k^p = min\{t > \sigma_k^p : x^\epsilon(t) \notin Q(\lambda_1)\}$, $\sigma_{k+1}^p = min\{t > \tau_k^p : x^p(t) \in Q(\lambda_0)\}$. Recall that if a stopping time is undefined at some ω, we set it equal to infinity there. By using the tightness of $\{x^p(\tau_k^p + \cdot), p, k\}$ and the estimates (1.2), (1.3), it can be shown that $\sigma_k^p \to \infty$, $\tau_k^p \to \infty$ w.p.1 as $k \to \infty$. Fix $T < \infty$. Define the set and indicator function:

$$J_k^p = [\tau_k^p, \tau_k^p + T), \quad C_k^p = I_{\{J_k^p - \cup_{i < k} J_i^p\}}(t).$$

For large N,

$$I_{\{|x^p(t)| \geq N\}} = \sum_k I_{\{|x^p(t)| \geq N\}} I_{[\tau_k^p, \sigma_{k+1}^p)}(t)$$

$$= \sum_k I_{\{|x^p(t)| \geq N\}} C_k^p + \sum_k I_{\{|x^p(t)| \geq N\}} I_{[\tau_k^p + T, \sigma_{k+1}^p)}(t). \tag{1.5}$$

The expectation of the second term on the right side of (1.5) goes to zero as $T \to \infty$ by (1.3). By (1.4),

$$EI_{\{|x^p(t)| \geq N\}} C_k^p \leq EP_k^p(\sup_{s \leq T} |x^p(\tau_k^p + s)| \geq N)C_k^p \leq \delta_T(N)EC_k^p,$$

where $\delta_T(N) \to 0$ as $N \to \infty$ for each T. The theorem follows from these facts.   Q.E.D.

2. Singularly Perturbed Deterministic Systems: Bounds on Paths

The Liapunov function method which we use to get tightness and weak convergence of $\{x^\epsilon(t), z^\epsilon(t), \epsilon > 0, t < \infty\}$ for the case (4.1.1), (4.1.2), is modelled on the approach used in [S1] for the deterministic problem. Thus, to provide the appropriate motivation, it is convenient to outline the method first for the simpler deterministic case:

$$\dot{x}^\epsilon = G(x^\epsilon, z^\epsilon), \qquad (2.1)$$

$$\epsilon \dot{z}^\epsilon = H(x^\epsilon, z^\epsilon), \qquad (2.2)$$

where the functions $G(\cdot)$ and $H(\cdot)$ are continuous.

Suppose that, for each parameter value x, the "fixed-x" and "stretched out" system

$$\dot{z}_0 = H(x, z_0) \qquad (2.3)$$

is asymptotically stable. Since $x^\epsilon(\cdot)$ varies "slowly" in comparison to $z^\epsilon(\cdot)$, it seems intuitively reasonable to expect that for small ϵ, and t bounded away from zero, the solution of (2.2) would be close to the "asymptotic manifold" set $\{z: H(x(t), z) = 0\}$. This might or might not be the case, but it is true under broad conditions, and we will informally develop a special case of the idea in [S1, K4] for proving it.

Suppose that there is a unique measurable function $h(\cdot)$ such that $z = h(x)$ is the unique solution to $H(x, z) = 0$, and $h(\cdot)$ is bounded on each bounded x-set. Suppose that the ODE

$$\dot{x} = G(x, h(x)) \qquad (2.4)$$

is asymptotically stable and has a Liapunov function $V(\cdot)$ with $V'_x(x)G(x, h(x))$ suitably negative for large values of $|x|$. The actual required properties are given in the next section. Now, let x be a *parameter* and $W(x, z_0)$ be a Liapunov function for the (parametrized by x) system (2.3). It is often the case that there is a function $\phi(\cdot)$ such that $\liminf_{|u| \to \infty} \phi^2(u) > 0$ and such that

$$\dot{W}(x, z_0) \leq -\phi^2(z_0 - h(x)) \qquad (2.5)$$

for large $|z_0 - h(x)|$. Inequality (2.5) is a quantification of a "uniform in x" stability property of (2.3), in that the rate of decrease of $W(x, z_0(t))$ is bounded above by a function of the distance between z_0 and the stable point $h(x)$.

The properties of $\dot{W}(x, \cdot)$ and $\phi(\cdot)$ and the stability of (2.4) are not usually sufficient by themselves to guarantee boundedness of $\{z^\epsilon(t), t < \infty, \epsilon > 0\}$, since the *variations* of $x^\epsilon(\cdot)$ were neglected. In fact,

$$\dot{W}(x^\epsilon, z^\epsilon) = W'_z(x^\epsilon, z^\epsilon)H(x^\epsilon, z^\epsilon)/\epsilon + W'_x(x^\epsilon, z^\epsilon)G(x^\epsilon, z^\epsilon). \qquad (2.6)$$

Some method of compensating for the fact that $x^\epsilon(\cdot)$ actually varies with time in (2.6) is needed. In [S1, K4], it is shown that, under appropriate conditions, a Liapunov function of the form

$$k\epsilon W(x, z) + V(x) \tag{2.7}$$

can be used, for an appropriate value of k. The form (2.7) represents a type of perturbed Liapunov function, although it is different from that used in [K20] and in Section 5 for the wide band noise driven stochastic problems. The use of $V(\cdot)$ in (2.7) and the stability of (2.4) allows as to account for the effects of the $W_x'G$ term in (2.6). The use of such perturbed Liapunov functions seems to go back to [K3]. The details of use of (2.7) are in [S1], but are also included in Section 3. The reference [S2] also contains some interesting applications.

3. Singularly Perturbed Itô Processes: Tightness

In this Section, we will use Liapunov functions similar to (2.7) to get (A4.1.6), (A4.4.1) and (A4.4.2), for the singularly perturbed system (4.1.1), (4.1.2). For the stochastic case, there is usually no simple analog of the asymptotic manifold $\{z: H(x, z) = 0\}$ defined for the $z_0(\cdot)$ of (2.3). The idea of a "*recurrence set*" for the fixed-x process $z_0(\cdot|x)$ given by (3.4) below is one natural substitute. Suppose, for purposes of discussion only, that there is a continuous $R^{r'}$-valued function $h(\cdot)$ on R^r and, for some small $\delta > 0$, a neighborhood $N(x)$ of $h(x)$ satisfying $\delta \geq \inf_x[\text{diameter } N(x)] > 0$ and such that $N(x)$ is a recurrence set for the fixed-x process $z_0(\cdot|x)$ of (3.4). Then a natural replacement for (2.5) is the condition that the process $y^\epsilon(t) = z^\epsilon(t) - h(x^\epsilon(t))$ have some stochastic stability property in the sense that if $|y^\epsilon(t)|$ is "large," then the "local mean drift" is towards the origin in some metric. Also, in analogy to the stability of (2.4), we might also need a stochastic stability property for $x^\epsilon(\cdot)$ for large $x^\epsilon(t)$, when $z^\epsilon(t)$ is near $h(x^\epsilon(t))$. These properties are implicit in the assumptions below. These assumptions are just "stochastic copies" of those used for the deterministic problem in [S1]. They present one possibility, and surely not the best or last. Often, we can use $h(x) = \int z\mu_x(dz)$, the mean value with respect to the invariant measure of the fixed-x process. There are other variations of the general idea, but the scheme below does cover many cases of interest. The details of the linear case in the next section can also help to motivate the assumptions of this section.

We will work with the model:

$$dx^\epsilon = G(x^\epsilon, z^\epsilon, \epsilon x^\epsilon, \epsilon z^\epsilon, t)dt + \sigma(x^\epsilon, z^\epsilon, \epsilon x^\epsilon, \epsilon z^\epsilon, t)dw$$

$$+ \int_{\Gamma_1} q_1(x^\epsilon, z^\epsilon, \gamma)N_1(d\gamma dt) \tag{3.1}$$

$$\epsilon dz^\epsilon = H(x^\epsilon, z^\epsilon, \epsilon x^\epsilon, \epsilon z^\epsilon, t)dt + \sqrt{\epsilon}v(x^\epsilon, z^\epsilon, \epsilon x^\epsilon, \epsilon z^\epsilon, t)dw$$

$$+ \int_{\Gamma_2} q(x^\epsilon, z^\epsilon, \gamma)N_2^\epsilon(d\gamma dt), \tag{3.2}$$

where $(w(\cdot), N_1(\cdot), N_2^\epsilon(\cdot))$ satisfies (A4.5.1). The ϵz^ϵ and ϵx^ϵ on the right sides of (3.1), (3.2), were not included in the models of the previous chapters, since they are usually unimportant in the weak convergence analysis under the conditions which were used. We include them here since they do often appear, and should be taken account of in the stability analysis. Let $A^\epsilon(t)$ denote the differential operator at t of $(x^\epsilon(\cdot), z^\epsilon(\cdot))$ and $A_0^\epsilon(t)$ that of the stretched out process $(x^\epsilon(\epsilon\cdot), z^\epsilon(\epsilon\cdot))$. The notation is different from that used previously. Let $A_h(t)$ denote the differential operator at t of the process defined by

$$dx = G(x, h(x), 0, 0, t)dt + \sigma(x, h(x), 0, 0, t)dw$$

$$+ \int_{\Gamma_1} q_1(x, h(x), \gamma)N_1(d\gamma dt). \tag{3.3}$$

Let $A_x^0(t)$ denote the differential operator at t of the *fixed-x system* $z^0(\cdot|x)$:

$$dz_0 = H(x, z_0, 0, 0, t)dt + v(x, z_0, 0, 0, t)dw + \int_{\Gamma_2} q_2(x, z_0, \gamma)N_2(d\gamma dt). \tag{3.4}$$

where the jump rate of $N_2(\cdot)$ is λ_2, and the jump distribution is $\Pi_2(\cdot)$.

All or some of the following conditions will be used. (A3.3) is the natural analog of the conditions for the deterministic case as given in [S1], and (A3.3a, b, f, g) make precise the stability requirements mentioned in the heuristic discussion above.

A3.1. $\sigma(\cdot)$ *and* $v(\cdot)$ *are bounded measurable functions.* $H(\cdot)$ *and* $G(\cdot)$ *are measurable functions which are bounded on each bounded* (x, z)-*set. Also* $(w(\cdot), N_1(\cdot), N_2^\epsilon(\cdot))$ *has independent increments.*

A3.2. $0 \le W(\cdot)$ *is in* $C^2(R^{r+r'})$, $0 \le V(\cdot)$ *is in* $C^2(R^r)$ *and the mixed partial derivatives of* $W(\cdot)$ *and* $V(\cdot)$ *of second order are bounded. Also,* $V(x) \to \infty$ *as* $|x| \to \infty$, *and* $W(x, z) \to \infty$ *as* $|z| \to \infty$, *uniformly on each bounded* x-*set.*

A3.3. *There are real valued non-negative measurable functions* $\phi(\cdot)$ *and* $\psi(\cdot)$, *continuous* $h(\cdot)$, *and constants* c_i *such that the following inequalities hold for all* t:

(a) $$A_x^0(t)W(x, z) \le -\phi^2(z - h(x)) + c_1,$$

(b) $$A_h(t)V(x) \le -\psi^2(x) + c_2,$$

(c) $W'_x(x,z)G(x,z,\epsilon x,\epsilon z,t) + \lambda_1 \int_{\Gamma_1} [W(x + q_1(x,z,\gamma)) - W(x,z)]\Pi_1(d\gamma)$

$$\leq c_3\phi^2(z - h(x)) + c_4\psi(x)\phi(z - h(x)) + c_5,$$

(d)
$W'_z(x,z)[H(x,z,\epsilon x,\epsilon z,t) - H(x,z,0,0,t)] + |W_{zz}(x,z)| \cdot |v(x,z,\epsilon x,\epsilon z,t)$

$-v(x,z,0,0,t)|^2 \leq \epsilon c_6\phi^2(z - h(x)) + \epsilon c_7\psi(x)\phi(z - h(x)) + \epsilon c_8,$

(e) $V'_x(x)[G(x,z,\epsilon x,\epsilon z,t) - G(x,h(x),0,0,t)] + \lambda_1 \int_{\Gamma_1} [V(x + q_1(x,z,\gamma)) -$

$-V(x + q_1(x,h(x),\gamma))]\Pi_1(d\gamma) \leq c_9\psi(x)\phi(z - h(x)) + c_{10}$

(f)
$$\liminf_{|u|\to\infty} \phi^2(u)/c_1 > 1$$

(g)
$$\lim_{|u|\to\infty} \psi(u) = \infty,$$

There are K and $K(x)$, bounded on each bounded x-set, such that

(h)
$$|H(x,z,\epsilon x,\epsilon z,t)| \leq K(x)(1 + |z|),$$

(i)
$$|G(x,z,\epsilon x,\epsilon z,t)| \leq K(1 + |x|).$$

Theorem 3.1. *Assume (A4.5.1), (A3.1), (A3.2) and (A3.3a,f,i). Let $\{x^\epsilon(0),$ $z^\epsilon(0)\}$ be tight. Let $T < \infty$ and for each $N < \infty$, let*

$$\overline{\lim_{|z|\to\infty}} \sup_{\substack{|x|\leq N \\ t\leq T}} |W'_x(x,z)G(x,z,\epsilon x,\epsilon z,t)|/\phi^2(z - h(x)) < \infty, \qquad (3.4)$$

$$\lim_{\epsilon\to 0} \overline{\lim_{|z|\to\infty}} \sup_{\substack{|x|\leq N \\ t\leq T}} \frac{|W'_z(x,z)(H(x,z,\epsilon x,\epsilon z,t) - H(x,z,0,0,t))|}{\phi^2(z - h(x))} = 0 \quad (3.5)$$

and similarly for $|v(x,z,\epsilon x,\epsilon z,t)|^2 - |v(x,z,0,0,t)|^2$ replacing the numerator. Let

$$\overline{\lim_{|z|\to\infty}} \sup_{\gamma} \sup_{\substack{|x|\leq N \\ t\leq T}} \frac{|W(x + q(x,z,\gamma),z) - W(x,z)|}{\phi^2(z - h(x))} < \infty. \qquad (3.6)$$

Then

$$\{z^\epsilon(t), \epsilon > 0, t \leq T\} \qquad (3.7)$$

is tight.

Proof. (A3.3i), (A4.5.1) and (A3.1) together with Theorem 2.3.3 guarantee that $\{x^\epsilon(\cdot), \epsilon > 0\}$ is tight. Thus, in order to prove (3.7) we can suppose that $x^\epsilon(\cdot)$ is uniformly bounded. We apply the Liapunov function $W(\cdot)$ to get

$$A^\epsilon(t)W(x,z) \leq -\phi^2(z - h(x))/\epsilon + c_1/\epsilon + \delta_\epsilon/\epsilon + \delta_\epsilon \phi(z - h(x))/\epsilon$$

$$+ W_x'(x,z)G(x,z,\epsilon x,\epsilon z,t) + \lambda_1 \int_{\Gamma_1} [W(x + q_1(x,z,\gamma),z) - W(x,z)]\Pi_1(d\gamma)$$

$$+ W_z'(x,z)[H(x,z,\epsilon x,\epsilon z,t) - H(x,z,0,0,t)]/\epsilon + O(1/\sqrt{\epsilon}),$$

where $\delta_\epsilon \xrightarrow{\epsilon} 0$. The $O(1/\sqrt{\epsilon})$ term is due to the contributions of the mixed (x,z) derivatives.

By our allowed boundedness hypothesis on $x^\epsilon(\cdot)$, the conditions of the theorem imply that

$$A^\epsilon(t)W(x,z) \leq -(1 - \delta_\epsilon'')\phi^2(z - h(x))/\epsilon + (c_1 + \delta_\epsilon')/\epsilon, \qquad (3.8)$$

where δ_ϵ' and δ_ϵ'' go to zero as $\epsilon \to 0$. Thus, working with the stretched out process $z_0^\epsilon(\cdot)$, $x_0^\epsilon(\cdot)$, we have

$$A_0^\epsilon(t)W(x,z) \leq -(1 - \delta_\epsilon'')\phi^2(z - h(x)) + (c_1 + \delta_\epsilon'). \qquad (3.9)$$

Now Theorem 1.1 applied to z_0^ϵ yields the tightness.   Q.E.D.

Theorem 3.2. *Assume* (A4.1.5), *and* (A3.1) *to* (A3.3), *and let* $\{x^\epsilon(0),$ $z^\epsilon(0)\}$ *be tight. Then*

$$\{x^\epsilon(t), z^\epsilon(t), t < \infty, (\text{small}) \ \epsilon > 0\} \qquad (3.10)$$

is tight.

Proof. Part 1. Applying the Liapunov function $V(\cdot)$ and using (A3.3e) and the definitions of $A^\epsilon(t)$ and $A_h(t)$, we have

$$A^\epsilon(t)V(x) \leq A_h(t)V(x) + c_9\psi(x)\phi(z - h(x)) + c_{10} + O(1).$$

By (A3.3b), for some constant c_1',

$$A^\epsilon(t)V(x) \leq -\psi^2(x) + c_9\psi(x)\phi(z - h(x)) + c_1'. \qquad (3.11)$$

Also, as in Theorem 3.1 (but not assuming boundedness of $x^\epsilon(\cdot)$)

$$A^\epsilon(t)W(x,z) \leq -\phi^2(z - h(x))/\epsilon + c_1/\epsilon + O(1/\sqrt{\epsilon})$$

$$+ O(1)[\phi^2(z - h(x)) + \psi(x)\phi(z - h(x))]. \qquad (3.12)$$

Next, use the inequality $\psi\phi \leq \psi^2/c + c\phi^2$ to get that

$$A^\epsilon(t)(k\epsilon W(x,z) + V(x)) \le -(1 - c_9/c)\psi^2(x)$$
$$- (k - cc_1 - \epsilon kc - \epsilon kO(1))\phi^2(z - h(x)) + (c_1' + kc_1) + kO(\sqrt{\epsilon}). \quad (3.13)$$

Using appropriate choices of c and k, we see that there is $\alpha_0 > 0$ and a compact set Q_0 such that the right side of (3.13) is less than $-\alpha_0$ for $(x, z) \notin Q_0$. Thus Q_0 is a "recurrence set" for $(x^\epsilon(\cdot), z^\epsilon(\cdot))$. Let Q_1 be a compact set whose interior contains Q_0. Define the stopping times : $\beta_1^\epsilon = \min\{t: (x^\epsilon(t), z^\epsilon(t)) \in Q_0\}$, and for $k \ge 1$, $\gamma_k^\epsilon = \min\{t \ge \beta_k^\epsilon: (x^\epsilon(t), z^\epsilon(t)) \notin Q_1\}$, $\beta_{k+1}^\epsilon = \min\{t \ge \gamma_k^\epsilon: (x^\epsilon(t), z^\epsilon(t)) \in Q_0\}$. By (3.13) and Theorem 1.1, there is $\alpha_1 < \infty$ such that (w.p.1)

$$\sup_{\epsilon,k} E_{\gamma_k^\epsilon}^\epsilon(\beta_{k+1}^\epsilon - \gamma_k^\epsilon) \le \alpha_1. \quad (3.14)$$

By (3.14) and the linear growth condition (A3.3i) and boundedness of $\sigma(\cdot)$ and $q_1(\cdot)$, the tightness of $\{x^\epsilon(t), t < \infty, \text{ (small) } \epsilon > 0\}$ follows from Theorem 1.2 (with our $x^\epsilon(\cdot)$ replacing the $x^p(\cdot)$ there).

Part 2. We now prove the tightness of

$$\{z_0^\epsilon(t) - h(x_0^\epsilon(t)) = y_0^\epsilon(t), t < \infty, \text{ (small) } \epsilon > 0\}. \quad (3.15)$$

Using (3.11), (3.12) and the definition of the differential operator A_0^ϵ of the "stretched out" process $(x_0^\epsilon(\cdot), z_0^\epsilon(\cdot))$, we have

$$A_0^\epsilon(t)(W(x,z) + V(x)) \le -(1 - \delta_\epsilon)\phi^2(z - h(x)) + c_1 + O(\sqrt{\epsilon}), \quad (3.16)$$

where $\delta_\epsilon \xrightarrow{\epsilon} 0$. If $\{x^\epsilon(t), t < \infty, \epsilon > 0\}$ could be assumed to be bounded, then the required tightness would follow from (3.16), similarly to what was done in Part 1. We will show that boundedness of the $\{x^\epsilon(\cdot), \epsilon > 0\}$ can be assumed, w.l.o.g.

Define $y = z - h(x)$. By (A3.3f), there are $\alpha_2 > 0$ and a compact set Λ_0, such that $-(1 - \delta_\epsilon)\phi^2(y) + c_1 + O(\sqrt{\epsilon}) \le -\alpha_2$ for $y \notin \Lambda_0$, and small ϵ. Let Λ_1 be a compact set whose interior contains Λ_0. The sets Λ_0 and Λ_1 can be used to define recurrence times analogous to the $\{\tau_k^\epsilon, \sigma_k^\epsilon\}$ of Theorem 1.2. We will work on each interval $[n/\epsilon, (n+2m)/\epsilon]$ for suitable m, and get bounds on $z_0^\epsilon(t)$ on $[(n+m)/\epsilon, (n+2m)/\epsilon]$ which are uniform in n. For each integer n, define $v_1^\epsilon(n) = \min\{t \ge n/\epsilon: y_0^\epsilon(t) \in \Lambda_0\}$, $u_k^\epsilon(n) = \min\{t \ge v_k^\epsilon(n): y_0^\epsilon(t) \notin \Lambda_1\}$, $v_{k+1}^\epsilon(n) = \min\{t \ge u_k^\epsilon(n): y_0^\epsilon(t) \in \Lambda_0\}$. From the results in Part 1, it can be shown that

$$P(|v_1^\epsilon(n) - n/\epsilon| \ge m/\epsilon) \to 0 \quad (3.17)$$

uniformly in n and (small) ϵ, as $m \to \infty$. Choose $\delta_1 > 0$ small and choose m large enough so that the left side of (3.17) is less than δ_1 for all (small) ϵ and all n.

By (3.16), and with $E_{u_k^\epsilon}^\epsilon$ denoting the expectation conditioned on the data up to time $u_k^\epsilon(n)$,

$$E^\epsilon_{u^\epsilon_k(n)}[v^\epsilon_{k+1}(n) - u^\epsilon_k(n)] \le [V(x^\epsilon_0(u^\epsilon_k(n)))$$

$$+ W(x^\epsilon_0(u^\epsilon_k(n)), z^\epsilon_0(u^\epsilon_k(n)))]/\alpha_2. \qquad (3.18)$$

By the hypotheses on $G(\cdot)$, $q_1(\cdot)$ and $\sigma(\cdot)$ and the tightness of the set of random variables $\{x^\epsilon(t), t < \infty, \text{(small)} \ \epsilon > 0\}$, the set of processes $\{x^\epsilon(t + \cdot), t < \infty, \text{(small)} \ \epsilon > 0\}$ is tight. Thus, with an arbitrary large probability (say, $1 - \delta_1$), $\{x^\epsilon_0(\cdot), \text{(small)} \ \epsilon > 0\}$ is bounded (by, say, N_1) on $[n/\epsilon, (n + 2m)/\epsilon]$ for each n. Consequently, for purposes of proving the tightness of (3.15), we can suppose that (3.18) reduces to (w.p.1)

$$E^\epsilon_{u^\epsilon_k(n)}[v^\epsilon_{k+1}(n) - u^\epsilon_k(n)] \le K(N_1) < \infty, \qquad (3.19)$$

where $K(N_1)$ does not depend on either k or n. The tightness of $\{z^\epsilon_0(t), t \in [(n + m)/\epsilon, (n + 2m)/\epsilon], \text{small } \epsilon\}$ for each n now follows by using the method of Part 1. Since δ_1 is arbitrary and does not depend on n and the bounds on the probability of large values of $|z^\epsilon_0(t)|$ on that interval do not depend on n, we have the tightness of $\{z^\epsilon(t), t < \infty, \text{(small)} \ \epsilon\}$.　　Q.E.D.

4. The Linear Case

Theorem 3.2 can be adjusted to be usable for the linear systems case. But a direct proof is instructive and helps explain the conditions in Theorem 3.2, since the Liapunov functions are known. Droping the jump term for simplicity, the linear systems analog of (3.1) and (3.2) is

$$dx^\epsilon = (A_0 x^\epsilon + A_1 z^\epsilon + \epsilon A_2 x^\epsilon + \epsilon A_3 z^\epsilon)dt + \sigma dw \qquad (4.1)$$

$$\epsilon dz^\epsilon = (H_0 x^\epsilon + H_1 z^\epsilon + \epsilon H_2 x^\epsilon + \epsilon H_3 z^\epsilon)dt + \sqrt{\epsilon} \ vdw. \qquad (4.2)$$

The v and σ can be state dependent, provided that they are bounded. Henceforth, we drop the "ϵ" terms on the right of (4.1) and (4.2) for notational simplicity, since they do not play an important role, and the described results are true even if these terms are included. Also, the addition of arbitrary bounded control terms does not affect the results. Define

$$\tilde{A}_0 = (A_0 - A_1 H_1^{-1} H_0).$$

Theorem 4.1. *Let $\tilde{A}_0$ and H_1 be stable (Hurwitz) matrices. Then (3.10) holds. Write*

$$dx^\epsilon = [\tilde{A}_0 x^\epsilon + A_1(z^\epsilon + H_1^{-1} H_0 x^\epsilon)]dt + \sigma dw \qquad (4.3)$$

$$\epsilon dz^\epsilon = H_1[z^\epsilon + H_1^{-1} H_0 x] + \sqrt{\epsilon} \ vdw. \qquad (4.4)$$

The function $h(\cdot)$ defined in Section 3 is $h(x) = -H_1^{-1} H_0 x$. Rewrite (4.3) and (4.4) in terms of x^ϵ and $y^\epsilon = (z^\epsilon + H_1^{-1} H_0 x) = (z^\epsilon - h(x^\epsilon))$ yielding

$$dx^\epsilon = (\tilde{A}_0 x^\epsilon + A_1 y^\epsilon)dt + \sigma dw \qquad (4.5)$$

$$dy^\epsilon = [\frac{1}{\epsilon}H_1 y^\epsilon + H_1^{-1} H_0 \tilde{A}_0 x^\epsilon + H_1^{-1} H_0 A_1 y^\epsilon] dt$$

$$+ \frac{1}{\sqrt{\epsilon}} v \, dw + H_1^{-1} H_0 \sigma \, dw. \tag{4.6}$$

Let $V(x) = x' P_1 x$ and $W(y) = y' P_2 y$ be Liapunov functions for the systems $\dot{x} = A_0 x$ and $\dot{y} = H_1 y$, resp., where $P_i > 0$ (positive definite symmetric matrices). Let $\tilde{A}_0' P_1 + P_1 \tilde{A}_0 = -C_1$, $H_1' P_2 + P_2 H_1 = -C_2$, where $C_i > 0$. Define $d_1 = $ trace $P_1 \cdot \sigma\sigma'$ and $d_2 = $ trace $P_2 \cdot vv'$ (if σ or v are not constant, then we let the d_i be an upper bound). Letting A^ϵ denote the differential operator of (4.5), (4.6), we can write

$$A^\epsilon V(x) \leq 2(\tilde{A}_0 x + A_1 y)' P_1 x + d_1 = -x' C_1 x + 2y' A_1' P_1 x + d_1,$$

$$A^\epsilon W(y) \leq 2[\frac{1}{\epsilon}H_1 y + H_1^{-1} H_0 \tilde{A}_0 x + H_1^{-1} H_0 A_1 y]' P_2 y + \frac{d_2}{\epsilon} + O(\frac{1}{\sqrt{\epsilon}})$$

$$= -\frac{1}{\epsilon}y' C_2 y + (H_1^{-1} H_0 \tilde{A}_0 x + H_1^{-1} H_0 A_1 y)' P_2 y + \frac{d_2}{\epsilon} + O(\frac{1}{\sqrt{\epsilon}}).$$

As in Theorem 3.2, given $\delta \in (0,1)$, there are $k < \infty$ and $\delta' < \infty$ (not depending on k) such that for small ϵ,

$$A^\epsilon[k\epsilon W(y) + V(x)] \leq -x'(C_1 - \delta I)x - ky' C_2 y/2 + kd_2 + \delta'. \tag{4.7}$$

By (4.7) and Itô's Formula, there is $d_3 > 0$ such that

$$EV(x^\epsilon(t)) \leq EV(x^\epsilon(t)) + k\epsilon EW(y^\epsilon(t)) \leq EV(x^\epsilon(0))$$

$$+ k\epsilon EW(y^\epsilon(0)) - d_3 \int_0^t EV(x^\epsilon(s))ds + (kd_2 + \delta')t. \tag{4.8}$$

By the Gronwall–Bellman inequality, (4.8) implies that

$$\overline{\lim_t} EV(x^\epsilon(t)) \leq (kd_2 + \delta')/d_3, \tag{4.9}$$

yielding the tightness result of $\{x^\epsilon(t), t < \infty, (\text{small}) \; \epsilon > 0\}$.

To get the tightness results for $\{y^\epsilon(t), t < \infty\}$, we follow the procedure of Theorem 3.2 and consider the stretched out process $y_0^\epsilon(\cdot)$. For some $c_i > 0$, we have

$$A_0^\epsilon W(y) \leq -c_1 W(y) + d_2 + \epsilon c_2 |x|^2 + O(\sqrt{\epsilon}). \tag{4.10}$$

Now, (4.9) and (4.10) yield that

$$\overline{\lim_\epsilon} \, \overline{\lim_t} \, EW(y_0^\epsilon(t)) \leq d_2/c_1. \quad \text{Q.E.D.} \tag{4.11}$$

5. Wide Bandwidth Noise

In this section, we give a brief survey of the perturbed Liapunov function method which was developed in [K20, Chapter 6] for systems with "wide bandwidth" noise disturbances. In the next section, there is a brief discussion of the singularly perturbed wide band noise driven system, where the methods of this Section and that of Section 3 need to be combined. For simplicity in the exposition we work with the model used in Chapter 7, namely:

$$\dot{x}^\rho = \overline{G}(x^\rho) + \tilde{G}(x^\rho, \xi^\rho) + F(x^\rho, \xi^\rho)/\rho, \ \rho > 0, x \in R^r, \qquad (5.1)$$

where $\xi^\rho(\cdot)$ is a bounded (uniformly in ρ) right continuous wide-band noise process. The functions $\overline{G}(\cdot), \tilde{G}(\cdot)$ and $F(\cdot)$ are continuous and satisfy a linear growth condition in x, uniformly in ξ ; i.e, $|\tilde{G}(x, \xi)| \leq K(1+|x|), K < \infty$, and similarly for $\overline{G}$ and F. The development will be entirely formal, since we are only concerned with illustrating the main idea and showing the types of conditions which are needed. The interested reader is referred to [K20].

Suppose that, if the initial conditions $x^\rho(0) \Rightarrow x$, we have the weak convergence, $x^\rho(\cdot) \Rightarrow x(\cdot)$, where $x(\cdot)$ satisfies (5.2):

$$dx = [\overline{G}(x) + \overline{F}(x)]dt + \sigma(x)dw, \ x(0) = x. \qquad (5.2)$$

See Section 7 for the details of the derivation and definition of the "correction term" $\overline{F}(\cdot)$. Let A denote the differential operator of $x(\cdot)$. Suppose that $V(\cdot)$ is a Liapunov function for $x(\cdot)$ in that $V(\cdot)$ satisfies $0 \leq V(x) \to \infty$ as $|x| \to \infty, V(\cdot) \in C^2(R^r)$ with the second order partial derivatives being bounded, and for some $\alpha_0 < 0, AV(x) \leq -\alpha_0$ for large $|x|$. It is reasonable to try to use $V(\cdot)$ as a Liapunov function for (5.1) in order to prove tightness of the set $\{x^\rho(t), t < \infty, (\text{small}) \rho > 0\}$. We will now see how this can be done.

Recall the definition of the operator $\hat{A}^\rho$ from Chapter 7. Formally calculate (writing $x^\rho = x^\rho(t)$):

$$\hat{A}^\rho V(x^\rho) = V_x'(x^\rho)\overline{G}(x^\rho) + V_x'(x^\rho)\tilde{G}(x^\rho, \xi^\rho) + V_x'(x^\rho)F(x^\rho, \xi^\rho)/\rho. \qquad (5.3)$$

Clearly, (5.3) can't be used directly to get the desired stability proof due to the $1/\rho$ and $\tilde{G}$ terms. We need to do an "averaging" in order to relate $\hat{A}^\rho$ to A and to take advantage of the fact that $AV(x) < 0$ for large x. In the formal development below, it is implicitly supposed that the stability of $x(\cdot)$ is due to the behavior of $\overline{G}(\cdot)$ for large $|x|$. If this is not the case, (i.e., if the diffusion term in (5.2) needs to be accounted for due to its strongly stabilizing or destabilizing effect) then a "second order" perturbed Liapunov function might need to be used [K20, Chapter 6].

The $1/\rho$ term in (5.3) is similar to what we dealt with in Chapter 7. The "averaging" of this $1/\rho$ term and of $\tilde{G}(x, \xi^\rho)$ is done by introducing a small perturbation to $V(\cdot)$, very similarly to the procedure that was followed in Chapter 7 in perturbing the test functions $f(\cdot)$. Define the Liapunov function perturbation $V_1^\rho(t) = V_1^\rho(x^\rho(t), t)$, where

$$V_1^\rho(x, t) = \int_t^\infty V_x'(x) E_t^\rho [\tilde{G}(x, \xi^\rho(s)) + F(x, \xi^\rho(s))/\rho] ds. \qquad (5.4)$$

The E_t^ρ, as defined in Chapter 7 is the expectation conditioned on $\{\xi^\rho(s), x^\rho(s), s \leq t\}$. Let $V^\rho(\cdot) = V(x^\rho(\cdot)) + V_1^\rho(\cdot)$. Suppose that the integral of the first integrand of (5.4) is bounded above by $O(\rho^2)(1 + |V_x(x)|^2)$ and the integral in the second integrand is bounded above by $O(\rho)(1+|V_x(x)|)$. Furthermore, let

$$|V_x(x)|^2 + |V_x'(x)\overline{G}(x)| \leq O(1)(1 + V(x)).$$

These conditions are satisfied under a variety of reasonable conditions on the growth of $\tilde{G}(\cdot)$, $V(\cdot)$, $F(\cdot)$ and "mixing" and scaling conditions on $\xi^\rho(\cdot)$. The reader can find ways in which they can be weakened by studying the derivation in [K20, Chapter 6].

Analogously to the calculations in Chapter 7, we have formally (write $x^\rho(t) = x^\rho$ for simplicity)

$$\hat{A}^\rho V^\rho(t) = V_x'(x^\rho)\overline{G}(x^\rho) + \int_t^\infty ds \left\{ V_x'(x^\rho) E_t^\rho \left[\tilde{G}(x^\rho, \xi^\rho(s)) \right. \right.$$

$$\left. \left. + \frac{F(x^\rho, \xi^\rho(s))}{\rho} \right] \right\}_x' \dot{x}^\rho(t). \qquad (5.5)$$

Often the right hand term of (5.5) is negligible in the sense that

$$[\text{r.h.s. of (5.5) with } x^\rho = x]/|V_x'(x)\overline{G}(x)| \qquad (5.6)$$

goes to zero as $|x| \to \infty$, uniformly in (ρ, t) (w.p.1). Suppose that this is true. Under all of the above conditions, there is $\alpha_1 > 0$ such that (again using the abbreviation $x^\rho(t) = x^\rho$)

$$V^\rho(t) \geq (1 - O(\rho))V(x^\rho) - O(\rho),$$

$$V^\rho(t) - V^\rho(0) - \int_0^t \hat{A}^\rho V^\rho(s) ds = \text{martingale}, \qquad (5.7)$$

$$\hat{A}^\rho V^\rho(t) \leq -\alpha_1 < 0 \text{ for large } |x^\rho|, \text{ and all small } \rho.$$

The relations in (5.7) imply that $x^\rho(\cdot)$ is "uniformly in ρ" recurrent. In particular, for large enough $\lambda_0, \hat{A}^\rho V^\rho(t) \leq -\alpha_1$ for small ρ if $x^\rho(t) \notin Q(\lambda_0) = \{x : V(x) \leq \lambda_0\}$, and $x^\rho(\cdot)$ will always (w.p.1) return to $Q(\lambda_0)$. For $\lambda_1 > \lambda_0$, the mean time (conditioned on the data up to the arrival time at $\partial Q(\lambda_1)$) required to go from $\partial Q(\lambda_1)$ to $\partial Q(\lambda_0)$ is bounded uniformly in ρ. A bound on the conditional mean time is given by $2\lambda_1/\alpha_1$, for small ρ.

To get the desired tightness, we need just one more condition: We need to be certain that $x^\rho(\cdot)$ can't move "very fast" (with a high probability) from $Q(\lambda_0)$ to the exterior of $Q(\lambda_1)$ as $\rho \to 0$. The following (usually unrestrictive) condition guarantees this. Let τ_n^ρ be a sequence of (finite w.p.1) stopping times for which $x^\rho(\tau_n^\rho) \in Q(\lambda_0)$. Suppose that the set $\{x^\rho(\tau_n^\rho +\cdot), \rho > 0, n < \infty\}$ is tight. Under this condition, and the conditions leading to (5.7),

$$\{x^\rho(t), t < \infty, \rho > 0\} \tag{5.8}$$

is tight. See [K20, Chapter 6] for the full details.

Such perturbed Liapunov function methods are adaptable for use on many problems with either a discrete or continuous time parameter.

6. Singularly Perturbed Wide Bandwidth Noise Driven Systems

Consider the system (8.1.1), (8.1.2), with the control term dropped for convenience, and where $\psi^\rho(\cdot)$ and $\xi^\beta(\cdot)$ are bounded (uniformly in ρ) and right continuous "wide-band" noise processes, namely:

$$\dot{x}^\gamma = G(x^\gamma, z^\gamma) + \tilde{G}(x^\gamma, z^\gamma, \psi^\rho) + F(x^\gamma, z^\gamma, \psi^\rho)/\rho \tag{6.1}$$

$$\epsilon \dot{z}^\gamma = H(x^\gamma, z^\gamma) + \tilde{H}(x^\gamma, z^\gamma, \xi^\beta) + \sqrt{\epsilon}J(x^\gamma, z^\gamma, \xi^\beta)/\beta. \tag{6.2}$$

As in Chapter 8, we write $\gamma = (\epsilon, \rho, \beta)$.

If $x^\gamma(\cdot) \Rightarrow x(\cdot)$, an "averaged" system as in Chapter 8, then for each $T < \infty$,

$$\{\sup_{t \le T} |x^\gamma(t)|, \ \gamma > 0\}$$

is bounded in probability. This fact can then be used to obtain tightness of $\{z_0^\gamma(t), \gamma > 0, t \le T/\epsilon\} = \{z^\gamma(t), \gamma > 0, t \le T\}$ by a perturbed Liapunov function method of the type used in Section 5.

The problem of showing tightness of $\{x^\gamma(t), z^\gamma(t), \gamma > 0, t < \infty\}$ is much harder. Since it is both singularly perturbed and driven by wide band noise, we will need to use perturbed Liapunov functions with *both* the perturbations of Sections 3 and 5. Our intention in this Section is simply to describe the general approach. We present a very brief outline here, and only for the case where $\tilde{G}, F, \tilde{H}$ and J do not depend on either x or z. The treatment of the general case follows the same lines, but with much more detail.

Let $W(\cdot)$ and $V(\cdot)$ satisfy (A3.2). Following what by now is the logical line of development (and using the abbreviations $x^\gamma = x^\gamma(t), z^\gamma = z^\gamma(t)$) we apply the $\hat{A}^\gamma$ operator to yield

$$\hat{A}^\gamma V(x^\gamma) = V'_x(x^\gamma)G(x^\gamma, z^\gamma) + V'_x(x^\gamma)[\tilde{G}(\psi^\rho) + F(\psi^\rho)/\rho], \tag{6.2}$$

$$\epsilon \hat{A}^\gamma W(x^\gamma, z^\gamma) = W_z'(x^\gamma, z^\gamma) H(x^\gamma, z^\gamma) + W_z'(x^\gamma, z^\gamma)[\tilde{H}(\xi^\beta) + \sqrt{\epsilon} J(\xi^\beta)/\beta]$$
$$+ \epsilon W_x'(x^\gamma, z^\gamma)[G(x^\gamma, z^\gamma) + \tilde{G}(\psi^\rho) + F(\psi^\rho)/\rho].$$

In order to use (6.2) to get the desired stability properties, the terms on the right side which contain the noise will have to be averaged out via use of perturbations $V_1^\gamma(\cdot)$ and $W_1^\gamma(\cdot)$ to the Liapunov functions. These will be of the type used in the last section and will be defined in the next theorem. Define the perturbed Liapunov functions $\epsilon W^\gamma(t) = \epsilon W(x^\gamma(t), z^\gamma(t)) + W_1^\gamma(t)$ and $V^\gamma(t) = V(x^\gamma(t)) + V_1^\gamma(t)$. Then the Liapunov functions $(V^\gamma + \epsilon k W^\gamma)$ and $(V^\gamma + W^\gamma)$ resp., are used just as the Liapunov functions $(V + \epsilon k W)$ and $(V + W)$, resp., were used in Section 3. We next state the required conditions. Condition (A6.2) is not very restrictive; e.g., it holds if $\xi^\beta(t) = \xi(t/\beta^2)$, $\psi^\rho(t) = \psi(t/\rho^2)$ under appropriate mixing conditions on $\xi(\cdot), \psi(\cdot)$.

A6.1. $F(\cdot), J(\cdot), \tilde{G}(\cdot)$ and $\tilde{H}(\cdot)$ *are continuous functions and do not depend on either x or z, and $\xi^\beta(\cdot), \psi^\rho(\cdot)$ are bounded and right continuous.*

A6.2. *Let τ denote a stopping time which is finite w.p.1, and E_τ^γ the expectation given all data up to time τ. Then*

$$\int_\tau^\infty E_\tau^\gamma \tilde{G}(\psi^\rho(s)) ds = O(\rho^2)$$

$$\int_\tau^\infty E_\tau^\gamma F(\psi^\rho(s)) ds = O(\rho^2)$$

$$\int_\tau^\infty E_\tau^\gamma \tilde{H}(\xi^\beta(s)) ds = O(\beta^2)$$

$$\int_\tau^\infty E_\tau^\gamma J(\xi^\beta(s)) ds = O(\beta^2).$$

where the $O(\rho^2)$ and $O(\beta^2)$ do not depend on τ or γ.

A6.3. *There is a constant K and a $K(x)$ which is bounded on each bounded x-set such that*

$$|G(x, z)| \leq K(1 + |x|),$$
$$|H(x, z)| \leq K(x)(1 + |z|).$$

A6.4. *There are constants d_i, a continuous function $h(\cdot)$, and continuous $\phi(\cdot), \psi(\cdot)$ where $\phi(y) \to \infty$ as $|y| \to \infty$, $\psi(x) \to \infty$ as $|x| \to \infty$, and*

$$V_x'(x) G(x, h(x)) \leq -\psi^2(x) + d_1,$$
$$W_z'(x, z) H(x, z) \leq -\phi^2(z - h(x)) + d_2,$$

See Section 3 for the motivation for the introduction of $h(\cdot), \psi(\cdot), \phi(\cdot)$.

A6.5.

$$|V'_x(x)(G(x,z) - G(x, h(x)))| \leq d_3 \psi(x) \phi(z - h(x)) + d_4.$$

A6.6.

$$|V_x(x)| \leq d_5(1 + V(x)),$$
$$|W_x(x,z)| + |W_z(x,z)| \leq d_6(1 + W(x,z) + V(x)).$$

A6.7.

$$|G(x,z)| + |H(x,z)| \leq d_7(1 + \phi^2(z - h(x)) + \psi^2(x)).$$

Theorem 6.1. *Assume* (A8.2.1), (A3.2) *and* (A6.1)–(A6.7). *Let* $\{x^\gamma(0), z^\gamma(0)\}$ *be tight. Then*

$$\{z^\gamma(t), x^\gamma(t), (\text{small}) \gamma > 0, t < \infty\}$$

is tight.

Proof. Only a few details will be given. Let $\{\tau^\gamma\}$ be a sequence of stopping times, each being finite w.p.1. First we note that (A8.2.1) and (A6.1)–(A6.3) imply that $\{x^\gamma(\cdot)\}$ is tight, and also that if $\{x^\gamma(\tau^\gamma), z^\gamma(\tau^\gamma)\}$ is tight, then so is $\{x^\gamma(\tau^\gamma + \cdot), z_0^\gamma(\tau^\gamma/\epsilon + \cdot)\}$. Define the Liapunov function perturbations $V_1^\gamma(t) = V_1^\gamma(x^\gamma(t), t)$, $W_1^\gamma(t) = W_1^\gamma(x^\gamma(t), z^\gamma(t), t)$, where

$$V_1^\gamma(x,t) = \int_t^\infty V'_x(x) E_t^\gamma [\tilde{G}(\psi^\rho(s)) + F(\psi^\rho(s))/\rho] ds$$

$$W_1^\gamma(x,z,t) = \int_t^\infty W'_z(x,z) E_t^\gamma [\tilde{H}(\xi^\beta(s)) + \sqrt{\epsilon} J(\xi^\beta(s))/\beta] ds$$

$$+ \epsilon \int_t^\infty W'_x(x,z) E_t^\gamma [\tilde{G}(\psi^\rho(s)) + F(\psi^\rho(s))/\rho] ds.$$

Then set $V^\gamma(t) = V(x^\gamma(t)) + V_1^\gamma(t)$ and $\epsilon W^\gamma(t) = \epsilon W(x^\gamma(t), z^\gamma(t)) + W_1^\gamma(t)$. Then, for $k \geq 1$ and some constants $c_i \geq 0$ and $c_i(k) \geq 0$

$$V^\gamma(t) \geq -c_1 ,$$

$$V^\gamma(t) + \epsilon k W^\gamma(t) \geq -c_2(k), \tag{6.3}$$

$$V^\gamma(t) + W^\gamma(t) \geq -c_3.$$

Also, writing $x^\gamma = x^\gamma(t)$ and $z^\gamma = z^\gamma(t)$, we have

$$\hat{A}^\gamma[V^\gamma(t) + \epsilon k W^\gamma(t)] \leq V'_x(x^\gamma) G(x^\gamma, h(x^\gamma)) + V'_x(x^\gamma)[G(x^\gamma, z^\gamma)$$

$$- G(x^\gamma, h(x^\gamma))] + k W'_z(x^\gamma, z^\gamma) H(x^\gamma, z^\gamma) +$$

$$+ (O(k\sqrt{\epsilon}\beta) + O(\rho))|G(x^\gamma, z^\gamma)| + k(O(\sqrt{\epsilon}\beta) + O(\rho))$$
$$+ k[O(\beta^2/\epsilon) + O(\beta\sqrt{\epsilon})]|H(x^\gamma, z^\gamma)| + O(1). \qquad (6.4)$$

As in Theorem 3.2, given arbitrarily small $\delta > 0$, there is $k < \infty$ such that the right side is bounded above by

$$-k\phi^2(z - h(x))/2 - (1 - \delta)\psi^2(x) + O(1 + k), \qquad (6.5)$$

for small γ.

From (6.3) and (6.5), we can get the tightness of $\{x^\gamma(t), t \geq 0, \text{(small)} \gamma\}$, analogously to what was done in Theorem 3.2 (with the adjustments for the "wideband noise" case outlined in Section 5). To get the tightness of $\{z^\gamma(t), t \geq 0, \text{(small)} \gamma > 0\}$ we work with the "stretched out" processes $x_0^\gamma(t) = x^\gamma(t/\epsilon)$ and $z_0^\gamma(t) = z^\gamma(t/\epsilon)$ and use the perturbed Liapunov function $[V(x_0^\gamma(t)) + V_1^\gamma(x_0^\gamma(t), t) + W(x_0^\gamma(t), z_0^\gamma(t)) + W_1^\gamma(x_0^\gamma(t), z_0^\gamma(t), t)]$, analogously to what was done in Part 2 of Theorem 3.2. The rest of the details are omitted.

10

Parametric Singularities

0. Outline of the Chapter

Multiple time scale systems can arise in applications due to the effects of small parameters, and this gives rise to a class of systems which are quite different from the models (4.1.1), (4.1.2), or their wide band noise driven analogs. Consider the following case:

$$\epsilon \ddot{y}^\epsilon + a_2 \dot{y}^\epsilon + a_1 y^\epsilon = u(t) + \text{ white noise.} \tag{0.1}$$

where $a_i > 0$, $\epsilon > 0$ and $u(t)$ is a control. The small parameter ϵ might represent, for example, a small inductance in an electric circuit. Define $x_1^\epsilon = y^\epsilon$ and $x_2^\epsilon = \dot{y}^\epsilon$. Then, we can write

$$dx_1^\epsilon = x_2^\epsilon dt$$
$$\epsilon dx_2^\epsilon = [-a_2 x_2^\epsilon - a_1 x_1^\epsilon + u(t)]dt + \sigma_2 dw, \tag{0.2}$$

for some constant $\sigma_2 > 0$. It will be seen in Section 1 that $x_2^\epsilon(\cdot)$ is a "wide band noise" process which converges to white noise in the sense that its integral converges to a Wiener process with a state and control dependent bias.

The methods for the convergence analysis are similar to those used for the wide band width noise case of Chapters 7 and 8. In the reference [E3] wide band noise driven versions of our systems model are dealt with and some interesting applications are discussed.

This Chapter is very brief, and contains only an outline of some weak convergence and stability results. In Section 1, we prove the weak convergence result for a class of systems which includes (0.2), and obtain the system which plays the role of the averaged limit system. We do not deal with the questions of near optimality of controls or of convergence of the cost functions. These questions can all be dealt with, but the detail is considerable, as can be seen from the nature of the proofs in Section 1.

Section 2 is concerned with a stability problem, which arises when the system is of interest over an arbitrarily long time interval. The subject of the chapter could easily be expanded into another monograph.

1. Singularly Perturbed Ito Processes: Weak Convergence

In this section, we will work with the controlled system

$$dx^\epsilon = [A_{11}(x^\epsilon) + A_{12}(x^\epsilon)z^\epsilon]dt + \int_{\mathcal{U}} B_1(x^\epsilon, \alpha)m_i^\epsilon(d\alpha)dt$$

$$+ \sigma_1(x^\epsilon)dw, \quad x \in R^r, \tag{1.1}$$

$$\epsilon dz^\epsilon = [A_{21}(x^\epsilon) + A_{22}z^\epsilon]dt + \int_{\mathcal{U}} B_2(x^\epsilon, \alpha)m_i^\epsilon(d\alpha)dt$$

$$+ \sigma_2(x^\epsilon)dw, \quad y \in R^{r'}, \tag{1.2}$$

under the assumption:

A1.1. A_{22} *is stable. The functions* $A_{12}(\cdot)$, $B_i(\cdot)$ *and* $\sigma_i(\cdot)$ *are bounded and continuous also* $A_{11}(\cdot)$ *and* $A_{21}(\cdot)$ *satisfy*

$$|A_{11}(x)| + |A_{21}(x)| \leq K(1 + |x|), K < \infty.$$

The $A_{ij}(\cdot)$ *are continuous and have bounded and continuous mixed partial derivatives up to third order, and the* $\sigma_i(\cdot)$ *up to second order. The control* α *takes value in a compact set* $\mathcal{U}$.

A centered "fast" process. It will be convenient to do the analysis with the $z^\epsilon(\cdot)$ process centered about the control term. To do this, define the centered process $\hat{z}^\epsilon(\cdot)$ by

$$\hat{z}^\epsilon(t) = z^\epsilon(t) - \int_0^t e^{A_{22}(t-u)/\epsilon}du \int_{\mathcal{U}} B_2(x^\epsilon(u), \alpha)m_u^\epsilon(d\alpha)/\epsilon. \tag{1.3}$$

We then can write

$$d(\epsilon\hat{z}^\epsilon) = [A_{21}(x^\epsilon) + A_{22}\hat{z}^\epsilon]dt + \sigma_2(x^\epsilon)dw \tag{1.4}$$

$$dx^\epsilon = [A_{11}(x^\epsilon) + A_{12}(x^\epsilon)\hat{z}^\epsilon]dt + \sigma_1(x^\epsilon)dw + \hat{B}_1^\epsilon(t)dt + \hat{B}_2^\epsilon(t)dt, \tag{1.5}$$

where

$$\hat{B}_2^\epsilon(t) = \frac{A_{12}(x^\epsilon(t))}{\epsilon} \int_0^t e^{A_{22}(t-u)/\epsilon}du \int_{\mathcal{U}} B_2(x^\epsilon(u), \alpha)m_u^\epsilon(d\alpha),$$

$$\hat{B}_1^\epsilon(t) = \int_{\mathcal{U}} B_1(x^\epsilon(t), \alpha)m_i^\epsilon(d\alpha).$$

Let A^ϵ denote the differential operater of (1.4) and (1.5).

The fixed-x uncontrolled "fast" process. In Chapter 4, the fast process defined by (4.1.2) behaves as a diffusion with a "squeezed" time scale. As

$\epsilon \to 0$, the "stretched out" fast process actually converged to a well defined process. Here, as $\epsilon \to 0$ the fast process converges in some sense to a "white noise" process. We now do some calculations which illustrate its asymptotic properties and which will be needed in Theorem 1.1 below. Let $\hat{z}^{\epsilon}(s|x,t)$ denote the fixed-x form of (1.3), starting at time t; i.e.,

$$\epsilon d\hat{z}^{\epsilon}(s|x,t) = [A_{21}(x) + A_{22}\hat{z}^{\epsilon}(s|x,t)]ds + \sigma_2(x)dw(s), \qquad (1.6)$$

where $s \geq t$, x is treated as a parameter and the initial condition is $\hat{z}^{\epsilon}(t|x,t) = \hat{z}^{\epsilon}(t)$. Then

$$\hat{z}^{\epsilon}(s|x,t) = e^{A_{22}(s-t)/\epsilon}\hat{z}^{\epsilon}(t) + \frac{1}{\epsilon}\int_t^s e^{A_{22}(s-u)/\epsilon}[A_{21}(x)du + \sigma_2(x)dw(u)].$$
$$(1.7)$$

Let E_t^{ϵ} denote the expectation conditioned on $\{x^{\epsilon}(s), z^{\epsilon}(s), w(s), m_s^{\epsilon}(\cdot), s \leq t\}$. Then

$$E_t^{\epsilon}\hat{z}^{\epsilon}(s|x,t) = e^{A_{22}(s-t)/\epsilon}\hat{z}^{\epsilon}(t) + \frac{1}{\epsilon}\int_t^s e^{A_{22}(s-u)/\epsilon}A_{21}(x)du$$

$$= e^{A_{22}(s-t)/\epsilon}\hat{z}^{\epsilon}(t) + A_{22}^{-1}[e^{A_{22}(s-t)/\epsilon} - 1]A_{21}(x). \qquad (1.8)$$

We will need the following expression for the conditional variances:

$$\epsilon E_t^{\epsilon}\hat{z}^{\epsilon}(s|t,x)(\hat{z}^{\epsilon}(s|t,x))' = \epsilon e^{A_{22}(s-t)/\epsilon}\hat{z}^{\epsilon}(t)\hat{z}^{\epsilon}(t)'e^{A_{22}'(s-t)/\epsilon}$$

$$+ O(\epsilon)e^{A_{22}(s-t)/\epsilon}\hat{z}^{\epsilon}(t) + \frac{1}{\epsilon}\int_t^s e^{A_{22}(s-u)/\epsilon}\sigma_2(x)\sigma_2'(x)e^{A_{22}'(s-u)/\epsilon}du$$

$$+ \frac{1}{\epsilon}\left(\int_t^s e^{A_{22}(s-u)/\epsilon}A_{21}(x)du\right)\left(\int_t^s e^{A_{22}(s-u)/\epsilon}A_{21}(x)du\right)'. \qquad (1.9)$$

There is a $\lambda_1 > 0$, such that the last term on the right of (1.9) can be written (and the function $g_1(x)$ defined) as

$$\epsilon\left(\int_0^{\infty} e^{A_{22}u}A_{21}(x)du\right)\left(\int_0^{\infty} e^{A_{22}u}A_{21}(x)du\right)' + O(\epsilon)e^{-\lambda_1(s-t)/\epsilon}$$

$$\equiv \epsilon g_1(x) + O(\epsilon)e^{-\lambda_1(s-t)/\epsilon}.$$

The next to last term on the right side of (1.9) equals, for some $\lambda_2 > 0$,

$$\int_0^{\infty} e^{A_{22}u}\sigma_2(x)\sigma_2'(x)e^{A_{22}'u}du + e^{-\lambda_2(s-t)/\epsilon}g_3(x, s-t)$$

$$\equiv g_2(x) + e^{-\lambda_2(s-t)/\epsilon}g_3(x, s-t),$$

where $g_3(\cdot)$ is a bounded function, and $g_2(\cdot)$ is defined in the obvious way. It can also be seen that the correlation function of $\hat{z}^{\epsilon}(s|x,t)$ converges to that of white noise as $\epsilon \to 0$; i.e., to a δ-function.

In the next theorem, it is proved that the sequence defined by (1.1) is tight, despite the "wild" behavior of the $z^\epsilon(\cdot)$ terms. The limit process is identified in Theorem 1.2. That limit process plays the same role that the averaged system did in Chapter 4—for purposes of approximations for the control problem. Note that the $z^\epsilon(\cdot)$ does not get "averaged out" of (1.1); in the limit it contributes to both the drift and diffusion terms.

Theorem 1.1. *Under* (A1.1) *and the tightness of* $\{x^\epsilon(0), \epsilon > 0\}$, *the set* $\{x^\epsilon(\cdot)\}$ *is tight.*

Proof. Recall from Chapter 7 that a N-truncation of $x^\epsilon(\cdot)$ is any process which is bounded by $N +$ constant and equals $x^\epsilon(\cdot)$ until at least the first time that $|x^\epsilon(t)| = N$. In order to prove the tightness of a sequence such as $\{x^\epsilon(\cdot)\}$, a usual method is to prove the tightness of N-truncations $\{x^{\epsilon,N}(\cdot)\}$ for each N. Suppose that the weak limits $x_N(\cdot)$ of $\{x^{\epsilon,N}(\cdot)\}$ are N-truncations of a process $x(\cdot)$ defined on $[0, \infty)$, or simply that for each $T_1 < \infty$, $P(\sup_{t \leq T_1} |x_N(t)| \geq M) \xrightarrow{M} 0$, uniformly in N. Then, as discussed in Chapter 7, $\{x^\epsilon(\cdot)\}$ is tight.

To get the N-truncation, let $q_N(\cdot)$ be a smooth non-negative real valued function which is bounded by unity, with value unity on $\{x : |x| \leq N\}$, and value zero on $\{x : |x| \geq N + 1\}$. We can define a suitable N-truncation $x^{\epsilon,N}(\cdot)$ by $x^{\epsilon,N}(t) = x^\epsilon(t) q_N(x^\epsilon(t))$. It will turn out that the weak limits of $\{x^{\epsilon,N}(\cdot)\}$ will be N-truncations of a process $x(\cdot)$ whose differential operator is the A^α of Theorem 1.2 below. Because of that, $\{x^\epsilon(\cdot)\}$ will also be tight. In order to save ourselves a lot or notation, we simply suppose that the $\{x^\epsilon(\cdot)\}$ is *uniformly bounded*, and ignore the $q_N(\cdot)$-truncation term. In the calculations below, the $q_N(\cdot)$ term would not affect the values of any of the expressions for $|x^\epsilon(t)| \leq N$.

By the truncation assumption on $\{x^\epsilon(\cdot)\}$, we have that for each $T_1 < \infty$,

$$\sup_{t \leq T_1} E|\epsilon \tilde{z}^\epsilon(t)|^2 = O(\epsilon),$$

$$E \sup_{t \leq T_1} |\epsilon \hat{z}^\epsilon(t)| \xrightarrow{\epsilon} 0.$$

(1.10)

Even though the system (1.4), (1.5) is driven by the white noise, the $\hat{z}^\epsilon(\cdot)$ process in (1.5) plays the role of a wide band noise, whose integral converges weakly to a Wiener process. We therefore use the perturbed test function method of Chapters 7 and 8 to prove the tightness; in particular, we use Theorem 7.3.3. It will be necessary to use a second order perturbed test function, rather than the first order one used in Chapter 8. Since the wide band noise $z^\epsilon(\cdot)$ taken together with $x^\epsilon(\cdot)$ is Markov, we will be able to use A^ϵ in lieu of the $\hat{A}^\epsilon$ operater of Chapter 7. We often write $x^\epsilon(t)$ as x^ϵ and $\hat{z}^\epsilon(t)$ as $\hat{z}^\epsilon$ for notational simplicity. Following the procedure used in

Chapter 8 or Chapter 7.4, let $f(\cdot) \in C_0^2(R^r)$ and write

$$A^\epsilon f(x^\epsilon) = f_x'(x^\epsilon)[A_{11}(x^\epsilon) + A_{12}(x^\epsilon)\hat{z}^\epsilon + \hat{B}_1^\epsilon(t) + \hat{B}_2^\epsilon(t)]$$

$$+ \frac{1}{2} \text{trace } f_{xx}(x^\epsilon) \cdot \sigma_1(x^\epsilon)\sigma_1'(x^\epsilon). \tag{1.11}$$

The $\hat{z}^\epsilon(\cdot)$ in (1.11) needs to be averaged out. To do this, we next define the test function perturbation $f_1^\epsilon(t) = f_1^\epsilon(x^\epsilon, \hat{z}^\epsilon)$, where

$$f_1^\epsilon(x^\epsilon, \hat{z}^\epsilon) = \int_t^\infty f_x'(x^\epsilon)A_{12}(x^\epsilon)E_t^\epsilon[\hat{z}^\epsilon(s|x^\epsilon, t) + A_{22}^{-1}A_{21}(x^\epsilon)]ds, \tag{1.12}$$

where $\hat{z}^\epsilon(s|x,t)$ is defined by (1.6) and $\hat{z}^\epsilon(t|x,t) = \hat{z}^\epsilon(t)$. The second term in the integrand in (1.12) asymptotially centers the integrand about the value zero (see (1.8)).

By (1.8), we have

$$f_1^\epsilon(x^\epsilon, \hat{z}^\epsilon) = -\epsilon f_x'(x^\epsilon)A_{12}(x^\epsilon)A_{22}^{-1}[\hat{z}^\epsilon + A_{22}^{-1}A_{21}(x^\epsilon)] \tag{1.13}$$

By (1.10) , for each $T_1 < \infty$

$$E \sup_{t \leq t_1} |f_1^\epsilon(t)| \xrightarrow{\epsilon} 0. \tag{1.14}$$

In order to evaluate $A^\epsilon f_1^\epsilon(x^\epsilon, \hat{z}^\epsilon)$, we can use either form (1.12) or (1.13). We use (1.13) and get

$$A^\epsilon f_1^\epsilon(x^\epsilon, \hat{z}^\epsilon) = -f_x'(x^\epsilon)A_{12}(x^\epsilon)A_{22}^{-1}[A_{21}(x^\epsilon) + A_{22}\hat{z}^\epsilon]$$

$$+ O(\epsilon) + O(\epsilon)\hat{z}^\epsilon - \epsilon[f_x'(x^\epsilon)A_{12}(x^\epsilon)A_{22}^{-1}\hat{z}^\epsilon]_x'A_{12}(x^\epsilon)\hat{z}^\epsilon \tag{1.15}$$

$$- E[dw'\sigma_2'(x^\epsilon)[f_x'(x^\epsilon)A_{12}(x^\epsilon)A_{22}^{-1}]_x'\sigma_1(x^\epsilon)dw]/dt,$$

where the expectation E in the last term on the right is over the "dw" only, and is the contribution of the "$dz \cdot dx$ component" of $A^\epsilon f_1^\epsilon(x^\epsilon, \hat{z}^\epsilon)$. If x^ϵ and z^ϵ process were driven by mutually independent Wiener processes, then this term would not appear. Define the operator A_1 acting on $f(\cdot) \in C^2(R^r)$ by defining the last term on the right of (1.15) *including the minus sign* to be $A_1 f(x^\epsilon)$.

The terms in (1.15) which include $\epsilon\hat{z}^\epsilon(t)$ are asymptotically negligible by (1.10). Since the next to last term on the right of (1.15) involves the square $\epsilon\hat{z}^\epsilon(\hat{z}^\epsilon)'$, it must be averaged out. This will be done via a second perturbation to the test function. Define the matrix valued operator $K_f(x)$ acting on $f(\cdot)$ by

$$-\epsilon[f_x'(x^\epsilon)A_{12}(x^\epsilon)A_{22}^{-1}\hat{z}^\epsilon]_x'A_{12}(x^\epsilon)\hat{z}^\epsilon$$

$$= \epsilon(\hat{z}^\epsilon)'K_f(x^\epsilon)\hat{z}^\epsilon = \text{trace } K_f(x^\epsilon) \cdot [\epsilon\hat{z}^\epsilon(\hat{z}^\epsilon)'].$$

Define the second perturbation $f_2^\epsilon(t) = f_2^\epsilon(x^\epsilon, \hat{z}^\epsilon)$ by

$$f_2^\epsilon(x^\epsilon, \hat{z}^\epsilon) = \text{ trace } K_f(x^\epsilon) \int_t^\infty E_t^\epsilon[\epsilon \hat{z}^\epsilon(s|x^\epsilon, t)\hat{z}^\epsilon(s|x^\epsilon, t)'$$

$$- \epsilon g_1(x^\epsilon) - g_2(x^\epsilon)]ds,$$

where the $g_i(\cdot)$ are defined below(1.9). By (1.9),

$$f_2^\epsilon(x^\epsilon, \hat{z}^\epsilon) = \text{ trace } K_f(x^\epsilon)[\epsilon^2 \int_0^\infty e^{A_{22}u}\hat{z}^\epsilon \hat{z}^{\epsilon\prime} e^{A_{22}'u}du$$

$$+ O(\epsilon^2)|\hat{z}^\epsilon| + O(\epsilon)], \tag{1.16}$$

where the $O(\epsilon^2)$ and $O(\epsilon)$ are functions of x^ϵ whose values and x-derivatives up to second order are uniformly $O(\epsilon^2)$ and $O(\epsilon)$, resp. (and similarly for the $O(\epsilon^2)$ which will appear below).

By these facts, we have

$$A^\epsilon f_2^\epsilon(x^\epsilon, \hat{z}^\epsilon) = O(\epsilon)\hat{z}^\epsilon + O(\epsilon) + O(\epsilon^2)|\hat{z}^\epsilon|^2 + O(\epsilon^2)|\hat{z}^\epsilon|^3$$

$$+ \text{ trace } K_f(x^\epsilon) \cdot A^\epsilon[\epsilon^2 \int_0^\infty e^{A_{22}u}\hat{z}^\epsilon \hat{z}^{\epsilon\prime} e^{A_{22}'u}du]. \tag{1.17}$$

In order to evaluate the right hand term in (1.17), let $P^\epsilon(t)$ denote the bracketed expression in that term and define $C^\epsilon(t) = \epsilon \hat{z}^\epsilon(t)\hat{z}^\epsilon(t)'$. Then, for each t, $P^\epsilon(t)$ satisfies the Liapunov equation

$$\frac{1}{\epsilon}[A_{22}P^\epsilon(t) + P^\epsilon(t)A_{22}'] = -C^\epsilon(t) \tag{1.18}$$

Now, apply Itô's formula to $P^\epsilon(t)$ and use the property (1.18) to get

$$A^\epsilon P^\epsilon(t) = -C^\epsilon(t) + \Sigma(x^\epsilon) + O(\epsilon)|\hat{z}^\epsilon(t)|,$$

where we define

$$\Sigma(x) = \int_0^\infty e^{A_{22}u}\sigma_2(x)\sigma_2'(x)e^{A_{22}'u}du.$$

We can now put all the estimates together. Define the perturbed test function $f^\epsilon(t) = f(x^\epsilon) + f_1^\epsilon(t) + f_2^\epsilon(t)$. For any $T_1 < \infty$ and $\delta > 0$ we have

$$P(\sup_{t \leq T_1} |f_1^\epsilon(t) + f_2^\epsilon(t)| \geq \delta) \xrightarrow{\epsilon} 0,$$

$$\tag{1.19}$$

$$\sup_{t \leq T_1} E|f_1^\epsilon(t) + f_2^\epsilon(t)| \xrightarrow{\epsilon} 0.$$

Also

$$A^\epsilon f^\epsilon(t) = f_x'(x^\epsilon)[A_{11}(x^\epsilon) + A_{12}(x^\epsilon)\hat{z}^\epsilon + \hat{B}_1^\epsilon(t) + \hat{B}_2^\epsilon(t)]$$

$$+ \frac{1}{2} \text{ trace } f_{xx}(x^\epsilon) \cdot \sigma_1(x^\epsilon)\sigma_1'(x^\epsilon)$$

$$- \epsilon f_x'(x^\epsilon)A_{12}(x^\epsilon)A_{22}^{-1}[A_{21}(x^\epsilon) + A_{22}\hat{z}^\epsilon] + \text{ trace } K_f(x^\epsilon)(\epsilon\hat{z}^\epsilon\hat{z}^{\epsilon\prime})$$

$$+ \text{ trace } K_f(x^\epsilon)[\Sigma(x^\epsilon) - \epsilon\hat{z}^\epsilon\hat{z}^{\epsilon\prime}] + A_1 f(x^\epsilon) + O(\epsilon) + O(\epsilon)\hat{z}^\epsilon + O(\epsilon^2)|\hat{z}^\epsilon|^3$$

$$= f_x'(x^\epsilon)[A_{11}(x^\epsilon) + \hat{B}_1^\epsilon(t) + \hat{B}_2^\epsilon(t)] + \frac{1}{2} \text{ trace } f_{xx}(x^\epsilon) \cdot \sigma_1(x^\epsilon)\sigma_1'(x^\epsilon)$$

$$+ \text{ trace } K_f(x^\epsilon) \cdot \Sigma(x^\epsilon) + A_1 f(x^\epsilon) + \text{ negligable error terms.} \quad (1.20)$$

The error terms are uniformly (in $\epsilon, t \leq T_1$) integrable and go to zero in the mean as $\epsilon \to 0$. Tightness of $\{x^\epsilon(\cdot)\}$ now follows from (1.19), (1.20) and Theorem 7.3.3   Q.E.D.

Theorem 2.2. *Under* (A1.1), *the limit* $(x(\cdot), m(\cdot))$ *of any weakly convergent subsequence of* $\{x^\epsilon(\cdot), m^\epsilon(\cdot)\}$ *solves the martingale problem for the operator* A^α *defined by*

$$A^\alpha f(x) = f_x'(x)A_{11}(x) + \frac{1}{2} \text{ trace } f_{xx}(x) \cdot \sigma_1(x)\sigma_1'(x)$$

$$+ \text{trace} K_f(x) \cdot \Sigma(x) + f_x'(x)B_1(x, \alpha) \quad (1.21)$$

$$- f_x'(x)A_{12}(x)A_{22}^{-1}B_2(x, \alpha) + A_1 f(x).$$

Proof. The proof follows from Theorems 1.1 and 7.3.1, where the perturbed test function $f^\epsilon(\cdot)$ of Theorem 1.1 is used. To deal with the limits of the $\hat{B}^\epsilon(\cdot)$ term in (1.20), note that the expression (1.22) goes to zero in the mean, uniformly on any interval $[0, T_1]$ as $\epsilon \to 0$:

$$\left| \int_0^t \hat{B}_2^\epsilon(s)ds + \int_0^t ds\, A_{12}(x^\epsilon(s))A_{22}^{-1} \int_U B_2(x^\epsilon(s), \alpha)m_s^\epsilon(d\alpha) \right|. \quad (1.22)$$

Q.E.D.

2. Stability

If the system (1.1), (1.2) is of interest over an infinite time interval, then we need to prove the stability or tightness of $\{x^\epsilon(t), t < \infty, \epsilon > 0\}$ in order to get results for the average cost per unit time problem. We will prove the stablity for the linear case only (the $\sigma_i(\cdot)$ and $B_i(\cdot)$ can be non-linear) since the pertubed Liapunov functions which are used are relatively simple in that case. The method of Chapter 8.4 cannot be used directly since our fast system is different, and plays the role of wide band noise. But we are able to employ a very similar method.

Theorem 2.1. *Suppose that the main part of the system* (1.1) (1.2) *is linear; i.e.,* $A_{11}(x) = A_{11}x, A_{12}(x) = A_{12}$ *a constant matrix, and* $A_{21}(x) = A_{21}x$. *Let* A_0 *and* A_{22} *be stable matrices, where* $A_0 = A_{11} - A_{12}A_{22}^{-1}A_{21}$. *Let* $B_i(\cdot)$ *and* $\sigma_i(\cdot)$ *be bounded and continuous. Then for small enough* ϵ,

$$\sup_t E|x^\epsilon(t)|^2 < \infty, \quad \sup_t E|\epsilon z^\epsilon(t)| < \infty. \tag{2.1}$$

Proof. Owing to the linearity, the control terms in (1.1), (1.2) do not affect the stability and we drop them for simplicity. The second inequality in (2.1) is a consequence of the first. Center the process $z^\epsilon(\cdot)$ by defining $y^\epsilon = z^\epsilon + A_{22}^{-1}A_{21}x^\epsilon$. We write y^ϵ and x^ϵ in lieu of $y^\epsilon(t)$ and $x^\epsilon(t)$, respectively. Then

$$dx^\epsilon = A_0 x^\epsilon dt + A_{12}y^\epsilon dt + \sigma_1(x^\epsilon)dw, \tag{2.2}$$

$$\epsilon \, dy^\epsilon = A_{22}y^\epsilon dt + \sigma_2(x^\epsilon)dw + \epsilon \, d(A_{22}^{-1}A_{21}x^\epsilon). \tag{2.3}$$

Let $V(\cdot)$ be a quadratic Liapunov function for the system $\dot{x} = A_0 x$ such that $V_x'(x)A_0 x \leq -c_1|x|^2$ for some $c_1 > 0$. We can write $A^\epsilon V(x^\epsilon) \leq V_x'(x^\epsilon)A_0 x^\epsilon + V_x'(x^\epsilon)A_{12}y^\epsilon +$ constant. We need to average out the y^ϵ-term, and this will be done by use of a perturbed Liapunov function, as in Chapter 8. For each x and t, define $y^\epsilon(s|x,t)$, $s \geq t$, by

$$\epsilon \, dy^\epsilon(s \mid x,t) = A_{22}y^\epsilon(s \mid x,t)dt + \sigma_2(x)dw,$$

where the initial condition is $y^\epsilon(t|x,t) = y^\epsilon(t)$. Define the Liapunov function perturbation $V_1^\epsilon(t) = V_1^\epsilon(x^\epsilon, y^\epsilon)$ where

$$\begin{aligned} V_1^\epsilon(x^\epsilon, y^\epsilon) &= \int_t^\infty V_x'(x^\epsilon)A_{12}E_t^\epsilon y^\epsilon(s|x^\epsilon,t)ds \\ &= -\epsilon V_x'(x^\epsilon)A_{12}A_{22}^{-1}y^\epsilon \\ &= O(\epsilon)|x^\epsilon| \cdot |y^\epsilon|. \end{aligned}$$

Then, we can write

$$A^\epsilon V_1^\epsilon(x^\epsilon, y^\epsilon) = -[\epsilon V_x'(x^\epsilon)A_{12}A_{22}^{-1}y^\epsilon]_x'(A_0 x^\epsilon + A_{12}y^\epsilon)$$

$$- [V_x'(x^\epsilon)A_{12}A_{22}^{-1}][A_{22}y^\epsilon + \epsilon(A_{22}^{-1}A_{21})(A_0 x^\epsilon + A_{12}y^\epsilon)]$$

$$= O(\epsilon)[|x^\epsilon|^2 + |y^\epsilon|^2 + 1] - V_x'(x^\epsilon)A_{12}y^\epsilon. \tag{2.4}$$

Thus

$$A^\epsilon[V(x^\epsilon) + V_1^\epsilon(x^\epsilon, y^\epsilon)] \leq V_x'(x^\epsilon)A_0 x^\epsilon + \text{ constant}$$

$$+ O(\epsilon)[|x^\epsilon|^2 + |y^\epsilon|^2 + 1]. \tag{2.5}$$

In order to eliminate the $O(\epsilon)|y^\epsilon|^2$ term in (2.5), we introduce another perturbation which is similar to that used in Chapter 9.4. Let $W(y)$ be a quadratic Liapunov function for the system $\dot{y} = A_{22}y$ and satisfying $W'_y(y)A_{22}y \le -d_1|y|^2$ for some $d_1 > 0$. We can write

$$A^\epsilon \epsilon^2 W(y^\epsilon) \le \epsilon W'_y(y^\epsilon)A_{22}y^\epsilon + O(\epsilon^2)[|y^\epsilon|^2 + |y^\epsilon||x^\epsilon|] + \text{constant}.$$

Define $V^\epsilon(x,y) = V(x) + V_1^\epsilon(x,y) + k\epsilon^2 W(y)$. Using the inequality $|x|\cdot|y| \le \alpha|x|^2/\epsilon + \epsilon|y|^2/\alpha$ for $\epsilon > 0$, $\alpha > 0$, we have

$$V_1^\epsilon(x^\epsilon, y^\epsilon) = O(\epsilon^2/\alpha)|y^\epsilon|^2 + O(\alpha)|x^\epsilon|^2.$$

Then for large enough k and appropriate α, there are constants $d_i > 0$ such that for small ϵ,

$$A^\epsilon V^\epsilon(x^\epsilon, y^\epsilon) \le \text{constant} - d_2|x^\epsilon|^2 - d_3\epsilon|y^\epsilon|^2$$

$$\le \text{constant} - d_4 V^\epsilon(x^\epsilon, y^\epsilon). \tag{2.6}$$

The theorem follows from (2.6) and the fact that there is $d_5 > 0$ such that $V^\epsilon(x,y) \ge d_5 V(x)$ for large k and small enough ϵ.   Q.E.D.

References

[A1] Arnold, L., 1974, *Stochastic Differential Equations*, Wiley, New York.

[B1] Benes, V.E., 1971, "Existence of optimal control laws," *SIAM J. on Control*, **9** 446–475.

[B2] Bensoussan, A., 1989, *Perturbation Methods in Optimal Control*, Wiley, New York.

[B3] Berkovitz, L., 1974, *Optimal Control Theory*, Appl. Math. Sci. Series, v. 12, Springer, Berlin.

[B4] Bielecki, T. and Stettner, L., 1989, "On ergodic control problems for singularly perturbed Markov process," *J. Applied Math. and Optimization*, **20**, 131–161.

[B5] Bielecki, T. and Stettner, L., 1988, "On some modelling problems arising in asymptotic analysis of Markov processes with singularly perturbed generators," *Stochastic Analysis and Applic.*, **6**, 129–168.

[B6] Billingsley, P., 1968, *Convergence of Probability Measures*, Wiley, New York.

[B7] Blankenship, G. and Papanicolaou, G.C., 1978, "Stability and control of systems with wide band noise disturbances," *SIAM J. Appl. Math.*, **34**, 437–476.

[B8] Borkar, V.S. and Ghosh, M.K., 1988, "Ergodic control of multidimensional diffusions: I, the existence results," *SIAM J. on Control and Optimization*, **26**, 112–126.

[B9] Breiman, L., 1968, *Probability Theory*, Addison-Wesley, Reading, Mass.

[C1] Clark, J.M.C., 1978, "The design of robust approximations to the stochastic differential equations of non linear filtering," in NATO Advanced Study Institute Series, (ed. J.K. Skwirzynski), Sitjhoff, Noordhoff, Alphen aan den Rijn, Amsterdam.

[C2] Cox, R.M. and Karatzas, I., 1985, "Stationary control of Brownian motion in several dimensions," *Adv. in Applied Probability*, **17**, 531–561.

[D1] Davenport, W.B. and Root, W.L., 1958, *Random Signals and Noise*, McGraw-Hill, New York.

[D2] Davis, J.M.C., 1982, "A pathwise solution to the equations of non linear filtering," *Theory of Prob. and Applic.*, **27**, 167–175.

[D3] DiMasi, G.B. and Runggaldier, W.J., 1982, "Approximations and bounds for discrete time non-linear filtering," *Lect. Notes in Control and Information Sciences*, **44**, Springer, Berlin.

[D4] Donsker, M. and Varadhan, S.R.S., 1975, "Asymptotic evaluation of certain Markov process expectations for large time," I, II, III, *Comm. Pure Appl. Math.*, **28**, 1–47, 279–301, **29**, 389–461.

[D5] Doob, J., 1953, *Stochastic Processes*, Wiley, New York.

[D6] Dynkin, E.B., 1965, *Markov Processes*, Springer, Berlin.

[E1] Elliot, R., 1982, *Stochastic Calculus and Applications*, Springer, Berlin.

[E2] Ethier, S.N. and Kurtz, T.G., 1986, *Markov Processes: Characterization and Convergence*, Wiley, New York.

[E3] El-Ansary, M. and Khalil, H., 1986, "On the interplay of singular perturbations and wide band stochastic fluctuations," *SIAM J. on Control and Optimization*, **24**, 83–93.

[F1] Fleming, W.H., 1977, "Generalized solutions in optimal stochastic control," Second Kingston Conference (URI) on Differential Games, Marcel Dekker.

[F2] Fleming, W.H. and Nisio, M., 1984, "On stochastic relaxed controls for partially observed diffusions," *Osaka Math. J.*, **93**, 71–108.

[F3] Fleming, W.H. and Rishel, R., 1975, *Deterministic and Stochastic Optimal Control*, Springer, Berlin.

[F4] Friedman, A., 1975, *Stochastic Differential Equations and Applications*, Academic Press, New York.

[F5] Fujisaki, H., Kallianpur, G. and Kunita, H., 1972, "Stochastic differential equations for the nonlinear filtering problem," *Osaka Math. J.*, 19–40.

[G1] Gihman, I.I. and Skorohod, A.V., 1972, *Stochastic Differential Equations*, Springer, Berlin.

[G2] Gihman, I.I. and Skorohod, A.V., 1965, *Introduction to the Theory of Random Processes*, Saunders, Philadephia.

[I1] Ikeda, N. and Watanabe, S., 1981, *Stochastic Differential Equations and Diffusion Processes*, North-Holland, Amsterdam.

[K1] Khalil, H. and Gajic, Z., 1984, "Near optimum regulators for stochastic linear singularly perturbed systems," *IEEE Trans. on Automatic Control*, **AC-29**, 531–541.

[K2] Khazminskii, R.Z., 1982, *Stochastic Stability of Differential Equations*, Sijthoff, Noordhoff, Alphen aan den Rijn, Amsterdam.

[K3] Klimushchev, A.I. and Krasovskii, N.N., 1962, "Uniform asymptotic stability of systems of differential equations with a small parameter in the derivative terms," *J. Appl. Math. Mech.*, **25**, 1011–1025.

[K4] Kokotovic, P., Bensoussan, A. and Blankenship, G., Eds., 1987, *Singular Perturbations and Asymptotic Analysis in Control Systems*, Vol. 90, Lect. Notes in Control and Information Science, Springer, Berlin.

[K5] Kokotovic, P., Khalil, H. and O'Reilly, J., 1986, *Singular Perturbation Methods in Control: Analysis and Design*, Academic Press, New York.

[K6] Krishnan, V., 1984, *Nonlinear Filtering and Smoothing*, Wiley, New York.

[K7] Kunita, H., 1971, "Asymptotic behavior of the nonlinear filtering errors of a Markov process," *J. Multivariate Anal.*, **1**, 365–393.

[K8] Kunita, H. and Watanabe, S., 1967, "On square integrable martingales," *Nagoya Math. J.*, 209–245.

[K9] Kurtz, T.G., 1981, *Approximation of Population Processes*, Vol. 36 of CBMS-NSF Regional Conf. Series in Appl. Math., SIAM, Philadelphia.

[K10] Kurtz, T.G., 1975, "Semigroups of conditional shifts and approximation of Markov processes," *Ann. Prob.*, **4**, 618–642.

[K11] Kushner, H.J. and Runggaldier, W., 1987, "Nearly optimal state feedback controls for stochastic systems with wideband noise disturbances," *SIAM J. on Control and Optimization*, **25**, 289–315.

[K12] Kushner, H.J., 1978, "Optimality conditions for the average cost per unit time problem with a diffusion model," *SIAM J. on Control and Optimization*, **16**, 330–346

[K13] Kushner, H.J., 1977, *Probability Methods for Approximations in Stochastic Control and for Elliptic Equations*, Academic Press, New York.

[K14] Kushner, H.J., 1967, "Dynamical equations for non linear filtering," *J. Diff. Equations*, **3**, 179–190.

[K15] Kushner, H.J., 1967, *Stochastic Stability and Control*, Academic Press, New York.

[K16] Kushner, H.J., 1972, "Stability of stochastic dynamical systems," *Lecture Notes in Math.*, **294**, Springer, Berlin.

[K17] Kushner, H.J., 1985, "Direct averaging and perturbed test function methods for weak convergence," *Lect. Notes in Control and Information Sciences*, **81**, 412–426.

[K18] Kushner, H.J., 1979, "A robust computable approximation to the optimal nonlinear filter," *Stochastics*, **3**, 75–83.

[K19] Kushner, H.J., 1979, "Jump-diffusion approximations for ordinary differential equations with wideband random right hand sides," *SIAM J. on Control and Optimization*, **17**, 729–744.

[K20] Kushner, H.J., 1984, *Approximation and Weak Convergence Methods for Random Processes, with Applications to Stochastic Systems Theory*, MIT Press, Cambridge, Mass.

[K21] Kushner, H.J. and Huang, H., 1986, "Approximation and limit results for nonlinear filters with wide bandwidth observation noise," *Stochastics*, **16**, 65–96.

[K22] Kushner, H.J., 1989, "Diffusion approximations and nearly optimal maintenance policies for system breakdown and repair problems," *Applied Math. and Optimization*, **20**, 33–53.

[K23] Kushner, H.J., 1989, "Approximations and optimal control for the pathwise average cost per unit time and discounted problems for wideband noise driven systems," *SIAM J. on Control and Optimization*, **27**, 546–562.

[K24] Kushner, H.J., 1988, Numerical methods for stochastic control in continuous time, LCDS Rept., Brown University, 1988, to appear *SIAM J. Control and Optimiz.*

[K25] Kushner, H.J. and DiMasi, G.B., 1978, "Approximations for functionals and optimal control problems on jump-diffusion processes," *J. Math. Anal. and Applic.*, **63**, 772–800.

[K26] Karlin, S. and Taylor, H.M., 1975, *A First Course in Stochastic Processes, Second Edition*, Academic Press, New York.

[K27] Kushner, H.J. and Martins, L.F., "Limit theorems for pathwise average cost per unit time problems for queues in heavy traffic," Sub. to *SIAM J. on Control and Optimization*, LCDS Report 89-18, Brown University.

[L1] Lions, P.L. and Sznitman, A.S., 1984, "Stochastic differential equations with reflecting boundary conditions," *Comm. Pure and Appl. Math.*, **37**, 511–553.

[L2] Liptser, R. and Shiryaev, A.N., 1977, *Statistics of Random Processes*, Springer, Berlin.

[M1] Meyer, P., 1966, *Probability and Potentials*, Blaisdell, Waltham, Mass.

[N1] Neveu, J., 1965, *Mathematical Foundations of the Calculus of Probability*, Holden-Day, San Francisco.

[P1] Papanicolaou, G.C., Stroock, D. and Varadhan, S.R.S., 1977, "Martingale approach to some limit theorems," Proc. of the 1977 Duke University Conference on Turbulence.

[P2] Papoulis, A., 1965, *Probability, Random Variables and Stochastic Processes*, McGraw-Hill, New York.

[R1] Rishel, R., 1970, "Necessary and sufficient dynamic programming conditions for continuous time stochastic optimal control," *SIAM J. on Control*, **8**, 559–571.

[S1] Saberi, A. and Khalil, H., 1984, "Quadratic-type Liapunov functions for singularly perturbed systems," *IEEE Trans. on Automatic Control*, **AC-29**, 542–550.

[S2] Saberi, A. and Khalil, H., 1985, "Stabilization and regulation of non-linearly singularly perturbed systems—composite control," *IEEE Trans. on Automatic Control*, **AC-30**, 739–747.

[S3] Saskena, V.R., O'Reilly, J. and Kokotovic, P., 1984, "Singular perturbations and time scale methods in control theory; a survey. 1976–1983," *Automatica*, **20**, 273–293.

[S4] Schuss, Z., 1980, *Theory and Applications of Stochastic Differential Equations*, Wiley, New York.

[S5] Shiryaev, A.N., 1966, "Stochastic equations of non linear filtering of jump Markov processes," *Problemy Peredachi Informatsii*, **3**, 3–22.

[S6] Shiryaev, A.N., 1978, *Optimal Stopping Rules*, Springer, Berlin.

[S7] Stroock, D.W. and Varadhan, S.R.S., 1972, "On degenerate elliptic and parabolic operators of second order and their associated diffusions," *Comm. Pure and Appl. Math.*, **25**, 651–713.

[S8] Stroock, D.W. and Varadhan, S.R.S., 1971, "Diffusion processes with boundary conditions," *Comm. Pure and Appl. Math.*, **24**, 147–225.

[S9] Stroock, D.W. and Varadhan, S.R.S., 1979, *Multidimensional Diffusion Processes*, Springer, Berlin.

[S10] Szpirglas, J., 1978, "Sur l'equivalence d'equations differentielles stochastiques a valeurs mesures intervenant dans le filtrage markovien non lineare," *Ann. Inst. H. Poincaré*, **XIV**, 33–59.

[V1] Varadhan, S.R.S., 1984, *Large Deviations and Applications*, CBMS-NSF Regional Conference Series in Mathematics, SIAM, Philadelphia.

[W1] Warga, J., 1962, "Relaxed variational problems," *J. Math. Anal. and Appl.*, 4, 111–128.

[W2] Wonham, W.M., 1965, "Some applications of stochastic differential equations to optimal nonlinear filtering," *SIAM J. on Control*, **2**, 347–369.

[Z1] Zakai, M., 1969, "On optimal filtering of diffusion processes," *Z. Wahrsch. verw. Gebeite*, **11**, 230–243.

List of Symbols

Index